Why This Book Will Help You

THIS BOOK WILL BE AN EFFECTIVE AID if you are serious about study, if you want to improve your grades, and if you want to prepare successfully for exams.

Designed to help you learn and review important ideas, definitions, theorems, and calculus, this book shows you how to achieve these goals by writing solutions carefully and precisely.

In the beginning most steps in these solutions are provided but, as your mathematical skills develop, you will be able to fill in more steps on your own. By the end of the manual only the major steps are provided.

As you progress you will become more independent of this manual and use it only when you are "stuck" and need a quick review to clarify a point. You will learn by *doing* and will gain confidence in your abilities to reach the correct solutions.

LOUIS A. GUILLOU
Saint Mary's College
Winona, Minnesota

STUDENT SOLUTIONS MANUAL FOR

FOURTH EDITION
CALCULUS
with Analytic Geometry

Edwin J. Purcell
Dale Varberg

Prentice-Hall, Inc., Englewood Cliffs, New Jersey 07632

Library of Congress Catalog Card Number: 83-13753

Computer art on cover: Compute-A-Slide, Inc.

© 1984 by Prentice-Hall, Inc., Englewood Cliffs, New Jersey 07632

All rights reserved. No part of this book may be
reproduced, in any form or by any means,
without permission in writing from the publisher.

Printed in the United States of America

10 9 8 7 6 5 4 3 2

ISBN 0-13-111824-2 01

Prentice-Hall International, Inc., *London*
Prentice-Hall of Australia Pty. Limited, *Sydney*
Editora Prentice-Hall do Brasil, Ltda., *Rio de Janeiro*
Prentice-Hall Canada Inc., *Toronto*
Prentice-Hall of India Private Limited, *New Delhi*
Prentice-Hall of Japan, Inc., *Tokyo*
Prentice-Hall of Southeast Asia Pte. Ltd., *Singapore*
Whitehall Books Limited, *Wellington, New Zealand*

CONTENTS

Chapter 1	Preliminaries	1
Chapter 2	Functions and Limits	21
Chapter 3	The Derivative	43
Chapter 4	Applications of the Derivative	79
Chapter 5	The Integral	113
Chapter 6	Applications of the Integral	135
Chapter 7	Transcendental Functions	155
Chapter 8	Techniques of Integration	179
Chapter 9	Indeterminate Forms and Improper Integrals	197
Chapter 10	Numerical Methods, Approximations	209
Chapter 11	Infinite Series	225
Chapter 12	Conics and Polar Coordinates	245
Chapter 13	Geometry in the Plane, Vectors	267
Chapter 14	Geometry in Space, Vectors	279
Chapter 15	The Derivative in n-Space	293
Chapter 16	The Integral in n-Space	311
Chapter 17	Vector Calculus	327
Chapter 18	Differential Equations	341

NOTE TO THE STUDENT

This manual contains solutions to problems whose numbers are divisible by three. It was prepared with the purpose of helping you to carefully write solutions that will in turn help you to learn and review important ideas, definitions, theorems, and applications of calculus.

In the early chapters most steps are provided. As you progress and your mathematical skills develop you will be expected to fill in more steps. By the end of the manual, only major steps are provided.

Make it your goal to become independent of this manual. Only use it after you have tried a problem and are "stuck," and then only use enough of the solution to get you started. It is important that you learn to "do," not just be able to see that someone else's solution is correct.

Finally, a request. Although much care to detail was given in the preparation of this manual, it is virtually impossible that there are no errors. If you find any errors, typographical or real, or if you have suggestions for improved solutions, I would appreciate hearing from you. This will be especially beneficial to subsequent users of the manual. Your comments can be sent to me at: Department of Mathematics and Statistics, Saint Mary's College, Winona, MN 55987. Thank you.

L.G.

ACKNOWLEDGEMENTS

The problems in the book were also solved by Walter Fleming, Hamline University. His work was an excellent aid for checking solutions as well as for providing improved or corrected solutions for some problems. In addition, many solutions were checked by one of the authors, Dale Varberg.

This solutions manual was prepared using a Xerox 820 Information Processor and a Xerox 630 Printer with a Qume Letter Gothic printwheel. Additional characters were added using Copyaid Symbol Transfers. Don Scott and Steve Cohen of Symbols International, 2817 Beverly Boulevard, Los Angeles, California, were most helpful in providing the transfer symbols needed.

Many of the graphs were drawn by David Rowlands and my sons, Michael, Joseph, Thomas, and Daniel Guillou, who also did much of the transfer symbols work. Their help with the graphs and many clerical details has been invaluable and much appreciated.

Special thanks is due my wife, Maureen, who gave up her desk so that this project could have a place of its own.

CHAPTER 1   PRELIMINARIES

Problem Set 1.1   The Real Number System

3. $-4[3(-6 + 13) - 2(5 - 9)] = -4[3(7) - 2(-4)] = -4[21 + 8] = -116$

6. $\dfrac{3}{4} - \left(\dfrac{7}{12} - \dfrac{2}{9}\right) = \dfrac{3}{4} - \dfrac{7}{12} + \dfrac{2}{9} = \dfrac{27}{36} - \dfrac{21}{36} + \dfrac{8}{36} = \dfrac{27 - 21 + 8}{36} = \dfrac{14}{36} \approx 0.3889$

9. $\dfrac{14}{33}\left(\dfrac{2}{3} - \dfrac{1}{7}\right)^2 = \dfrac{14}{33}\left(\dfrac{14}{21} - \dfrac{3}{21}\right)^2 = \dfrac{14}{33}\left(\dfrac{14 - 3}{21}\right)^2 = \dfrac{14}{33} \cdot \dfrac{11}{21} \cdot \dfrac{11}{21} = \dfrac{22}{189} \approx 0.1164$

12. $\dfrac{\frac{1}{2} - \frac{3}{4} + \frac{7}{8}}{\frac{1}{2} + \frac{3}{4} - \frac{7}{8}} = \dfrac{\left(\frac{1}{2} - \frac{3}{4} + \frac{7}{8}\right)8}{\left(\frac{1}{2} + \frac{3}{4} - \frac{7}{8}\right)8} = \dfrac{4 - 6 + 7}{4 + 6 - 7} = \dfrac{5}{3} \approx 1.6667$

15. $(\sqrt{2} + \sqrt{3})(\sqrt{2} - \sqrt{3}) = \sqrt{4} - \sqrt{9} = 2 - 3 = -1$

18. $2\sqrt[3]{4}(\sqrt[3]{2} + \sqrt[3]{16}) = 2\sqrt[3]{8} + 2\sqrt[3]{64} = 2 \cdot 2 + 2 \cdot 4 = 12$

21. $(2x - 3)(2x + 3) = 2x(2x + 3) - 3(2x + 3) = 4x^2 + 6x - 6x - 9 = 4x^2 - 9$

24. $(3x+11)(2x-4) = 3x(2x-4) + 11(2x-4) = 6x^2 - 12x + 22x - 44 = 6x^2 + 10x - 44$

27. $\dfrac{x^2 - 4}{x - 2} = \dfrac{(x + 2)(x - 2)}{x - 2} = x + 2 \quad (x \neq 2)$

30. $\dfrac{2x - 2x^2}{x^3 - 2x^2 + x} = \dfrac{2x(1 - x)}{x(x^2 - 2x + 1)} = \dfrac{-2x(x - 1)}{x(x - 1)^2} = \dfrac{-2}{x - 1} \quad (x \neq 0)$

33. $\dfrac{x^2+x-6}{x^2-1} \cdot \dfrac{x^2+x-2}{x^2+5x+6} = \dfrac{(x+3)(x-2)}{(x+1)(x-1)} \cdot \dfrac{(x+2)(x-1)}{(x+3)(x+2)} = \dfrac{x-2}{x+1} \quad (x \neq -3, -2, 1)$

36. If $0/0 = b$, then $0 = 0 \cdot b$, which is no contradiction (as did occur in the $a \neq 0$ case). However, $0 \cdot b$ equals 0 for every real number b. Thus $0/0$, if defined in the same way as division by nonzero numbers, would not yield a unique real number. Hence, division by zero is undefined.

39. Suppose $\sqrt{2}$ = p/q where p and q are natural numbers. Neither p nor q can equal 1. (If q = 1, then $\sqrt{2}$ = p, a natural number. If p = 1, then $\sqrt{2}$ would be less than 1; neither is true so p ≠ 1 and q ≠ 1.)

$$\sqrt{2} = p/q \implies 2 = p^2/q^2 \implies p^2 = 2q^2.$$

Therefore, p^2 has one more prime factor than q^2 (an extra factor of 2). Then, either p^2 or q^2 has an odd number of prime factors (the other having an even number). This contradicts the result obtained in Problem 38. We conclude that $\sqrt{2}$ is not a rational number.

42. Let r be a rational number and s be an irrational number.
    (a) Let the sum of r and s be t; i.e., r + s = t. Then s = -r + t. If t were rational, then s (which equals -r + t) would be rational (Problem 41). But s is an irrational number, so t (the sum of r and s) must be irrational too.
    (b) Assume r ≠ 0. Let the product of r and s be u; i.e., rs = u. Then s = u/r. If u were rational, then s (which equals u/r) would be rational. But s is an irrational number, so u (the product of r and s) must be irrational too.

45. (a) False since (-20) - (-2) = -18 which is not positive.
    (b) True   since (1) - (-39) = 40 which is positive.
    (c) True   since (5/9) - (-3) = 32/9 which is positive.
    (d) True   since (-4) - (-16) = 12 which is positive.
    (e) True   since (34/39) - (6/7) = 4/273 which is positive.
    (f) False since (-44/59) - (-5/7) = -13/413 which is not positive.

48. (a) Always correct. [Added -4 to each side]
    (b) Not always correct. [For example, 3 < 7 but -3 > -7.]
    (c) Not always correct. [For example, -5 < 3 but $(-5)^2$ > (-5)(3).]
    (d) Always correct. [Multiplied each side by a^2, a nonnegative number]

Problem Set 1.2 Decimals, Denseness, Calculators

3.
```
        .15
   20 )3.00
       2 0
       1 00
       1 00
```

6.
```
      1.571428···
   7 ⟌ 11.000000···
       7
       ──
       40
       35
       ──
        50
        49
        ──
         10
          7
         ──
         30
         28
         ──
          20
          14
          ──
          60
          56
          ──
           40
```
Remainder repeats so digits in quotient will begin to repeat beginning with the digit 5. Thus, the repeating decimal representation of 11/7 is 1.571428 571428 ···.

9. Let x = 2.56 56 56···   [Cycle has a 2-digit length so multiply each
    $100x$ = 256.56 56 56···   side by 10^2.]
    Therefore, $99x = 254$, so $x = 254/99$.

12. Let x = 0.39999···   [Cycle has a 1-digit length so multiply each
    $10x$ = 3.99999···   side by 10.]
    Therefore, $9x = 3.6$, so $x = 3.6/9 = 36/90$.

15. 0.000001 is a positive rational number less than 0.00001.
    0.00000101001000100001··· is an irrational number less than 0.00001.
    [Even though there is a pattern, it is not a repeating decimal.]

18. $\frac{22}{7}$ = 3.142857 142857 ··· (Obtain by long division.)

    π = 3.141592 ··· (Given in Problem 17)

    Therefore, $\pi - \frac{22}{7}$ is negative since $\frac{22}{7} > \pi$.

21. Irrational. It is not a repeating decimal.

24. 0.010205144336

27. 12.433227831

30. 0.010273214286

Problem Set 1.2

33. Using four-decimal-place accuracy:

   (a) 25.4828   (b) 9.1692   (c) 2046.9136

Problem Set 1.3   Inequalities

3. $4x-7 < 3x+5$
   $x < 12$
   Solution Set: $(-\infty, 12)$ Graph: ——————————)——————
                                              12   x

6. $6x-10 \geq 5x-16$
   $x \geq -6$
   Solution Set: $[-6, \infty]$ Graph: ———————[——————————
                                              -6   x

9. $-6 < 2x+3 < -1$
   $-9 < 2x < -4$
   $-9/2 < x < -2$
   Solution Set: $(-9/2, -2)$ Graph: ———(———)———————
                                          -9/2 -2   x

12. $4 < 5-3x < 7$
    $-1 < -3x < 2$
    $1/3 > x > -2/3$
    $-2/3 < x < 1/3$
    Solution Set: $(-2/3, 1/3)$ Graph: ———(———)———————
                                            -2/3 1/3   x

15. $x^2 + x - 12 < 0$
    $(x + 4)(x - 3) < 0$
    Split points: $-4, 3$ Test points: $-5, 0, 4$

    Auxiliary axis for $(x+4)(x-3)$: (+)   (0)   (-)   (0)   (+)
                                      ——————•—————————•—————————
                                            -4   3   x

    Solution Set: $(-4, 3)$ Graph: ———(—————————)———————
                                          -4   3   x

18. $2x^2 + 7x - 15 \geq 0$

$(2x - 3)(x + 5) \geq 0$

Split points: $-5, 3/2$ Test points: $-6, 0, 2$

Auxiliary axis for $(2x-3)(x+5)$: $\quad \underline{\quad (+) \quad \overset{(0)}{\underset{-5}{\bullet}} \quad (-) \quad \overset{(0)}{\underset{3/2}{\bullet}} \quad (+) \quad}$ x

Solution set: $(-\infty, -5] \cup [3/2, \infty)$ Graph: $\underline{\quad\quad\quad]_{-5} \quad\quad\quad [_{3/2} \quad\quad\quad}$ x

21. $\dfrac{x + 5}{2x - 1} \leq 0$

Split points: $-5, 1/2$ Test points: $-6, 0, 1$

Auxiliary axis for $\dfrac{x + 5}{2x - 1}$: $\quad \underline{\quad (+) \quad \overset{(0)}{\underset{-5}{\bullet}} \quad (-) \quad \overset{(u)}{\underset{1/2}{\bullet}} \quad (+) \quad}$ x

Solution set: $[-5, 1/2)$ Graph: $\underline{\quad\quad [_{-5} \quad\quad\quad)_{1/2} \quad\quad}$ x

24. $\dfrac{7}{2x} < 3$

$\dfrac{7}{2x} - 3 < 0$

$\dfrac{7 - 6x}{2x} < 0$

Split points: $0, 7/6$ Test points: $-1, 1, 2$

Auxiliary axis for $\dfrac{7 - 6x}{x} < 0$: $\quad \underline{\quad (-) \quad \overset{(u)}{\underset{0}{\bullet}} \quad (+) \quad \overset{(0)}{\underset{7/6}{\bullet}} \quad (-) \quad}$ x

Solution set: $(-\infty, 0) \cup (7/6, \infty)$ Graph: $\underline{\quad\quad)_{0} \quad\quad (_{7/6} \quad\quad}$ x

27. $\dfrac{x - 2}{x + 4} < 2$

$\dfrac{x - 2}{x + 4} - 2 < 0$

$\dfrac{x - 2 - 2(x + 4)}{x + 4} < 0$

Problem Set 1.3

27. (continued)

$\frac{-x - 10}{x + 4} < 0$

Split points: -10,-4   Test points: -11,-6,0

Auxiliary axis for $\frac{-x - 10}{x + 4} < 0$: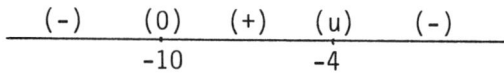

Solution set: $(-\infty,-10) \cup (-4,\infty)$ Graph:

30. $(2x + 3)(3x - 1)(x - 2) < 0$

Split points: -3/2,1/3,2   Test points: -2,0,1,3

Auxiliary axis for $(2x+3)(3x-1)(x-2)$:

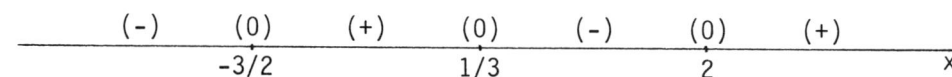

Solution set: $(-\infty,-3/2) \cup (1/3,2)$

Graph: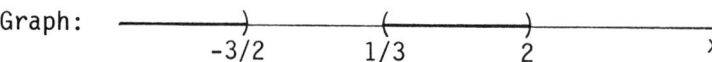

33. $x^3 - 5x^2 - 6x < 0$
$x(x^2 - 5x - 6) < 0$
$x(x + 1)(x - 6) < 0$

Split points: -1,0,6   Test points: -2,-1/2,1,7

Auxiliary axis for $x(x + 1)(x - 6)$:

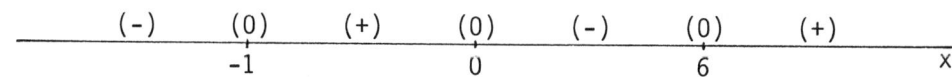

Solution set: $(-\infty,-1) \cup (0,6)$

Graph:

|)   (   ) |
| -1   0   6   x |

6   Problem Set 1.3

36. (a) $3x + 7 > 1$ or $2x + 1 < -5$ (b) $3x + 7 \leq 1$ or $2x + 1 < -8$
       $3x > -6$ or $2x < -6$ $3x \leq -6$ or $2x < -9$
       $x > -2$ or $x < -3$ $x \leq -2$ or $x < -9/2$
       Solution set: $(-\infty,-3)\cup(-2,\infty)$ $x \leq -2$
                                                                                         Solution set: $(-\infty,-2]$

     (c) $3x + 7 \leq 1$ or $2x + 1 > -8$
         $3x \leq -6$ or $2x > -9$
         $x \leq -2$ or $x > -9/2$
         Solution set: R (set of real numbers)

Problem Set 1.4   Absolute Values, Square Roots, Squares

3. $|3x+4| < 8$
   $-8 < 3x+4 < 8$
   $-12 < 3x < 4$
   $-4 < x < 4/3$ Solution Set: $(-4, 4/3)$

6. $\left|\dfrac{3x}{5} + 1\right| \leq 4$

   $-4 \leq \dfrac{3x+5}{5} \leq 4$
   $-20 \leq 3x+5 \leq 20$
   $-25 \leq 3x \leq 15$
   $-25/3 \leq x \leq 5$ Solution Set: $[-25/3, 5]$

9. $|4x+2| \geq 10$
   $4x+2 \leq -10$ or $4x+2 \geq 10$
   $4x \leq -12$ or $4x \geq 8$
   $x \leq -3$ or $x \geq 2$ Solution Set: $(-\infty, -3]\cup[2, \infty)$

12. $\left|\dfrac{1}{x} - 3\right| > 6$

    $\dfrac{1}{x} - 3 < -6$ or $\dfrac{1}{x} - 3 > 6$

    $\dfrac{1}{x} + 3 < 0$ or $\dfrac{1}{x} - 9 > 0$

    $\dfrac{1+3x}{x} < 0$ (*)   or $\dfrac{1-9x}{x} > 0$ (**)

12. (continued)

Split points for (*): -1/3, 0   Test points for (*): -1, -1/4, 1
         for (**): 0, 1/9   for (**): -1, 1/10, 1

Auxiliary axis for (*):
$$\underset{\underset{-1/3}{\bullet}\qquad\underset{0}{\bullet}}{\underline{\quad(+)\qquad(0)\qquad(-)\qquad(u)\qquad(+)\quad}}\; x$$

Auxiliary axis for (**):
$$\underset{\underset{0}{\bullet}\qquad\underset{1/9}{\bullet}}{\underline{\quad(-)\qquad(u)\qquad(+)\qquad(0)\qquad(-)\quad}}\; x$$

Solution set: $(-1/3, 0) \cup (0, 1/9)$

15. $4x^2 + x - 2 > 0$

$x = \dfrac{-1 \pm \sqrt{33}}{8} \approx -0.8, 0.6$ if $4x^2 + x - 2 = 0$

Split points: -0.8, 0.6   Test points: -1, 0, 1

Auxiliary axis for $4x^2 + x - 2$:
$$\underset{\underset{-0.8}{\bullet}\qquad\underset{0.6}{\bullet}}{\underline{\quad(+)\qquad(0)\qquad(-)\qquad(0)\qquad(+)\quad}}\; x$$

Solution set: $(-\infty, -0.8) \cup (0.6, \infty)$ [Approximately]

18. $|x+2| < 0.3 \Rightarrow 4|x+2| < 1.2$ [Multiply each side by 4.]
$\Rightarrow |4||x+2| < 1.2$ $[|4| = 4]$
$\Rightarrow |4(x+2)| < 1.2$ $[|a||b| = |ab|]$
$\Rightarrow |4x+8| < 1.2$ [Distributive property]

21. $|x-5| < \delta \Rightarrow 3|x-5| < 3\delta$
$\Rightarrow |3(x-5)| < 3\delta$
$\Rightarrow |3x-15| < 3\delta$
$\Rightarrow |3x-15| < \epsilon$ if $\delta = \epsilon/3$

24. $|x+5| < \delta \Rightarrow 5|x+5| < 5\delta$
$\Rightarrow |5(x+5)| < 5\delta$
$\Rightarrow |5x+25| < 5\delta$
$\Rightarrow |5x+25| < \epsilon$ if $\delta = \epsilon/5$

27. $2|2x-3| < |x+10|$
   $4(2x-3)^2 < (x+10)^2$ [Squaring each side]
   $4(4x^2-12x+9) < x^2+20x+100$
   $16x^2-48x+36 < x^2+20x+100$
   $15x^2-68x-64 < 0$
   $(3x-16)(5x+4) < 0$

   Split points: $-4/5, 16/3$ Test points: $-1, 0, 6$

   Auxiliary axis for $(3x-16)(5x+4)$: $\underset{}{(+)}\quad\underset{-4/5}{(0)}\quad(-)\quad\underset{16/3}{(0)}\quad(+)\qquad x$

   Solution set: $(-4/5, 16/3)$

30. $0 < a < b \Rightarrow (\sqrt{a})^2 < (\sqrt{b})^2$ [Since $(\sqrt{x})^2 = x$ for all $x \geq 0$]
    $\Rightarrow \sqrt{a} < \sqrt{b}$ [By Problem 29 in the $\Leftarrow$ direction]

33. $\left|\dfrac{x-2}{x^2+9}\right| = \dfrac{|x-2|}{|x^2+9|} \leq \dfrac{|x|+|-2|}{x^2+9} \leq \dfrac{|x|+2}{9}$

36. (a) (i) If $x < 0$, then $x^2 > 0$. Therefore, $x^2 > x$.

    (ii) If $x > 1$, then $x^2 > x$. [Multiplying each side by the positive number x]

    (b) If $0 < x < 1$, then $x^2 < x$. [Multiplying each side of $x < 1$ by the positive number x]

Problem Set 1.5 The Rectangular Coordinate System

3. $\sqrt{(2-4)^2 + (4-2)^2} = \sqrt{8} \approx 2.8284$

6. $\sqrt{(5.16-2.71)^2 + (4.33-[-3.42])^2} = \sqrt{(2.45)^2 + (7.75)^2} \approx 8.13$

9. See the diagram following Case II.

   (Case I) $(3,-1)$ and $(3,3)$ are diagonally opposite vertices. The diagonal is vertical, its midpoint is $(3,1)$, and its length is 4. The other pair of vertices is $(1,1)$ and $(5,1)$.

9. (continued)

(Case II) (3,-1) and (3,3) are end points of a side. There are two possible pairs of other vertices: (-1,-1) and (-1,3), and (7,-1) and (7,3).

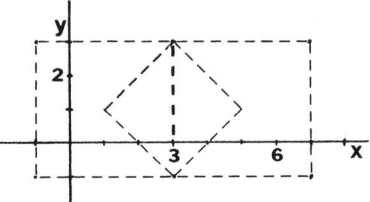

12. The midpoint of AB is $M_{AB} = \left(\frac{1+2}{2}, \frac{3+6}{2}\right) = \left(\frac{3}{2}, \frac{9}{2}\right)$

The midpoint of CD is $M_{CD} = \left(\frac{4+3}{2}, \frac{7+4}{2}\right) = \left(\frac{7}{2}, \frac{11}{2}\right)$

$d(M_{AB}, M_{CD}) = \sqrt{\left(\frac{7}{2} - \frac{3}{2}\right)^2 + \left(\frac{11}{2} - \frac{9}{2}\right)^2} = \sqrt{5} \approx 2.2361$

15. The radius is $d[(2,-1),(5,3)] = \sqrt{(5-2)^2 + (3-[-1])^2} = 5$

Equation of circle: $(x-2)^2 + (y+1)^2 = 25$

18. The radius is 4 [since the absolute value of the y-coordinate of the center is the distance of the center from the x-axis].

Equation of circle: $(x-3)^2 + (y-4)^2 = 16$

21. $x^2 + y^2 + 2x - 10y + 25 = 0$

$x^2 + 2x \qquad + y^2 - 10y \qquad = -25$

$(x^2 + 2x + 1) + (y^2 - 10y + 25) = -25 + 1 + 25$

$(x+1)^2 + (y-5)^2 = 1$

Center: (-1,5) Radius: 1

24. $x^2 + y^2 - 10x + 10y = 0$

$x^2 - 10x \qquad + y^2 + 10y \qquad = 0$

$(x^2 - 10x + 25) + (y^2 + 10y + 25) = 25 + 25$

$(x-5)^2 + (y+5)^2 = 50$

Center: (5,-5) Radius: $\sqrt{50} \approx 7.0711$

27. Length of each side is 4. Therefore, the radius of the inscribed circle is 2. Its center is (4,1), the midpoint of the diagonals. Then the equation is $(x - 4)^2 + (y - 1)^2 = 4$.

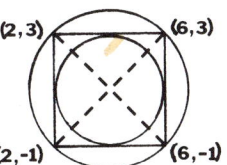

Length of a diameter is $\sqrt{16 + 16} = \sqrt{32}$. Therefore, the radius of the circumscribed circle is $\sqrt{32}/2 = \sqrt{8}$. Its center is also (4,1). The equation is $(x - 4)^2 + (y - 1)^2 = 8$.

30. Mary's route ----------
Distance to run = 6 miles

Distance to swim = $\sqrt{(4)^2 + (1/2)^2} = \sqrt{16.25}$

The formula to use is Time = $\dfrac{\text{Distance}}{\text{Rate}}$

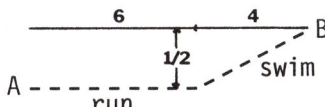

Total time = Time running + Time swimming

$= \dfrac{6 \text{ mi}}{8 \text{ mi/hr}} + \dfrac{\sqrt{16.25} \text{ mi}}{3 \text{ mi/hr}} \approx 2.0937$ hr [or 2 hr, 5 min, 37 sec]

33. $x^2 + y^2 - 4x - 2y - 11 = 0$ $x^2 + y^2 + 20x - 12y + 72 = 0$
    $(x - 2)^2 + (y - 1)^2 = 16$ $(x + 10)^2 + (y - 6)^2 = 64$
    Center: (2,1)   Radius: 4   Center: (-10,6)   Radius: 8

The circles do not intersect since the distance between the centers is $\sqrt{(-10 - 2)^2 + (6 - 1)^2} = 13$ which is greater than 12, the sum of the radii.

Problem Set 1.6 The Straight Line

3. $\dfrac{0 - 2}{3 - (-4)} = -\dfrac{2}{7}$

6. $\dfrac{6 - 0}{0 - (-6)} = 1$

9. $y - 3 = 4(x - 2)$ [point-slope form]
   $y - 3 = 4x - 8$
   $4x - y - 5 = 0$ [general form]

12. $y = 5$ [The line is horizontal.]
    $y - 5 = 0$ [general form]

15. $x = 2$ [The slope is undefined so the line is vertical.]
    $x - 2 = 0$ [general form]

18. $2y = 5x + 2$
    $y = (5/2)x + 1$ [slope-intercept form]
    Slope: $5/2$ y-intercept: 1

21. $y = 2x + 5$ has a slope of 2.

   (a) Slope is 2 [Parallel lines have the same slope.]
       Equations: $y - (-3) = 2(x - 3)$ [point-slope form]
                  $y + 3 = 2x - 6$
                  $y = 2x - 9$ [slope-intercept form]

   (b) Slope is $-1/2$ [Perpendicular lines have slopes that are negative reciprocals of each other.]
       Equations: $y - (-3) = (-1/2)(x - 3)$ [point-slope form]
                  $y + 3 = (-1/2)x + 3/2$
                  $y = (-1/2)x - 3/2$ [slope-intercept form]

   $2x + 3y = 6$ [or $y = (-2/3)x + 2$] has a slope of $-2/3$

   (c) Slope is $-2/3$ [Parallel lines have the same slope.]
       Equations: $y - (-3) = (-2/3)(x - 3)$ [point-slope form]
                  $y + 3 = (-2/3)x + 2$
                  $y = (-2/3)x - 1$ [slope-intercept form]

   (d) Slope is $3/2$ [Perpendicular lines have slopes that are negative reciprocals of each other.]
       Equations: $y - (-3) = (3/2)(x - 3)$ [point-slope form]
                  $y + 3 = (3/2)x - (9/2)$
                  $y = (3/2)x - (15/2)$ [slope-intercept form]

   Slope of the line through $(-1,2)$ and $(3,-1)$ is $\frac{-1 - 2}{3-(-1)} = -3/4$

   (e) Slope is $-3/4$ [Parallel lines have the same slope.]
       Equations: $y - (-3) = (-3/4)(x - 3)$ [point-slope form]
                  $y + 3 = (-3/4)x + (9/4)$
                  $y = (-3/4)x - (3/4)$ [slope-intercept form]

   $x = 8$ is a vertical line and has no slope.

   (f) The line is vertical. An equation of the line is $x = 3$.
   (g) The line is horizontal. An equation of the line is $y = -3$.

24. $kx - 3y = 10$ [or $y = (k/3)x - (10/3)$] has a slope of $k/3$.

$y = 2x + 4$ has a slope of 2.

(a) $k/3 = 2$ [Parallel lines have the same slope.]
$k = 6$

(b) $k/3 = -1/2$ [Perpendicular lines have slopes that are negative reciprocals of each other.]
$k = -3/2$

$2x + 3y = 6$ [or $y = (-2/3)x + 2$] has a slope of $-2/3$.

(c) $k/3 = 3/2$ [Perpendicular lines have slopes that are negative reciprocals of each other.]
$k = 9/2$

27. $\begin{bmatrix} 2x + 3y = 4 \\ -3x + y = 5 \end{bmatrix}$ <=> $\begin{bmatrix} 2x + 3y = 4 \\ 9x - 3y = -15 \end{bmatrix}$ => $[11x = -11]$ <=> $[x = -1]$

Set x equal to -1 in $-3x+y=5$, and obtain that y equals 2.
Point of intersection: (-1,2)

$2x + 3y = 4$ [or $y = (-2/3)x + (4/3)$] has slope of $-2/3$.
Therefore, the slope of a line perpendicular to $2x + 3y = 4$ is $3/2$.

Equations: $y - 2 = (3/2)(x - [-1])$ [point-slope form]
$y - 2 = (3/2)x + (3/2)$
$y = (3/2)x + (7/2)$ [slope-intercept form]

30. $\begin{bmatrix} 5x - 2y = 5 \\ 2x + 3y = 6 \end{bmatrix}$ <=> $\begin{bmatrix} 15x - 6y = 15 \\ 4x + 6y = 12 \end{bmatrix}$ => $[19x = 27]$ <=> $[x = 27/19]$

Set x equal to 27/19 in $2x+3y=6$, and obtain that y equals 20/19.
Point of intersection: (27/19, 20/19)

$5x - 2y = 5$ [or $y = (5/2)x - (5/2)$] has slope of $5/2$.
Therefore, the slope of a line perpendicular to $5x - 2y = 5$ is $-2/5$.

Equations: $y - 20/19 = (-2/5)(x - [27/19])$ [point-slope form]
$y - 20/19 = (-2/5)x + (54/95)$
$y = (-2/5)x + (154/95)$ [slope-intercept form]

33. Express $5y = 12x + 1$ as $12x - 5y + 1 = 0$. [general form]

Then the distance is $\dfrac{|(12)(-2) + (-5)(-1) + (1)|}{\sqrt{(12)^2 + (-5)^2}} = \dfrac{18}{13} \approx 1.3846$

Problem Set 1.6

36. Let y = 1 and solve for x. Thus, (-1,1) is one point on 5x + 12y = 7. The distance between the two lines is then the same as the distance from the point (-1,1) to the line 5x + 12y = 2.

Express 5x + 12y = 2 as 5x + 12y - 2 = 0. [general form]

Then the distance is $\dfrac{|(5)(-1) + (12)(1) + (-2)|}{\sqrt{(5)^2 + (12)^2}} = \dfrac{5}{13} \approx 0.3846$

39. Two points on the line are (0, 700 000) and (10, 820 000).

The slope is $\dfrac{820{,}000 - 700{,}000}{10 - 0} = 12{,}000$; N-intercept is 700,000.

Therefore, the formula is N = 12,000n + 700,000.
For the year 2000, n = 40; so N = 12,000(40) + 700,000 = 1,180,000.
That is, the prediction is that 1,180,000 cases of eggs will be produced in Matlin County in the year 2000.

42. The slope is 0.75. [The equation is in slope-intercept form.] It indicates that the cost increases by 0.75 dollars (75 cents) for each unit increase in production. [Note: it does not indicate that the cost per item is 75 cents, due to the 200 dollar term which indicates an initial cost of 200 dollars.]

45. First, note that for each real number k, 2x - y + 4 + k(x + 3y - 6) = 0 is equivalent to (2 + k)x + (-1 + 3k)y + (4 - 6k) = 0, which is an equation of a line since there is no value of k that will make both (2 + k) and (-1 + 3k) equal to zero. [See Problem 43.]

Second, note that the lines intersect since their slopes are different.

Finally, note that (x_0, y_0) is the point of intersection
=> $2x_0 - y_0 + 4 = 0$ and $x_0 + 3y_0 - 6 = 0$
=> $(2x_0 - y_0 + 4) + k(x_0 + 3y_0 - 6) = 0 + k \cdot 0 = 0$
=> (x_0, y_0) is on the line 2x-y+4 + k(x+3y-6) = 0

Problem Set 1.7 Graphs of Equations

3. $3x^2 + 4y = 0$
Symmetry: With respect to the y-axis since $3(-x)^2 + 4y = 0$ is equivalent to $3x^2 + 4y = 0$.
x-intercept: 0   y-intercept: 0

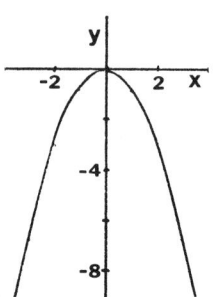

3. (continued)

   Other Points:

x	y
1	-0.75
2	-3
3	-6.75

6. The graph of $(x - 2)^2 + y^2 = 4$ is the circle with center $(2,0)$ and radius 2. [See Section 1.5.]

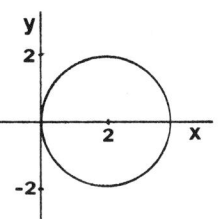

9. $y = x^3 - 3x = x(x^2 - 3) = x(x+\sqrt{3})(x-\sqrt{3})$

   Symmetry: With respect to origin since $(-y) = (-x)^3 - 3(-x)$ is equivalent to $y = x^3 - 3x$.

   x-intercepts: $-\sqrt{3}, 0, \sqrt{3}$ y-intercept: 0

   Other Points:

x	y
1	-2
2	2
3	18

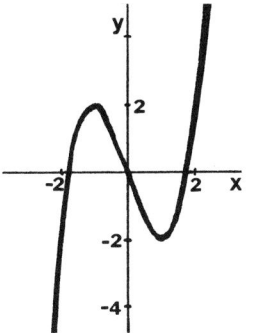

12. $y = \dfrac{x}{x^2 + 1}$

    Symmetry: With respect to origin since $(-y) = \dfrac{-x}{(-x)^2 + 1}$ is equivalent to $y = \dfrac{x}{x^2 + 1}$.

    x-intercept: 0   y-intercept: 0

    Other Points:

x	y
1	0.5
2	0.4
3	0.3

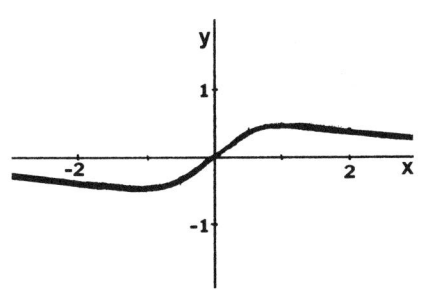

Problem Set 1.7

15. $y = (x - 2)(x + 1)(x + 3)$
Symmetry: None of those discussed
x-intercept: $-3, -1, 2$
y-intercept: -6

Other Points:

x	y
-4	-18
-2	4
1	-8
3	24

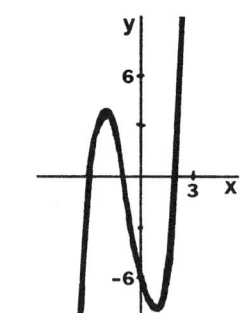

18. $|x| + |y| = 4$
Symmetry: With respect to all three since $|-a| = |a|$ so the required equivalences all exist.
x-intercepts: $-4, 4$
y-intercepts: $-4, 4$

For $x > 0$ and $y > 0$ (points in the first quadrant), the equation can be expressed as $x + y = 4$, whose graph is the first quadrant part of the line with slope -1 and y-intercept 4. The rest of the graph can then be obtained from the symmetry.

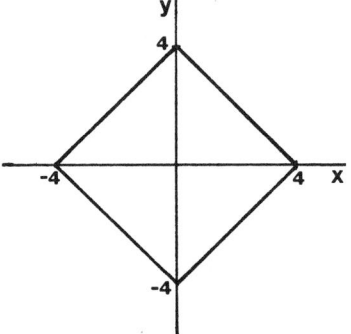

21. $y = -2x + 1$ [a line with slope -2 and y-intercept 1]
$y = -x^2 - x + 3$

$-2x+1 = -x^2-x+3$ at points of intersection.
$x^2 - x - 2 = 0$
$(x - 2)(x + 1) = 0$
$x = 2$ or $x = -1$
If $x = 2$, $y = -3$; if $x = -1$, $y = 3$.

Intersection points: $(2,-3), (-1,3)$

Other Points:

x	y
-3	-3
-2	1
-1	3
1	1
2	-3

For $y = -x^2 - x + 3$
Symmetry: None of those discussed
x-intercepts: Let $y = 0$

Solve $x^2 + x - 3 = 0$

$$x = \frac{-(1) \pm \sqrt{(1)^2 - 4(1)(-3)}}{2(1)} = \frac{-1 \pm \sqrt{13}}{2}$$

$x \approx -2.30, 1.30$
y-intercept: 3

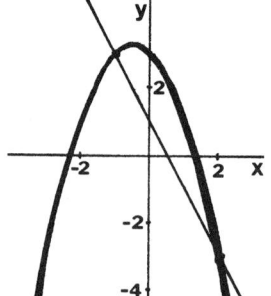

24. $y = 2.1x - 6.4$ [a line with slope 2.1 and y-intercept -6.4]
$y = -1.2x^2 + 4.3$

Therefore, $2.1x - 6.4 = -1.2x^2 + 4.3$ at points of intersection.
$$1.2x^2 + 2.1x - 10.7 = 0$$
$$x = \frac{-(2.1) \pm \sqrt{(2.1)^2 - 4(1.2)(-10.7)}}{2(1.2)}$$

Therefore, $x = \frac{-2.1 \pm \sqrt{55.77}}{2.4}$

$$x \approx 2.2366, -3.9866$$

The corresponding approximate values of y are -1.7031 and -14.7719.

For $y = -1.2x^2 + 4.3$
Symmetry: With respect to y-axis
x-intercepts: Let $y = 0$
$$1.2x^2 - 4.3 = 0$$
$$x^2 = \frac{4.3}{1.2}$$
$$x \approx -1.89, 1.89$$

y-intercept: 4.3

Other Points:

x	y
1	3.1
2	-0.5
3	-6.5
4	-14.9

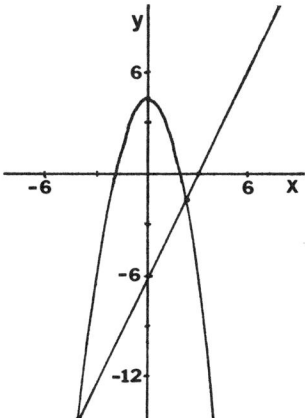

27. $y = 3x^4 - 2x + 1$
$x = -1 \Rightarrow y = 3 + 2 + 1 = 6$ so one point is $(-1, 6)$
$x = 1 \Rightarrow y = 3 - 2 + 1 = 2$ so the other point is $(1, 2)$

The distance between $(-1, 6)$ and $(1, 2)$ is $\sqrt{4 + 16} \approx 4.4721$

30. $y = 2^x - 2^{-x}$
Symmetry: With respect to origin
since $(-y) = 2^{(-x)} - 2^{-(-x)}$
is equivalent to
$y = 2^x - 2^{-x}$.
x-intercept: 0   y-intercept: 0
Other Points:

x	y
1	1.5
2	3.75
3	7.88

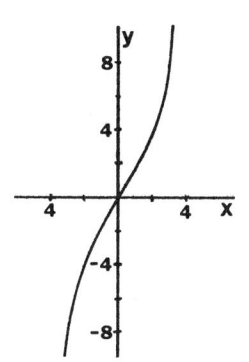

Problem Set 1.7

Problem Set 1.8 Chapter Review Problems

True-False Quiz

Note: In True-False quizzes, if a statement involving variables is true for some values of the variables and false for others, we classify the statement as **False**, meaning that it is not always, or not generally, true.

A counterexample is an example which shows that some statement involving variables is false for some values of the variables. Hence, a simple way to show that a statement is **False** is to provide one particular counterexample. "c^{ex}" will be used as an abbreviation for "Counterexample."

3. False. c^{ex}: $\sqrt{2}$ is irrational, but $\sqrt{2} - \sqrt{2} = 0$, which is rational.

6. True. $(a^m)^n = a^{mn} = a^{nm} = (a^n)^m$ [assuming m and n are integers].

9. True. Assume $x \neq 0$. Then $\frac{|x|}{2}$ is positive and $\frac{|x|}{2} < |x|$.

   This contradicts that $|x|$ is less than every positive number, so our assumption that x is not equal to zero must have been false.

12. True. For example, [1,3] and [3,7] have only the number 3 in common.

15. False. c^{ex}: $|-2| < |-7|$ but $-2 > -7$.

18. True. For example, 2 is the only solution for $|x-2| \leq 0$.

21. True. $ab > 0 \Rightarrow [a > 0 \text{ and } b > 0]$ or $[a < 0 \text{ and } b < 0]$
    $\Rightarrow (a,b)$ is in the 1st or in the 3rd quadrant.

24. True. The points are on the horizontal line y=a. Then, the distance between them is $|(a+b) - (a-b)| = |2b|$.

27. False. If each line has a positive slope, the product of their slopes is positive, not -1, so they are not perpendicular.

30. True. The equation can be written as $(3+2m)x + (-2+6m)y + (4-2m) = 0$ with $(3+2m)$ and $(-2+6m)$ not both zero for any real number m. [See Problem 43 of Problem Set 1.6.]

Miscellaneous Problems

3. Let a/b and c/d be rational numbers. That is, a,b,c,d are integers and bd ≠ 0. The average of a/b and c/d is

$$\frac{\frac{a}{b} + \frac{c}{d}}{2} = \frac{\left(\frac{a}{b} + \frac{c}{d}\right) \cdot bd}{2 \cdot bd} = \frac{ad + bc}{2bd} \quad \text{which is a rational number since}$$

(ad + bc) and (2bd) are integers, and 2bd ≠ 0.

6. 545.38680037

9. $2x^2 + 5x - 3 < 0$
   $(2x - 1)(x + 3) < 0$
   Split points: $-3, 1/2$ Test points: $-4, 0, 1$

   Auxiliary axis for (2x-1)(x+3): (+) (0) (-) (0) (+)
                                    ─────●────────●───── x
                                        -3   1/2

   Solution set: $(-3, 1/2)$ Graph: ──(─────)── x
                                                -3   1/2

12. $|3x-4| < 6$
    $-6 < 3x-4 < 6$
    $-2 < 3x < 10$
    $-2/3 < x < 10/3$

    Solution set: $(-2/3, 10/3)$ Graph: ──(─────────)── x
                                                -2/3   10/3

15. $\left|\dfrac{2x^2 + 3x + 2}{x^2 + 2}\right| = \dfrac{|2x^2 + 3x + 2|}{|x^2 + 2|} \leq \dfrac{2x^2 + 3|x| + 2}{x^2 + 2} \leq \dfrac{2(4) + 3(2) + 2}{2} = 8$

18. The center of the circle (midpoint of AB) is $M = \left(\dfrac{2+10}{2}, \dfrac{0+4}{2}\right) = (6,2)$.
    The radius of the circle is $d(A,M) = \sqrt{(6-2)^2 + (2-0)^2} = \sqrt{20}$.
    Therefore, the equation of the circle is $(x-6)^2 + (y-2)^2 = 20$.

Problem Set 1.8

21. (a) Slope: $\frac{3-1}{7-(-2)} = \frac{2}{9}$

   Equations: $y - 1 = (2/9)(x + 2)$ [point-slope form]
   $y - 1 = (2/9)x + (4/9)$
   $y = (2/9)x + (13/9)$ [slope-intercept form]

   (b) The slope of $3x - 2y = 5$ [or $y = (3/2)x - (5/2)$] is $3/2$ so the slope of a line parallel to it is also $3/2$.
   Equations: $y - 1 = (3/2)(x + 2)$ [point-slope form]
   $y - 1 = (3/2)x + 3$
   $y = (3/2)x + 4$ [slope-intercept form]

   (c) The slope of $3x + 4y = 9$ [or $y = (-3/4)x - 9/4$] is $-3/4$ so the slope of a line perpendicular to it is $4/3$.
   Equations: $y - 1 = (4/3)(x + 2)$ [point-slope form]
   $y - 1 = (4/3)x + (8/3)$
   $y = (4/3)x + (11/3)$ [slope-intercept form]

   (d) $y = 4$ is a horizontal line so each line perpendicular to it is a vertical line. The vertical line through $(-2,1)$ is $x = -2$.

   (e) Slope: $\frac{1-3}{-2-0} = 1$ [since $(0,3)$ and $(-2,1)$ are on the line]

   Then an equation of the line is $y = x + 3$.   [slope-intercept form]

24. $x^2 - 2x + y^2 = 3$

   $(x^2 - 2x + 1) + y^2 = 3 + 1$

   $(x - 1)^2 + y^2 = 4$ (a circle)

   Center: $(1,0)$
   Radius: 2

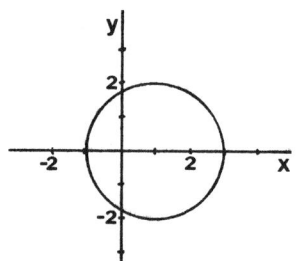

27. $y = x^2 - 2x + 4$
   $y - x = 4$ [or $y = x + 4$]

   Therefore, $x^2 - 2x + 4 = x + 4$ at points of intersection.
   $x^2 - 3x = 0$
   $x(x - 3) = 0$
   $x = 0$ or $x = 3$

   The corresponding values of y are $0 + 4 = 4$, and $3 + 4 = 7$, so the points of intersection are $(0,4)$ and $(3,7)$.

CHAPTER 2 FUNCTIONS AND LIMITS

Problem Set 2.1 Functions and Their Graphs

3. (a) $G(0) = \dfrac{1}{(0) - 1} = -1$ (b) $G(0.999) = \dfrac{1}{(0.999) - 1} = -1000$

   (c) $G(1.01) = \dfrac{1}{(1.01) - 1} = 100$ (d) $G(y^2) = \dfrac{1}{(y^2) - 1} = \dfrac{1}{y^2 - 1}$

   (e) $G(-x) = \dfrac{1}{(-x) - 1} = \dfrac{-1}{x + 1}$ (f) $G(1/x^2) = \dfrac{1}{(1/x^2) - 1} = \dfrac{x^2}{1 - x^2}$

6. (a) $g(-1.71) \approx 23.3323$ (b) $g(3.01) \approx 121.1186$ (c) $g(\sqrt{3}) \approx -8.1017$

9. (a) $x^2 + y^2 = 4$

   $y^2 = 4 - x^2$

   $y = \sqrt{4-x^2}$ or $y = -\sqrt{4-x^2}$

   Therefore, $x^2 + y^2 = 4$ does not determine a function with formula $y = f(x)$.

   (b) $xy + y + 3x = 4$
   $y(x + 1) = 4 - 3x$

   $y = \dfrac{4 - 3x}{x + 1}$

   Therefore, $xy + y + 3x = 4$ does determine a function with formula $y = f(x) = \dfrac{4-3x}{x+1}$.

   (c) $x = \sqrt{3y+1}$

   $x^2 = 3y+1$, $x \geq 0$

   $y = \dfrac{x^2-1}{3}$, $x \geq 0$

   Therefore, $x = 3y+1$ does determine a function with formula $y = f(x) = \dfrac{x^2-1}{3}$ and with domain $[0,\infty)$.

   (d) $3x = \dfrac{y}{y+1}$

   $3x(y+1) = y$
   $3xy+3x = y$
   $3xy-y = -3x$
   $(3x-1)y = -3x$

   $y = \dfrac{-3x}{3x-1}$

   Thus, $3x = \dfrac{y}{y+1}$ does determine a function with formula $y = f(x) = \dfrac{-3x}{3x-1}$.

Problem Set 2.1

12. $\dfrac{F(a+h) - F(a)}{h} = \dfrac{4(a+h)^3 - 4(a)^3}{h} = \dfrac{4(a^3 + 3a^2h + 3ah^2 + h^3) - 4a^3}{h}$

$= \dfrac{4a^3 + 12a^2h + 12ah^2 + 4h^3 - 4a^3}{h} = \dfrac{h(12a^2 + 12ah + 4h^2)}{h}$

$= 12a^2 + 12ah + 4h^2 \quad [h \neq 0]$

15. (a) The radicand must be nonnegative.
$(2z+3 \geq 0)$ iff $(2z \geq -3)$ iff $(x \geq -1.5)$, so the natural domain is $[-1.5, \infty)$.

(b) The denominator must be nonzero.
$(4v-1 \neq 0)$ iff $(4v \neq 1)$ iff $(v \neq 1/4)$, so the natural domain is $\{v : v \neq 1/4\}$.

(c) The radicand must be nonnegative.
$(x^2-9 \geq 0)$ iff $(x^2 \geq 9)$ iff $(x \leq -3$ or $x \geq 3)$, so the natural domain is $(-\infty, -3] \cup [3, \infty)$.

(d) The radicand must be nonnegative.
$(625-y^4 \geq 0)$ iff $(625 \geq y^4)$ iff $(25 \geq y^2)$ iff $(-5 \leq y \leq 5)$, so the natural domain is $[-5, 5]$.

18. $f(x) = 3x$

Then $f(-x) = 3(-x)$
$= -3x$
$= -f(x)$

Therefore, f is an odd function and the graph of f is symmetric with respect to the origin.

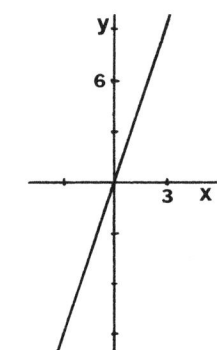

21. $g(x) = 3x^2 + 2x - 1$

Since $g(-1) = 0$ but $g(1) = 4$, for example, g is neither an even nor an odd function.

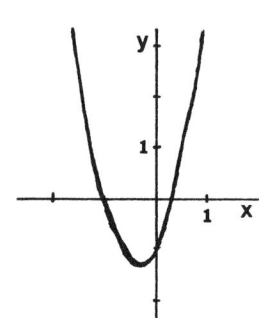

24. $\phi(z) = \frac{2z+1}{z-1}$

Since $\phi(-2) = 1$ but $\phi(2) = 5$, for example, ϕ is neither an even function nor an odd function.

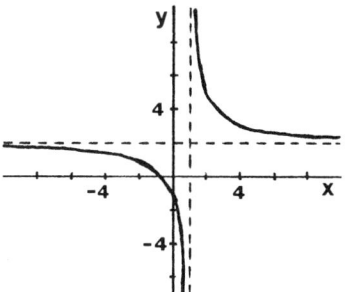

27. $f(x) = |2x|$

Then $f(-x) = |2(-x)|$
$= |2x|$
$= f(x)$.

Therefore, f is an even function and the graph of f is symmetric with respect to the y-axis. Moreover, for $x \geq 0$ the graph of $f(x) = 2x$ is part of the line through the origin with slope 2.

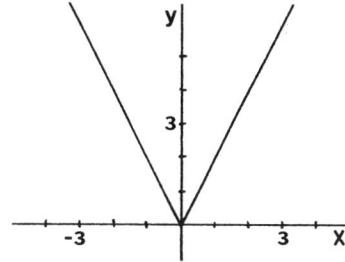

30. $G(x) = [\![2x-1]\!]$

Since $G(-2) = -5$, but $G(2) = 3$, for example, G is neither an even nor an odd function.

Note that when x is a multiple of 1/2, 2x-1 is an integer. Therefore, the graph of G and the graph of $y = 2x-1$ coincide at those points where x is a multiple of 1/2. And between those points, G(x) is a constant (the graph of G is horizontal.)

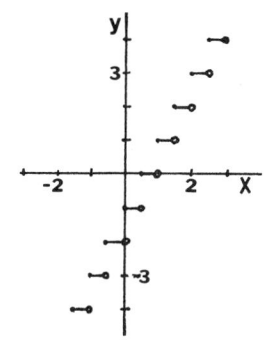

33. T(x) = (direct cost for x refrigerators) + (overhead cost).
= 151x + 2200, and $D_T = \{0,1,2,3,\ldots,100\}$.

$u(x) = \frac{(\text{Total Cost})}{(\text{No. units})} = \frac{T(x)}{x} = \frac{151x + 2200}{x}$, and $D_u = \{1,2,3,\ldots,100\}$.

36. The length of each side is p/3. Let h be the altitude. By the Pythagorean theorem,

$h = \sqrt{\left(\frac{p}{3}\right)^2 - \left(\frac{p}{6}\right)^2} = \sqrt{\frac{p^2}{9} - \frac{p^2}{36}} = \sqrt{\frac{3p^2}{36}} = \frac{\sqrt{3}p}{6}$

$A = \frac{1}{2}(\text{base})(\text{altitude}) = \frac{1}{2}\left(\frac{p}{3}\right)\left(\frac{\sqrt{3}p}{6}\right) = \frac{\sqrt{3}p^2}{36}$

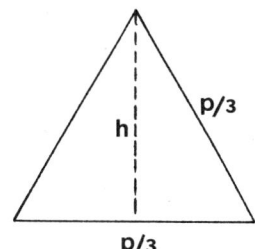

Problem Set 2.1

Problem Set 2.2   Operations on Functions

3. $f(x) = x^3+2$, $D_f = R$; and $g(x) = \frac{2}{x-1}$, $D_g = \{x : x \neq 1\}$.

(a) $(f+g)(x) = f(x) + g(x) = (x^3 + 2) + \frac{2}{x-1} = \frac{(x^3+2)(x-1) + 2}{x-1}$

$= \frac{x^4 - x^3 + 2x}{x-1}$, $D_{f+g} = D_f \cap D_g = \{x : x \neq 1\}$.

(b) $\left(\frac{g}{f}\right)(x) = \frac{g(x)}{f(x)} = \frac{\frac{2}{x-1}}{x^3+2} = \frac{2}{(x-1)(x^3+2)}$.

$D_{f/g} = \{x : x \neq 1, -\sqrt[3]{2}\}$.

(c) $(f \circ g)(x) = f[g(x)] = f\left(\frac{2}{x-1}\right) = \left(\frac{2}{x-1}\right)^3 + 2 = \frac{2x^3 - 6x^2 + 6x + 6}{(x-1)^3}$.

$D_{f \circ g} = \{x : x \neq 1\}$.

(d) $(g \circ f)(x) = g[f(x)] = g[x^3+2] = \frac{2}{(x^3+2) - 1} = \frac{2}{x^3+1}$.

$D_{g \circ f} = \{x : x \neq -1\}$.

6. $g(x) = x^2+1$, $D_g = R$.

$g^3(x) = [g(x)]^3 = [x^2+1]^3 = x^6 + 3x^4 + 3x^2 + 1$.

$(g \circ g \circ g)(x) = g \circ g[g(x)] = g \circ g(x^2+1) = g[g(x^2+1)]$

$= g[(x^2+1)^2 + 1] = g[x^4 + 2x^2 + 2]$

$= (x^4 + 2x^2 + 2)^2 + 1$

$= x^8 + 4x^6 + 8x^4 + 8x^2 + 5$.

9. Sequence of keys: 3.46 $\boxed{1/x}$ $\boxed{x^2}$ $\boxed{+}$ 4 $\boxed{x}$ 3.46 $\boxed{1/x}$ $\boxed{=}$ $\boxed{\sqrt{x}}$

Answer: Approximately 1.1134.

12. (a) Two valid answers are given. There are others.

   (i) Let $g(x) = x^2 + 2x + 1$; let $f(x) = \dfrac{2}{x^3}$.

   (ii) Let $g(x) = (x^2 + x + 1)^3$; let $f(x) = \dfrac{2}{x}$.

   (b) Let $g(x) = x^3 + 3x$; let $f(x) = \log x$. (There are other valid answers.)

15. Translate the graph of g two units right and three units down.

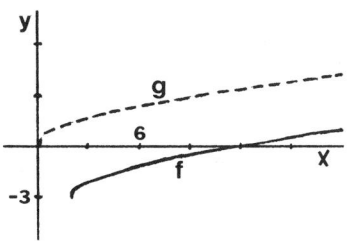

18. Graph $f(x) = x^3$. Then translate the graph of f one unit left and three units down.

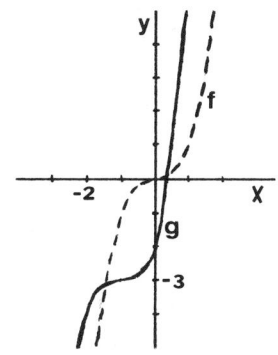

21. $F(t) = \begin{cases} \dfrac{t}{t} = 1, & \text{if } t > 0 \\ \dfrac{-t}{t} = -1, & \text{if } t < 0 \end{cases}$

   Note that 0 is not in the domain of F.

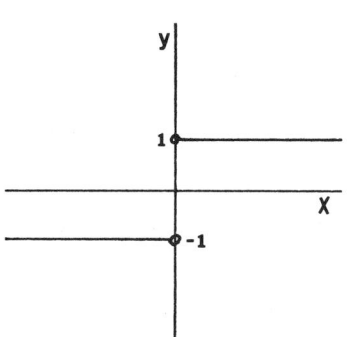

Problem Set 2.2

24. (a) Let $G(x) = F(x) - F(-x)$
Then $G(-x) = F(-x) - F(-[-x])$
$= F(-x) - F(x) = -[F(x) - F(-x)] = -G(x).$
Therefore, G is an odd function.

(b) Let $H(x) = F(x) + F(-x)$
Then $H(-x) = F(-x) + F(-[-x])$
$= F(-x) + F(x) = F(x) + F(-x) = G(x).$
Therefore, G is an even function.

(c) $G(x) + H(x) = [F(x) - F(-x)] + [F(x) + F(-x)] = 2F(x)$
Therefore, $F(x) = \dfrac{G(x) + H(x)}{2}$

Hence, F is the sum of (1/2)G [an odd function] and (1/2)H [an even function].

27. (a) $P = \sqrt{29 - 3D + D^2} = \sqrt{29 - 3(2+\sqrt{t}) + (2+\sqrt{t})^2}$
$= \sqrt{29 - 6 - 3\sqrt{t} + 4 + 4\sqrt{t} + t} = \sqrt{27 + \sqrt{t} + t}$

(b) $P(15) = \sqrt{27 + \sqrt{15} + 15} = \sqrt{42 + \sqrt{15}} \approx 6.7730$

30. $D(t) = 100\sqrt{25t^2 - 18t + 9}$ if $t > 1$ [from Problem 29]
At 2:30 p.m., $t = 2.5$

$D(2.5) = 100\sqrt{25(2.5)^2 - 18(2.5) + 9} \approx 1096.5856$ [miles]

Problem Set 2.3 The Trigonometric Functions

3. (a) $33.3° = 33.3° \times \dfrac{\pi \text{ rad}}{180°} = \dfrac{33.3\pi}{180}$ rad ≈ 0.5812 rad

(b) $471.5° = 471.5° \times \dfrac{\pi \text{ rad}}{180°} = \dfrac{471.5\pi}{180}$ rad ≈ 8.2292 rad

(c) $-391.4° = -391.4° \times \dfrac{\pi \text{ rad}}{180°} = \dfrac{-391.4\pi}{180}$ rad ≈ -6.8312 rad

(d) $14.9° = 14.9° \times \dfrac{\pi \text{ rad}}{180°} = \dfrac{14.9\pi}{180}$ rad ≈ 0.2601 rad

(e) $4.02° = 4.02° \times \dfrac{\pi \text{ rad}}{180°} = \dfrac{4.02\pi}{180}$ rad ≈ 0.0702 rad

(f) $-1.52° = -1.52° \times \dfrac{\pi \text{ rad}}{180°} = \dfrac{-1.52\pi}{180}$ rad ≈ -0.0265 rad

6. (a) 0.9425 (b) 0.8080 (c) 48.08
   (d) $\sin(-1.23) = -\sin(1.23) = -0.9425$ [sin is an odd function].
   (e) $\cos(-0.63) = \cos(0.63) = 0.8080$ [cos is an even function].
   (f) $\tan(-1.55) = -\tan(1.55) = -48.08$ [tan is an odd function].

9. The following two triangles along with the triangle definitions of the trigonometric functions are helpful for evaluations involving $\pi/6$, $\pi/4$, and $\pi/3$.

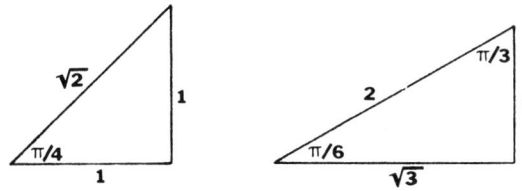

   (a) $\tan(\pi/6) = \frac{1}{\sqrt{3}}$ $\left[\frac{OPP}{ADJ}\right]$

   (b) $\sec \pi = \frac{1}{\cos \pi} = \frac{1}{-1} = -1$

   (c) $\sec(3\pi/4) = -\sec(\pi/4) = -\frac{\sqrt{2}}{1} = -\sqrt{2}$ $\left[\frac{HYP}{ADJ}\right]$

   (d) $\csc(\pi/2) = \frac{1}{\sin(\pi/2)} = \frac{1}{1} = 1$

   (e) $\cot(\pi/4) = \frac{1}{1} = 1$ $\left[\frac{ADJ}{OPP}\right]$

   (f) $\tan(-\pi/4) = -\tan(\pi/4) = -\frac{1}{1} = -1$ $\left[\frac{OPP}{ADJ}\right]$

12. (a) $\frac{\sin u}{\csc u} + \frac{\cos u}{\sec u} = \frac{\sin u}{1/(\sin u)} + \frac{\cos u}{1/(\cos u)} = \sin^2 u + \cos^2 u = 1$
    [if $\sin u \neq 0$ and $\cos u \neq 0$; i.e., u is not a multiple of $\pi/2$]

   (b) $(1-\cos^2 x)(1+\cot^2 x) = (\sin^2 x)(\csc^2 x) = \sin^2 x \cdot \frac{1}{\sin^2 x} = 1$
    [if $\sin x \neq 0$; i.e., x is not a multiple of π]

   (c) $\sin t (\csc t - \sin t) = \sin t \csc t - \sin^2 t = 1 - \sin^2 t = \cos^2 t$
    [if $\sin t \neq 0$; i.e., t is not a multiple of π]

   (d) $\frac{1 - \csc^2 t}{\csc^2 t} = \frac{1}{\csc^2 t} - 1 = \sin^2 t - 1 = -(1 - \sin^2 t) = -\cos^2 t = \frac{-1}{\sec^2 t}$
    [if $\sin t \neq 0$; i.e., t is not a multiple of π]

   (e) $\frac{1}{\sin t \cos t} - \frac{\cos t}{\sin t} = \frac{1 - \cos^2 t}{\sin t \cos t} = \frac{\sin^2 t}{\sin t \cos t} = \frac{\sin t}{\cos t} = \tan t$
    [if $\sin t \neq 0$ and $\cos t \neq 0$; i.e., t is not a multiple of $\pi/2$]

Problem Set 2.3

15. (a) 5.97 divided by $\pi/2$ is about 3.8. Thus, an angle of 5.97 radians can be obtained by rotating the positive x-axis counterclockwise through about 3.8 quadrants. Therefore, $P(x,y)$ is in the 4th quadrant, so $\cos(5.97)$ is positive.

(b) 9.34 divided by $\pi/2$ is about 5.9. Thus, an angle of 9.34 radians can be obtained by rotating the positive x-axis counterclockwise through about 5.9 quadrants. Therefore, $P(x,y)$ is in the 2nd quadrant, so $\cos(9.34)$ is negative.

(c) -16.1 divided by $\pi/2$ is about -10.2. Thus, an angle of -16.1 radians can be obtained by rotating the positive x-axis clockwise through about 10.2 quadrants. Therefore, $P(x,y)$ is in the 2nd quadrant, so $\cos(-16.1)$ is negative.

18. (a) $\sin(x-y) = \sin(x+[-y])$
$= \sin x \cos(-y) + \cos x \sin(-y)$
$= \sin x \cos y + \cos x [-\sin y]$
$= \sin x \cos y - \cos x \sin y.$

(b) $\cos(x-y) = \cos(x+[-y])$
$= \cos x \cos(-y) - \sin x \sin(-y)$
$= \cos x \cos y - \sin x [-\sin y]$
$= \cos x \cos y + \sin x \sin y.$

(c) $\tan(x-y) = \tan(x+[-y])$
$= \dfrac{\tan x + \tan(-y)}{1 - \tan x \tan(-y)} = \dfrac{\tan x - \tan y}{1 - \tan x [-\tan y]} = \dfrac{\tan x - \tan y}{1 + \tan x \tan y}.$

21. Formula: $s = rt$. [s is length of arc; r is radius of circle; t is central angle measured in radians.]

(a) $s = (2.5)(6) = 15$ cm.

(b) $225° = \dfrac{5\pi}{4}$ radians. Therefore, $s = (2.5)\dfrac{5\pi}{4} \approx 9.8175$ cm.

24. The formula for the circumference of wheel is $c = 2\pi r$.

Large wheel: The belt travels along the large wheel $2\pi(8) = 16\pi$ inches per revolution. The number of revolutions per second is 21. Therefore, it travels $16\pi(21) = 336\pi$ inches per second along the large wheel.

Small wheel: Let x denote the number of revolutions per second. The belt travels along the small wheel $2\pi(6) = 12\pi$ inches per revolution. Therefore, it travels $12\pi x$ inches per second along the small wheel.

Therefore, $12\pi x = 336\pi$ (The belt travels at the same rate along the two wheels.)

Therefore, $x = 28$ revolutions per second.

27. If m is the slope and α is the angle of inclination, $m = \tan\alpha$. [From Problem 26]

   (a) $y = \sqrt{3}x - 7$ => $m = \sqrt{3}$ => $\tan\alpha = \sqrt{3}$ => $\alpha = \pi/3$.

   (b) $\sqrt{3}x + 3y = 6$ => $y = (-\sqrt{3}/3)x + 2$ => $m = -\sqrt{3}/3$ => $\tan\alpha = -\sqrt{3}/3$ => $\alpha = 5\pi/6$. [Reference angle is $\pi/6$.]

30. The area of a sector is proportional to the vertex angle.

   When the vertex angle is 2π (full revolution), the area of the corresponding sector is πr^2 (area of full circle).

   Therefore, $\dfrac{A}{\theta} = \dfrac{\pi r^2}{2\pi}$ [using the proportionality]. Hence, $A = \dfrac{1}{2}r^2\theta$.

Problem Set 2.4 Introduction to Limits

3. $(-2)^2 - 3(-2) + 1 = 11$.

6. $\dfrac{5(1) - (1)^2}{(1)^2 + 2(1) - 4} = \dfrac{4}{-1} = -4$.

9. $\lim\limits_{x \to -3} \dfrac{2x^2+5x-3}{x+3} = \lim\limits_{x \to -3} \dfrac{(2x-1)(x+3)}{x+3} = \lim\limits_{x \to -3} (2x-1) = 2(-3) - 1 = -7$.

12. $\lim\limits_{x \to 2} \dfrac{x^3-8}{x-2} = \lim\limits_{x \to 2} \dfrac{(x-2)(x^2+2x+4)}{x-2} = \lim\limits_{x \to 2} (x^2+2x+4) = (2)^2 + 2(2) + 4 = 12$.

15. $\lim\limits_{t \to 2} \dfrac{t^2-5t+6}{t^2-t-2} = \lim\limits_{t \to 2} \dfrac{(t-2)(t-3)}{(t-2)(t+1)} = \lim\limits_{t \to 2} \dfrac{t-3}{t+1} = \dfrac{(2)-3}{(2)+1} = -\dfrac{1}{3}$.

18. Note that $\dfrac{1 - \cos x}{x}$ is an odd function since it is the ratio of an even function and an odd function. Therefore, we only need to observe what happens as x approaches zero from one side.

18. (continued)

x	$\dfrac{1-\cos x}{x}$
0.5	0.24483
0.1	0.04996
0.01	0.00500
0.001	0.00050

It seems that $\lim\limits_{x \to 0} \dfrac{1-\cos x}{x} = 0$

21. Note that $\dfrac{\sin t}{t^2}$ is an odd function since it is the ratio of an odd function and an even function. Therefore, we only need to observe what happens as x approaches zero from one side.

t	$\dfrac{\sin t}{t^2}$
0.5	1.9177
0.1	9.9833
0.01	99.9983

It seems that $\lim\limits_{t \to 0} \dfrac{\sin t}{t^2}$ doesn't exist.

24. Note that $\dfrac{\tan x - \sin x}{x^3}$ is an even function since it is the ratio of two odd functions. Therefore, we only need to observe what happens as x approaches zero from one side.

x	$\dfrac{\tan x - \sin x}{x^3}$
0.5	0.53502
0.1	0.50126
0.01	0.50001

It seems that $\lim\limits_{x \to 0} \dfrac{\tan x - \sin x}{x^3} = \dfrac{1}{2}$.

27. (a) 2   (b) 1   (c) doesn't exist   (d) 2.6
(e) 2   (f) doesn't exist   (g) 2   (h) 1

30. (a) 0

(b) doesn't exist

(c) 1

(d) 1

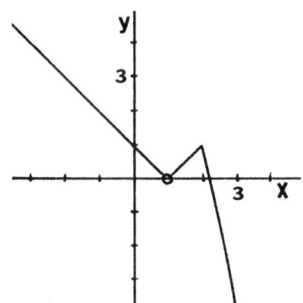

33. $\lim\limits_{x \to 1^+} \dfrac{x^2-1}{|x-1|} = \lim\limits_{x \to 1^+} \dfrac{(x+1)(x-1)}{x-1} = \lim\limits_{x \to 1^+} (x+1) = 2$

$\lim\limits_{x \to 1^-} \dfrac{x^2-1}{|x-1|} = \lim\limits_{x \to 1^-} \dfrac{(x+1)(x-1)}{-(x-1)} = \lim\limits_{x \to 1^-} \dfrac{(x+1)}{-1} = -2$

Therefore, $\lim\limits_{x \to 1} \dfrac{x^2-1}{|x-1|}$ doesn't exist.

36. There are infinitely many that will satisfy all the conditions. Here is one that does.

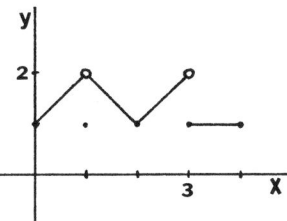

Problem Set 2.5 Rigorous Study of Limits

3. $\lim\limits_{z \to d} h(z) = P$ means that for each $\varepsilon > 0$, there is a corresponding $\delta > 0$ such that $0 < |z-d| < \delta \Rightarrow |h(z)-P| < \varepsilon$.

6. $\lim\limits_{t \to a^+} g(t) = D$ means that for each $\varepsilon > 0$, there is a corresponding $\delta > 0$ such that $0 < t-a < \delta \Rightarrow |g(t)-D| < \varepsilon$.

9. Note: In proving $\lim\limits_{x \to 5} \dfrac{x^2-25}{x-5} = 10$, we are not concerned with x being 5.

Therefore, it will be sufficient to prove that $\lim\limits_{x \to 5} (x+5) = 10$

since $\dfrac{x^2-25}{x-5} = \dfrac{(x-5)(x+5)}{x-5} = x+5$, if $x \neq 5$.

Proof: Let $\varepsilon > 0$ be given. Choose $\delta = \varepsilon$.

Then $0 < |x-5| < \delta \Rightarrow |x-5| < \varepsilon$

$\Rightarrow |(x+5) - 10| < \varepsilon$

Therefore, $\lim\limits_{x \to 5}(x+5) = 10$; hence, $\lim\limits_{x \to 5} \dfrac{x^2-25}{x-5} = 10$.

12. $\dfrac{3x^2-4x}{x} = \dfrac{x(3x-4)}{x} = 3x-4$, if $x \neq 0$. [See Note in Problem 9 Solution.]

Proof: Let $\varepsilon > 0$ be given. Choose $\delta = \varepsilon/3$.

Then $0 < |x-0| < \delta \implies |x| < \varepsilon/3$
$\implies |3x| < \varepsilon$
$\implies |(3x-4) - (-4)| < \varepsilon$

Therefore, $\lim\limits_{x \to 0} (3x-4) = -4$; hence, $\lim\limits_{x \to 0} \dfrac{3x^2-4x}{x} = -4$

15. $\dfrac{2x^3+3x^2-2x-3}{x^2-1} = \dfrac{(2x+3)(x^2-1)}{x^2-1} = 2x+3$, if $x \neq -1, 1$.

Proof: Let $\varepsilon > 0$ be given. Choose $\delta = \{\min \varepsilon/2, 2\}$ [to avoid $x = -1$].

Then $0 < |x-1| < \delta \implies |x-1| < \varepsilon/2$
$\implies |2x-2| < \varepsilon$
$\implies |(2x+3) - 5| < \varepsilon$

Therefore, $\lim\limits_{x \to 1} (2x+3) = 5$; hence, $\lim\limits_{x \to 1} \dfrac{2x^3+3x^2-2x-3}{x^2-1} = 5$

18. Proof: Let $\varepsilon > 0$ be given. Choose $\delta = \sqrt[4]{\varepsilon}$.

Then $0 < |x-0| < \delta \implies |x| < \sqrt[4]{\varepsilon}$
$\implies x^4 < \varepsilon$
$\implies |x^4 - 0| < \varepsilon$

Therefore, $\lim\limits_{x \to 0} x^4 = 0$

21. $0 \leq \sin^2(1/x) \leq 1 \implies 0 \leq x^4 \sin^2(1/x) \leq x^4$ [since $x^4 > 0$].

Therefore, $\lim\limits_{x \to 0} x^4 \sin^2(1/x) = 0$ [using Problems 20 and 18].

24. Proof: Let $\varepsilon > 0$ be given. Then there is a $\delta > 0$ such that

$$0 < |x-a| < \delta \implies |g(x) - 0| < \varepsilon/B \quad [\varepsilon/B \text{ is positive}]$$
$$\implies |g(x)| < \varepsilon/|f(x)|, \text{ if } f(x) \neq 0$$
$$[\text{since } |f(x)| < B \text{ so } \varepsilon/B < \varepsilon/|f(x)|]$$
$$\implies |f(x)g(x)| < \varepsilon \quad [\text{true even if } f(x) = 0]$$
$$\implies |f(x)g(x) - 0| < \varepsilon$$

Therefore, $\lim_{x \to a} f(x)g(x) = 0$.

Problem Set 2.6 Limit Theorems

3. $\lim_{x \to 2}[(x^2+1)(3x-1)] \overset{[6]}{=} \lim_{x \to 2}(x^2+1) \cdot \lim_{x \to 2}(3x-1)$

$\overset{[4,5]}{=} \left[\lim_{x \to 2}(x^2) + \lim_{x \to 2}(1)\right]\left[\lim_{x \to 2}(3x) - \lim_{x \to 2}(1)\right]$

$\overset{[8,1,3,1]}{=} \left[\left(\lim_{x \to 2}(x)\right)^2 + 1\right]\left[3\lim_{x \to 2}(x) - 1\right]$

$\overset{[2,2]}{=} [(2)^2 + 1][3(2) - 1] = 25$

6. $\lim_{x \to -2} \dfrac{3x^4-8}{x^3+24} \overset{[7]}{=} \dfrac{\lim_{x \to -2}(3x^4-8)}{\lim_{x \to -2}(x^3+24)} \overset{[5,4]}{=} \dfrac{\lim_{x \to -2}(3x^4) - \lim_{x \to -2}(8)}{\lim_{x \to -2}(x^3) + \lim_{x \to -2}(24)}$

$\overset{[3,1,8,1]}{=} \dfrac{3\lim_{x \to -2}(x^4) - 8}{\left(\lim_{x \to -2}(x)\right)^3 + 24} \overset{[8,2]}{=} \dfrac{3\left(\lim_{x \to -2}(x)\right)^4 - 8}{(-2)^3 + 24}$

$\overset{[2]}{=} \dfrac{3(-2)^4 - 8}{-8 + 24} = \dfrac{5}{2} = 2.5$

9. $\lim\limits_{t \to -2}(2t^3+15)^{13} \stackrel{[8]}{=} \left(\lim\limits_{t \to -2}(2t^3+15)\right)^{13} \stackrel{[4]}{=} \left(\lim\limits_{t \to -2}(2t^3) + \lim\limits_{t \to -2}(15)\right)^{13}$

$\stackrel{[3,1]}{=} \left(2\lim\limits_{t \to -2}(t^3) + 15\right)^{13} \stackrel{[8]}{=} \left[2\left(\lim\limits_{t \to -2}(t)\right)^3 + 15\right]^{13}$

$\stackrel{[2]}{=} [2(-2)^3 + 15]^{13} = (-1)^{13} = -1.$

12. $\lim\limits_{w \to 5}(2w^4-9w^3+19)^{-1/2} \stackrel{[7,9]}{=} \dfrac{\lim\limits_{w \to 5}(1)}{\sqrt{\lim\limits_{w \to 5}(2w^4-9w^3+19)}}$

$\stackrel{[1,5,4]}{=} \left(\lim\limits_{w \to 5}(2w^4) - \lim\limits_{w \to 5}(9w^3) + \lim\limits_{w \to 5}(19)\right)^{-1/2}$

$\stackrel{[3,3,1]}{=} \left(2\lim\limits_{w \to 5}(w^4) - 9\lim\limits_{w \to 5}(w^3) + 19\right)^{-1/2}$

$\stackrel{[8,8]}{=} \left[2\left(\lim\limits_{w \to 5}w\right)^4 - 9\left(\lim\limits_{w \to 5}w\right)^3 + 19\right]^{-1/2}$

$\stackrel{[2,2]}{=} [2(5)^4 - 9(5)^3 + 19]^{-1/2}$

$= (144)^{-1/2} = \dfrac{1}{12} \approx 0.0833.$

15. $\lim\limits_{x \to 4}\dfrac{x^2+2x-24}{x-4} = \lim\limits_{x \to 4}\dfrac{(x+6)(x-4)}{x-4} = \lim\limits_{x \to 4}(x+6) = 4+6 = 10.$

18. $\lim\limits_{u \to 2}\dfrac{u^2-2u}{u^2-4} = \lim\limits_{u \to 2}\dfrac{u(u-2)}{(u-2)(u+2)} = \lim\limits_{u \to 2}\dfrac{u}{u+2} = \dfrac{2}{2+2} = \dfrac{1}{2} = 0.5.$

21. $\lim\limits_{y \to 1}\dfrac{(y-1)(y^2+2y-3)}{(y^2-2y+1)} = \lim\limits_{y \to 1}\dfrac{(y-1)(y-1)(y+3)}{(y-1)^2} = \lim\limits_{y \to 1}(y+3) = 1+3 = 4.$

24. $\lim\limits_{x \to a}\dfrac{2f(x) - 3g(x)}{f(x) + g(x)} = \dfrac{2\lim\limits_{x \to a}f(x) - 3\lim\limits_{x \to a}g(x)}{\lim\limits_{x \to a}f(x) + \lim\limits_{x \to a}g(x)} = \dfrac{2(3) - 3(-1)}{(3) + (-1)} = \dfrac{9}{2} = 4.5.$

27. $\lim\limits_{t \to a} [f(t) + (t-a)g(t)] = \lim\limits_{t \to a} f(t) + \lim\limits_{t \to a}(t-a) \lim\limits_{t \to a} g(t)$

$= 3 + \left(\lim\limits_{t \to a}(t) - \lim\limits_{t \to a}(a)\right)(-1) = 3 + (a-a)(-1) = 3.$

30. $\lim\limits_{x \to 2} \dfrac{f(x) - f(2)}{x-2} = \lim\limits_{x \to 2} \dfrac{(3x^2-5) - (7)}{x-2} = \lim\limits_{x \to 2} \dfrac{3x^2-12}{x-2} = \lim\limits_{x \to 2} \dfrac{3(x+2)(x-2)}{x-2}$

$= \lim\limits_{x \to 2} 3(x+2) = 3(2+2) = 12.$

33. Let $\varepsilon > 0$ be given.
Choose δ_1 such that $0 < |x-c| < \delta_1 \Rightarrow |g(x) - M| < 1$
$\Rightarrow |g(x)| - |M| < 1$
$\Rightarrow |g(x)| < |M| + 1$

Choose δ_2 such that $0 < |x-c| < \delta_2 \Rightarrow |f(x) - L| < \dfrac{\varepsilon}{2(|M| + 1)}$

Choose δ_3 such that $0 < |x-c| < \delta_3 \Rightarrow |g(x) - M| < \dfrac{\varepsilon}{2(|L| + 1)}$

Now let $\delta = \min\{\delta_1, \delta_2, \delta_3\}$ and use the hint.

Then $0 < |x-c| < \delta \Rightarrow |f(x)g(x) - LM| \leq |g(x)||f(x)-L| + |L||g(x)-M|$

$\leq (|M| + 1) \dfrac{\varepsilon}{2(|M| + 1)} + |L| \dfrac{\varepsilon}{2(|L| + 1)}$

$\leq \dfrac{\varepsilon}{2} + \dfrac{\varepsilon}{2} = \varepsilon.$

Therefore, $\lim\limits_{x \to c} f(x)g(x) = LM.$

36. ($\Rightarrow$) Note: $|f(x) - 0| = |f(x)| = ||f(x)|| = ||f(x)| - 0|.$

Let $\varepsilon > 0$ be given.
Then there is a $\delta > 0$ such that $0 < |x-c| < \delta \Rightarrow |f(x) - 0| < \varepsilon$
$\Rightarrow ||f(x)| - 0| < \varepsilon.$

Therefore, $\lim\limits_{x \to c} |f(x)| = 0.$

($\Leftarrow$) Prove by exchanging the roles of $f(x)$ and $|f(x)|$ in ($\Rightarrow$) above.

Problem Set 2.6

39. (a) Let $f(x) = \frac{1}{x}$; $g(x) = \frac{-1}{x}$. Then $f(x)+g(x) = 0$ $[x \neq 0]$.

Thus, $\lim_{x \to 0}[f(x)+g(x)] = 0$, but $\lim_{x \to 0} f(x)$ and $\lim_{x \to 0} g(x)$ do not exist.

(b) Let $f(x) = \begin{cases} 1, & \text{if } x \text{ is rational} \\ 2, & \text{if } x \text{ is irrational} \end{cases}$; $g(x) = \begin{cases} 2, & \text{if } x \text{ is rational} \\ 1, & \text{if } x \text{ is irrational} \end{cases}$.

Then $\lim_{x \to 5}[f(x)g(x)] = 2$, but $\lim_{x \to 5} f(x)$ and $\lim_{x \to 5} g(x)$ do not exist.

42. $\lim_{x \to 3^-} \sqrt{9-x^2} = \sqrt{9-(3)^2} = 0$.

45. $\lim_{x \to 2^+} \frac{(x^2+1)[\![x]\!]}{(3x-1)^2} = \lim_{x \to 2^+} \frac{(x^2+1)2}{(3x-1)^2} = \frac{[(2)^2+1]2}{[3(2)-1]^2} = \frac{10}{25} = 0.4$.

48. $\lim_{x \to 3^+} [\![x^2+2x]\!] = \lim_{x \to 3^+}(15) = 15$.

Problem Set 2.7 Continuity of Functions

3. g is not continuous at 2 since g(2) is not defined.

6. h is continuous at 2 since h is the composite of a polynomial function and the absolute value function. [Theorems A, B, D]

9. g is not continuous at 2 since g(2) is not defined.

12. h is not continuous at 2 since h(2) = 2,
but $\lim_{t \to 2} h(t) = \lim_{t \to 2} \frac{4t-8}{t-2} = \lim_{t \to 2} \frac{4(t-2)}{t-2} = \lim_{t \to 2}(4) = 4$.

15. $\lim_{x \to 3} f(x) = \lim_{x \to 3} \frac{x^2-9}{x-3} = \lim_{x \to 3} \frac{(x+3)(x-3)}{x-3} = \lim_{x \to 3}(x+3) = 6$, so define $f(3) = 6$.

18. $\lim_{t \to 0} g(t) = \lim_{t \to 0} \frac{\sin t}{t} = 1$ [Section 2.6, Example 8], so define $g(0) = 1$.

21. f is discontinuous only at -2 and 3 since f is a rational function whose denominator is zero only at -2 and 3 [Theorem A].

24. f is discontinuous on $(-\infty,1]$ since f is undefined on $(-\infty,1]$. f is continuous everywhere else [Theorems A, C].

27. F is continuous on **R** since $4 + x^2$ is always positive [Theorems A, C].

30. g is continuous on $(-\infty,0)$, on $(0,1)$, and on $(1,\infty)$ since g is a polynomial on each of those intervals [Theorem A]. The only points then at which g might be discontinuous are 0 and 1.

    x=0: (i) $g(0) = 0$

    (ii) $\lim_{x \to 0^-}[g(x)] = \lim_{x \to 0^-}(x^2) = 0$ and $\lim_{x \to 0^+}[g(x)] = \lim_{x \to 0^+}(-x) = 0$

    Therefore, $\lim_{x \to 0} g(x) = 0$.

    (iii) Therefore, $\lim_{x \to 0} g(x) = g(0)$, so g is continuous at 0.

    x=1: (i) $g(1) = -1$

    (ii) $\lim_{x \to 1^-}[g(x)] = \lim_{x \to 1^-}(-x) = -1$ and $\lim_{x \to 1^+}[g(x)] = \lim_{x \to 1^+}(x) = 1$

    Therefore, $\lim_{x \to 1} g(x)$ does not exist.

    (iii) Therefore, $\lim_{x \to 1} g(x) \neq g(1)$, so g is discontinuous at 1.

    Therefore, g is discontinuous only at 1.

33. There are infinitely many. The following is one that satisfies all the conditions.

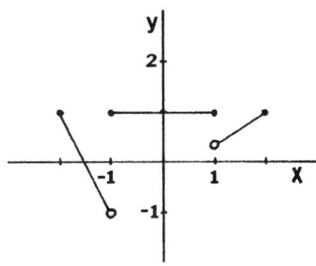

Problem Set 2.7

36. The function defined by $f(x) = x^5 + 4x^3 - 7x + 14$ is continuous on the interval $[-2,0]$.

$f(-2) = -36$ and $f(0) = 14$; and 0 is a number between $f(-2)$ and $f(0)$.

Therefore, by the Intermediate Value Theorem, there is at least one number c between -2 and 0 such that $f(c) = 0$.

Therefore, $c^5 + 4c^3 - 7c + 14 = 0$, so c is a solution of $x^5 + 4x^3 - 7x + 14 = 0$.

39. [Intuitive] If f is continuous on $[0,1]$ and all the $f(x)$ values lie between 0 and 1, then the graph of f has to touch the line $y=x$ at an end point or cross it somewhere inbetween. And $f(x) = x$ at such a point of intersection.

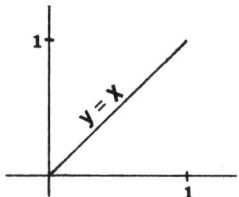

[Formal] I. If $f(0) = 0$ or $f(1) = 1$, then 0 or 1 is a fixed point. (This is the end point possibility.)

II. If $f(0) > 0$ and $f(1) < 1$, define $g(x) = x - f(x)$. Then g is continuous on $[0,1]$; $g(0) = 0 - f(0) < 0$; $g(1) = 1 - f(1) > 0$; so 0 is between $g(0)$ and $g(1)$.

Therefore, by the Intermediate Value Theorem, there is at least one number c between 0 and 1 such that $g(c) = 0$.

Thus, $c - f(c) = 0$ [since $g(c) = c - f(c)$].

Therefore, $f(c) = c$, so c is a fixed point of f.

Problem Set 2.8 Chapter Review Problems

True-False Quiz

3. True. Auxiliary axis for $\frac{x}{4-x}$: $\dfrac{(-) \quad (0) \quad (+) \quad (u) \quad (-)}{ 0 4 x}$

The radicand must be nonnegative. Thus, the natural domain of f is $[0,4)$.

6. False. c^{ex}: Let $f(x) = x^3$ and $g(x) = x$. [Each is an odd function.]

Then $(fg)(x) = f(x)g(x) = x^4$. [fg is an even function.]

9. False. cex: Consider $f(x) = 6$. The natural domain is **R** which contains at least two numbers, but the range is $\{6\}$.

12. False. cex: $f(x) = x^2$ and $g(x) = x-2$ each have **R** for natural domain, but the natural domain of f/g is $\{x : x \neq 2\}$.

15. False. cex: $\pi/2$ is not in the natural domain of the tan function.

18. False. cex: Let $f(x) = \frac{x^2-1}{x-1}$. $f(1)$ is not defined but $\lim_{x \to 1} f(x) = 2$.

21. False. cex: [The function defined in Number 18 will do.]

24. False. cex: [See Solution for Problem 39(a) of Problem Set 2.6.]

27. False. cex: Let $f(x) = \begin{cases} x, & \text{if } x \neq 0 \\ 1, & \text{if } x = 0 \end{cases}$ and let $g(x) = -x$

Then $f(x) \neq g(x)$ for all x, but $\lim_{x \to 0} f(x) = \lim_{x \to 0} g(x) = 0$.

30. True. For each number y between $1/f(a)$ and $1/f(b)$, $1/y$ is between $f(a)$ and $f(b)$ [since $1/f(a)$ and $1/f(b)$ are positive]. Then, by the Intermediate Value Theorem, $f(c) = 1/y$ for some c between a and b. Therefore, $(1/f)(c) = 1/f(x) = y$ for some c between a and b.

Miscellaneous Problems

3. (a) The denominator must be nonzero. $[x^2 - 1 \neq 0]$ iff $[x \neq -1, 1]$. Therefore, the natural domain is $\{x : x \neq -1, 1\}$.

   (b) The radicand must be nonnegative.

   $[4-x^2 \geq 0]$ iff $[4 \geq x^2]$ iff $[2 \geq |x|]$ iff $[-2 \leq x \leq 2]$, so the natural domain is $[-2, 2]$.

   (c) The denominator must be nonzero. $[|2x+3| \neq 0]$ iff $[x \neq -3/2]$. Therefore, the natural domain is $\{x : x \neq -3/2\}$.

6. For $0 \leq x \leq 4$, translate the graph of $y = \sqrt{x}$ 1 unit down. Then the part of the graph to the left of the y-axis can be obtained by symmetry with respect to the y-axis [since f is an even function].

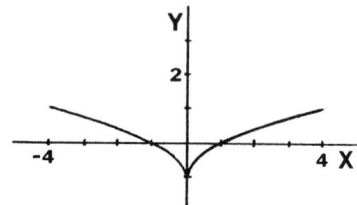

9. (a) $y = (1/4)x^2$

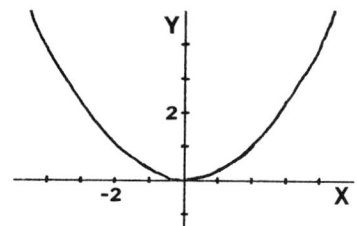

(b) $y = (1/4)(x+2)^2$
    Translate the graph in (a) 2 units left.

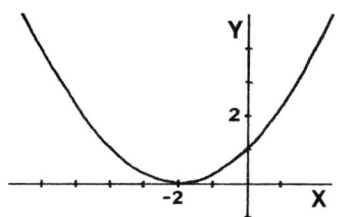

(c) $y = -1 + (1/4)(x+2)^2$
    Translate the graph in (b) 1 unit down.

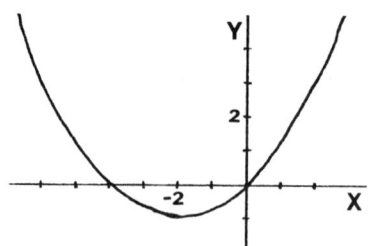

12. (a) $\sin(570°) = \sin(360° + 180° + 30°)$
    $= -\sin(30°)$
    $= -1/2.$

(b) $\cos(9\pi/2) = \cos(4\pi + \pi/2)$
    $= \cos(\pi/2)$
    $= 0.$

(c) $\sin^2(5) + \cos^2(5) = 1.$ [$\sin^2 x + \cos^2 x = 1$ for all real numbers x.]

(d) $\cos(-13\pi/6) = \cos(-2\pi - \pi/6)$
    $= \cos(-\pi/6)$
    $= \cos(\pi/6)$
    $= \sqrt{3}/2.$

15. The circumference of the circle is $2\pi \cdot 9$ inches = 18π inches. Now transform "revolutions per minute" to "inches per second."

$$\frac{20 \text{ rev}}{1 \text{ min}} \times \frac{18\pi \text{ in}}{1 \text{ rev}} \times \frac{1 \text{ min}}{60 \text{ sec}} = \frac{360\pi \text{ in}}{60 \text{ sec}} = \frac{6\pi \text{ in}}{1 \text{ sec}}.$$

In 1 second the fly travels 6π inches ≈ 18.8496 inches.

18. $\lim_{u \to 1} \frac{u+1}{u^2-1}$ does not exist since the numerator approaches 2 and the denominator approaches 0, so the fraction (in absolute value) gets larger without bound.

21. $\lim_{x \to 0} \frac{\tan x}{\sin 2x} = \lim_{x \to 0} \frac{(\sin x)/(\cos x)}{2 \sin x \cos x} = \lim_{x \to 0} \frac{1}{2 \cos^2 x} = \frac{1}{2(1)^2} = 0.5.$

24. $\lim_{x \to 0} \frac{\cos x}{x}$ does not exist since the numerator approaches 1 and the denominator approaches 0, so the fraction (in absolute value) gets larger without bound.

27. $\lim_{t \to 2^-} (\llbracket t \rrbracket - t) = \lim_{t \to 2^-} \llbracket t \rrbracket - \lim_{t \to 2^-} (t) = 1 - 2 = -1.$

30. (a) $\lim_{u \to a} g(u) = M$ means that for each $\varepsilon > 0$, there is a corresponding $\delta > 0$ such that $0 < |u-a| < \delta \Rightarrow |g(u)-M| < \varepsilon.$

(b) $\lim_{x \to a^-} f(x) = L$ means that for each $\varepsilon > 0$, there is a corresponding $\delta > 0$ such that $0 < a-x < \delta \Rightarrow |f(x)-L| < \varepsilon.$

Problem Set 2.8

33. The only possible points of discontinuity are 0 and 1.

At x=0: $f(0) = -1$; $\lim_{x\to 0^+} f(x) = \lim_{x\to 0^+} (ax+b) = b$; $\lim_{x\to 0^-} f(x) = \lim_{x\to 0^-} (-1) = -1$

Let $b = -1$. Then f is continuous at 0.

At x=1: $f(1) = 1$; $\lim_{x\to 1^-} f(x) = \lim_{x\to 1^-} (ax-1) = a-1$; $\lim_{x\to 1^+} f(x) = \lim_{x\to 1^+} (1) = 1$

Let $a-1 = 1$ (i.e. $a = 2$). Then f is continuous at 1.

Conclusion: a=2, b=-1

CHAPTER 3 THE DERIVATIVE

Problem Set 3.1 Two Problems with One Theme

3. The tangent line seems to go through points (1,8) and (4,0). Therefore, the slope is estimated to be

$$\frac{0 - 8}{4 - 1} = \frac{-8}{3} \approx -2.6667.$$

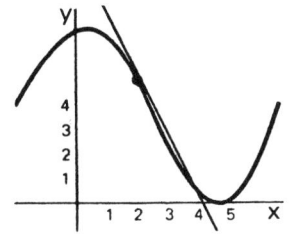

6. The tangent line seems to go through points (-2,3) and (0.6,0). Therefore, the slope is estimated to be

$$\frac{0 - 3}{0.6 - (-2)} = \frac{-3}{2.6} \approx -1.1538.$$

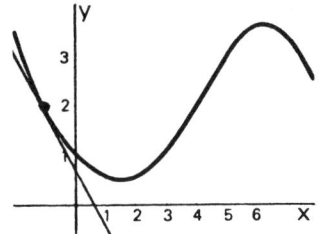

9. Let $f(x) = x^2 - 3x + 2$. Then the slope of the tangent line at $x=c$ is

$$m_{tan} = \lim_{h \to 0} \frac{f(c+h) - f(c)}{h} = \lim_{h \to 0} \frac{[(c+h)^2 - 3(c+h) + 2] - [c^2 - 3c + 2]}{h}$$

$$= \lim_{h \to 0} \frac{c^2 + 2ch + h^2 - 3c - 3h + 2 - c^2 + 3c - 2}{h} = \lim_{h \to 0} \frac{h(2c+h-3)}{h}$$

$$= \lim_{h \to 0} (2c+h-3) = 2c-3. \text{ See table below for slopes at } x=c.$$

c	-2	1.5	2	5
2c-3	-7	0	1	7

12. Let $f(x) = \frac{2}{x-2}$.

Then $m_{tan} = \lim_{h \to 0} \frac{f(0+h) - f(0)}{h} = \lim_{h \to 0} \frac{\frac{2}{h-2} - \frac{2}{-2}}{h} = \lim_{h \to 0} \frac{\left(\frac{2}{h-2} + 1\right)(h-2)}{h \quad (h-2)}$

$= \lim_{h \to 0} \frac{2 + (h-2)}{h(h-2)} = \lim_{h \to 0} \frac{h}{h(h-2)} = \lim_{h \to 0} \frac{1}{h-2} = -1/2.$

Equation of tangent at $(0,-1)$: $y-(-1) = (-1/2)(x-0)$ [Point-slope form]
$y = (-1/2)x - 1$ [Slope-intercept form]

15. (a) Let $s(t) = \sqrt{t}$. The instantaneous velocity at time $t=c$ ($c>0$) is

$v(c) = \lim_{h \to 0} \frac{s(c+h) - s(c)}{h} = \lim_{h \to 0} \frac{\sqrt{c+h} - \sqrt{c}}{h} = \lim_{h \to 0} \frac{(\sqrt{c+h} - \sqrt{c})(\sqrt{c+h} + \sqrt{c})}{h \quad (\sqrt{c+h} + \sqrt{c})}$

$= \lim_{h \to 0} \frac{c+h - c}{h(\sqrt{c+h} + \sqrt{c})} = \lim_{h \to 0} \frac{1}{\sqrt{c+h} + \sqrt{c}} = \frac{1}{2\sqrt{c}}$ [ft/sec]

(b) Solve $\frac{1}{2\sqrt{c}} = \frac{1}{6}$ for c and obtain $c = 9$.

It will reach a velocity of 1/6 feet per second in 9 seconds.

18. Let $P(t) = 1000t^2$ [dollars] be the total (accumulated) profit function.

(a) $P(2) = 4000$ [dollars] is the profit after 2 years.
$P(3) = 9000$ [dollars] is the profit after 3 years.
Therefore, during the third year the business made 5000 dollars.

(b) Average rate of profit during the first half of the third year is

$\frac{P(2.5) - P(2)}{2.5 - 2} = \frac{1000(2.5)^2 - 1000(2)^2}{0.5} = 4500$ [dollars/year].

(c) Instaneous rate of profit (marginal profit) at the end of the second year ($t=2$) is

$\lim_{h \to 0} \frac{P(2+h) - P(2)}{h} = \lim_{h \to 0} \frac{1000(2+h)^2 - 1000(2)^2}{h}$

$= \lim_{h \to 0} \frac{4000 + 4000h + 1000h^2 - 4000}{h}$

$= \lim_{h \to 0} \frac{1000h(4+h)}{h} = \lim_{h \to 0} 1000(4+h) = 4000$ [dollars/year]

21. The rate of growth when $t = 10$ [weeks] is

$$\lim_{h \to 0} \frac{W(10+h) - W(10)}{h} = \lim_{h \to 0} \frac{[0.2(10+h)^2 - 0.09(10+h)] - [0.2(10)^2 - 0.09(10)]}{h}$$

$$= \lim_{h \to 0} \frac{20 + 4h + 0.2h^2 - 0.9 - 0.09h - 20 + 0.9}{h}$$

$$= \lim_{h \to 0} \frac{h(4 + 0.2h - 0.09)}{h} = \lim_{h \to 0} (3.91 + 0.2h)$$

$$= 3.91 \text{ [grams/week]}$$

Problem Set 3.2 The Derivative

3. $f(x) = x^3 + 2x^2$

$$f'(-1) = \lim_{h \to 0} \frac{f(-1+h) - f(-1)}{h} = \lim_{h \to 0} \frac{[(-1+h)^3 + 2(-1+h)^2] - [(-1)^3 + 2(-1)^2]}{h}$$

$$= \lim_{h \to 0} \frac{[(-1+3h-3h^2+h^3) + 2(1-2h+h^2)] - [-1+2]}{h}$$

$$= \lim_{h \to 0} \frac{-h - h^2 + h^3}{h} = \lim_{h \to 0} \frac{h(-1-h+h^2)}{h} = \lim_{h \to 0} (-1-h+h^2) = -1$$

6. $f(x) = ax + b$

$$f'(x) = \lim_{h \to 0} \frac{f(x+h) - f(x)}{h} = \lim_{h \to 0} \frac{[a(x+h)+b] - [ax+b]}{h}$$

$$= \lim_{h \to 0} \frac{ax + ah + b - ax - b}{h} = \lim_{h \to 0} \frac{ah}{h} = \lim_{h \to 0} (a) = a$$

[Note that $f'(x) = a$ is indeed the slope of the line $f(x) = ax+b$.]

9. $f(x) = ax^2 + bx + c$

$$f'(x) = \lim_{h \to 0} \frac{f(x+h) - f(x)}{h} = \lim_{h \to 0} \frac{[a(x+h)^2 + b(x+h) + c] - [ax^2 + bx + c]}{h}$$

$$= \lim_{h \to 0} \frac{[ax^2 + 2axh + ah^2 + bx + bh + c] - [ax^2 + bx + c]}{h}$$

$$= \lim_{h \to 0} \frac{h(2ax + ah + b)}{h} = \lim_{h \to 0} (2ax + ah + b) = 2ax + b$$

12. $f(x) = x^4$

$$f'(x) = \lim_{h\to 0} \frac{f(x+h) - f(x)}{h} = \lim_{h\to 0} \frac{(x+h)^4 - x^4}{h}$$

$$= \lim_{h\to 0} \frac{[x^4 + 4x^3h + 6x^2h^2 + 4xh^3 + h^4] - x^4}{h}$$

$$= \lim_{h\to 0} \frac{h(4x^3 + 6x^2h + 4xh^2 + h^3)}{h} = \lim_{h\to 0}(4x^3 + 6x^2h + 4xh^2 + h^3) = 4x^3$$

15. $F(x) = \dfrac{6}{x^2+1}$

$$F'(x) = \lim_{h\to 0} \frac{F(x+h) - F(x)}{h} = \lim_{h\to 0} \frac{\dfrac{6}{(x+h)^2+1} - \dfrac{6}{x^2+1}}{h}$$

$$= \lim_{h\to 0} \frac{\left(\dfrac{6}{x^2+2xh+h^2+1} - \dfrac{6}{x^2+1}\right)(x^2+2xh+h^2+1)(x^2+1)}{h\,(x^2+2xh+h^2+1)(x^2+1)}$$

$$= \lim_{h\to 0} \frac{6(x^2+1) - 6(x^2+2xh+h^2+1)}{h(x^2+2xh+h^2+1)(x^2+1)}$$

$$= \lim_{h\to 0} \frac{h(-12x-6h)}{h(x^2+2xh+h^2+1)(x^2+1)} = \lim_{h\to 0} \frac{-12x-6h}{(x^2+2xh+h^2+1)(x^2+1)} = \frac{-12x}{(x^2+1)^2}$$

18. $G(x) = \dfrac{2x}{x^2-x} = \dfrac{2x}{x(x-1)} = \dfrac{2}{x-1} \quad [x \neq 0]$

$$G'(x) = \lim_{h\to 0} \frac{G(x+h) - G(x)}{h} = \lim_{h\to 0} \frac{\dfrac{2}{(x+h)-1} - \dfrac{2}{x-1}}{h}$$

$$= \lim_{h\to 0} \frac{\left(\dfrac{2}{x+h-1} - \dfrac{2}{x-1}\right)(x+h-1)(x-1)}{h\,(x+h-1)(x-1)} = \lim_{h\to 0} \frac{2(x-1) - 2(x+h-1)}{h(x+h-1)(x-1)}$$

$$= \lim_{h\to 0} \frac{-2h}{h(x+h-1)(x-1)} = \lim_{h\to 0} \frac{-2}{(x+h-1)(x-1)} = \frac{-2}{(x-1)^2} \quad [x \neq 0]$$

21. $H(x) = \dfrac{3}{\sqrt{x-2}}$

$$H'(x) = \lim_{h\to 0} \frac{H(x+h) - H(x)}{h} = \lim_{h\to 0} \frac{\dfrac{3}{\sqrt{(x+h)-2}} - \dfrac{3}{\sqrt{x-2}}}{h}$$

21. (continued)

$$= \lim_{h \to 0} \frac{\left(\frac{3}{\sqrt{x+h-2}} - \frac{3}{\sqrt{x-2}}\right)\sqrt{x+h-2}\sqrt{x-2}}{h\sqrt{x+h-2}\sqrt{x-2}}$$

$$= \lim_{h \to 0} \frac{3\sqrt{x-2} - 3\sqrt{x+h-2}}{h\sqrt{x+h-2}\sqrt{x-2}} = 3 \lim_{h \to 0} \frac{(\sqrt{x-2} - \sqrt{x+h-2})}{h\sqrt{x+h-2}\sqrt{x-2}} \cdot \frac{(\sqrt{x-2} + \sqrt{x+h-2})}{(\sqrt{x-2} + \sqrt{x+h-2})}$$

$$= 3 \lim_{h \to 0} \frac{(x-2) - (x+h-2)}{h\sqrt{x+h-2}\sqrt{x-2}(\sqrt{x-2} + \sqrt{x+h-2})} = 3 \lim_{h \to 0} \frac{-h}{h\sqrt{x+h-2}\sqrt{x-2}(\sqrt{x-2} + \sqrt{x+h-2})}$$

$$= 3 \lim_{h \to 0} \frac{-1}{\sqrt{x+h-2}\sqrt{x-2}(\sqrt{x-2} + \sqrt{x+h-2})} = \frac{-3}{\sqrt{x-2}\sqrt{x-2}(\sqrt{x-2} + \sqrt{x-2})}$$

$$= \frac{-3}{(x-2) \cdot 2\sqrt{x-2}} = \frac{-3}{2(x-2)^{3/2}}$$

24. $f(x) = x^3 + 5x$

$$f'(x) = \lim_{t \to x} \frac{f(t) - f(x)}{t - x} = \lim_{t \to x} \frac{(t^3 + 5t) - (x^3 + 5x)}{t - x}$$

$$= \lim_{t \to x} \frac{(t^3 - x^3) + (5t - 5x)}{t - x} = \lim_{t \to x} \frac{(t-x)(t^2 + tx + x^2) + 5(t - x)}{t - x}$$

$$= \lim_{t \to x}[(t^2 + tx + x^2) + 5] = x^2 + x^2 + x^2 + 5 = 3x^2 + 5$$

27. Derivative of $f(x) = 2x^3$ at $x = 5$

30. Derivative of $f(x) = x^3 + x$ at $x = 3$

33. Derivative of $f(x) = \frac{2}{x}$ at $x = t$

36. Derivative of $f(x) = \tan x$ at $x = t$

39. Use the values obtained in Problem 37 as well as those in the following table obtained by estimating the slope of tangents.

x	f'(x)
0.7	0.0 (horizontal tangent)
5.8	0.0 (horizontal tangent)
-1.0	-2.0
3.4	1.7 (steepest tangent line)

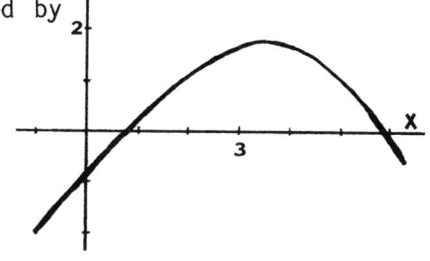

Problem Set 3.2

Problem Set 3.3 Rules for Finding Derivatives

3. $y = \pi x^2 \Rightarrow Dy = \pi \cdot 2x^1 = 2\pi x$

6. $y = 4x^{-2} \Rightarrow Dy = 4\cdot(-2)x^{-3} = -8x^{-3}$

9. $y = \dfrac{3}{5x^5} = \dfrac{3}{5}x^{-5} \Rightarrow Dy = \dfrac{3}{5}(-5)x^{-6} = \dfrac{-3}{x^6}$

12. $y = 2x^4 - 3x \Rightarrow Dy = 2\cdot 4x^3 - 3 = 8x^3 - 3$

15. $y = 5x^6 - 3x^5 + 11x - 9 \Rightarrow Dy = 30x^5 - 15x^4 + 11$

18. $y = 2x^{-6} + x^{-1} \Rightarrow Dy = 2(-6)x^{-7} + (-1)x^{-2} = -12x^{-7} - x^{-2}$

21. $y = \dfrac{1}{2x} + 2x = \dfrac{1}{2}x^{-1} + 2x \Rightarrow Dy = \dfrac{1}{2}(-1)x^{-2} + 2 = \dfrac{-1}{2x^2} + 2 = \dfrac{4x^2 - 1}{2x^2}$

24. $y = 3x(x^3 - 1) = 3(x^4 - x) \Rightarrow Dy = 3(4x^3 - 1)$

27. $y = (x^2 + 2)(x^3 + 1) = x^5 + 2x^3 + x^2 + 2 \Rightarrow Dy = 5x^4 + 6x^2 + 2x$.
    [Also try it using the Product Rule.]

30. $y = (x^4 + 2x)(x^3 + 2x^2 + 1) = x^7 + 2x^6 + 3x^4 + 4x^3 + 2x \Rightarrow Dy = 7x^6 + 12x^5 + 12x^3 + 12x^2 + 2$

33. $y = \dfrac{1}{3x^2 + 1} \Rightarrow Dy = \dfrac{(3x^2+1)\,D(1) - (1)D(3x^2+1)}{(3x^2+1)^2}$

$= \dfrac{(3x^2+1)(0) - (6x)}{(3x^2+1)^2} = \dfrac{-6x}{(3x^2+1)^2}$

36. $y = \dfrac{4}{2x^3-3x} \Rightarrow Dy = \dfrac{(2x^3-3x)\,D(4) - 4\,D(2x^3-3x)}{(2x^3-3x)^2} = \dfrac{(2x^3-3x)(0) - 4(6x^2-3)}{(2x^3-3x)^2}$

$= \dfrac{-4(6x^2-3)}{(2x^3-3x)^2} = \dfrac{-12(2x^2-1)}{[x(2x^2-3)]^2} = \dfrac{-12(2x^2-1)}{x^2(2x^2-3)^2}$

39. $y = \dfrac{2x^2-1}{3x+5} \Rightarrow Dy = \dfrac{(3x+5)\,D(2x^2-1) - (2x^2-1)\,D(3x+5)}{(3x+5)^2}$

$= \dfrac{(3x+5)(4x) - (2x^2-1)(3)}{(3x+5)^2} = \dfrac{6x^2+20x+3}{(3x+5)^2}$

42. $y = \dfrac{5x^2+2x-6}{3x-1} \Rightarrow Dy = \dfrac{(3x-1)\,D(5x^2+2x-6) - (5x^2+2x-6)\,D(3x-1)}{(3x-1)^2}$

$= \dfrac{(3x-1)(10x+2) - (5x^2+2x-6)(3)}{(3x-1)^2} = \dfrac{15x^2-10x+16}{(3x-1)^2}$

45. (a) $(f \cdot g)'(0) = f(0)g'(0) + g(0)f'(0) = (4)(5) + (-3)(-1) = 23$

(b) $(f+g)'(0) = f'(0) + g'(0) = (-1) + (5) = 4$

(c) $(f/g)'(0) = \dfrac{g(0)f'(0) - f(0)g'(0)}{[g(0)]^2} = \dfrac{(-3)(-1) - (4)(5)}{(-3)^2} = \dfrac{-17}{9} \approx -1.8889$

48. $D[f(x)g(x)h(x)] = [f(x)g(x)]Dh(x) + h(x)D[f(x)g(x)]$
$= f(x)g(x)Dh(x) + h(x)[f(x)Dg(x) + g(x)Df(x)]$
$= f(x)g(x)Dh(x) + h(x)f(x)Dg(x) + h(x)g(x)Df(x)$

An easy way to remember it is as $(fgh)' = f'gh + fg'h + fgh'$.

51. $y = x^3 - x^2 \Rightarrow Dy = 3x^2 - 2x$

Horizontal lines have slope zero so set $3x^2 - 2x = 0$ and solve for x.
$x(3x-2) = 0$ iff $x = 0$ or $x = 2/3$.

If $x = 0$, $y = 0$; if $x = 2/3$, $y = -4/27$. Therefore, the points at which the tangents are horizontal are $(0,0)$ and $(2/3, -4/27)$.

Problem Set 3.3

54. $s(t) = 4.5t^2 + 2t \Rightarrow v(t) = s'(t) = 9t + 2$.

Set $v(t) = 30$ and solve for t.
$9t + 2 = 30$ iff $t = 28/9 \approx 3.11$.

Thus, the instantaneous velocity will be 30 feet per second about 3.11 seconds after the ball starts to roll.

Problem Set 3.4 Derivatives of Sines and Cosines

3. $y = \cot x = \dfrac{\cos x}{\sin x} \Rightarrow Dy = \dfrac{(\sin x) D(\cos x) - (\cos x) D(\sin x)}{(\sin x)^2}$

$= \dfrac{(\sin x)(-\sin x) - (\cos x)(\cos x)}{(\sin x)^2} = \dfrac{-\sin^2 x - \cos^2 x}{\sin^2 x}$

$= \dfrac{-(\sin^2 x + \cos^2 x)}{\sin^2 x} = \dfrac{-1}{\sin^2 x} = -\csc^2 x.$

6. $y = \sin^2 x = (\sin x)(\sin x).$

$Dy = (\sin x) D(\sin x) + (\sin x) D(\sin x)$
$= (\sin x)(\cos x) + (\sin x)(\cos x) = 2 \sin x \cos x = \sin 2x.$

9. $y = \sin^2 x + \cos^2 x = 1 \Rightarrow Dy = D(1) = 0.$

12. $y = \dfrac{x^2 + 1}{x \sin x}.$

$Dy = \dfrac{(x \sin x) D(x^2+1) - (x^2+1) D(x \sin x)}{(x \sin x)^2}$ [Quotient Rule]

$= \dfrac{(x \sin x)(2x) - (x^2+1)[x D(\sin x) + (\sin x) D(x)]}{x^2 \sin^2 x}$ [Product Rule]

$= \dfrac{2x^2 \sin x - (x^2+1)[x(\cos x) + (\sin x)(1)]}{x^2 \sin^2 x}$

$= \dfrac{2x^2 \sin x - (x^3 \cos x + x^2 \sin x + x \cos x + \sin x)}{x^2 \sin^2 x}$

$= \dfrac{x^2 \sin x - x^3 \cos x - x \cos x - \sin x}{x^2 \sin^2 x}.$

15. The rate at which the seat is moving horizontally is the derivative of the horizontal coordinate, 30cos2t. [See Example 4.]

$$D(30\cos 2t) = 30\, D(\cos 2t)$$
$$= 30\, D(1-2\sin^2 t)$$
$$= 30[D(1) - 2\, D(\sin^2 t)]$$
$$= 30[0 - 2\sin 2t] \quad \text{[Use the result from Problem 6.]}$$
$$= -60\sin 2t.$$

The derivative evaluated at $t = \pi/4$ is $-60\sin(\pi/2) = -60$, so the seat is moving horizontally to the left at 60 feet per second when t is $\pi/4$ seconds.

18. $\lim\limits_{x \to 0} \dfrac{\sin(x/2)}{3x} = \lim\limits_{x \to 0} \dfrac{\sin(x/2)}{(6)(x/2)} = \dfrac{1}{6} \lim\limits_{x \to 0} \dfrac{\sin(x/2)}{(x/2)} = \dfrac{1}{6}(1) = \dfrac{1}{6}.$

21. $\lim\limits_{x \to 0} \dfrac{\tan x - \sin x}{x \cos x} = \lim\limits_{x \to 0} \dfrac{\dfrac{\sin x}{\cos x} - \sin x}{x \cos x} = \lim\limits_{x \to 0} \dfrac{\sin x - \sin x \cos x}{x \cos^2 x}$

$= \lim\limits_{x \to 0} \dfrac{(\sin x)(1 - \cos x)}{x \cos^2 x} = \lim\limits_{x \to 0} \left(\dfrac{\sin x}{x} \cdot \dfrac{1 - \cos x}{\cos^2 x} \right)$

$= \left(\lim\limits_{x \to 0} \dfrac{\sin x}{x} \right) \left(\lim\limits_{x \to 0} \dfrac{1 - \cos x}{\cos^2 x} \right) = (1)\left(\dfrac{1 - 1}{1^2} \right) = 0.$

24. If we set up a coordinate system as at the right, then $y = 2\sin t$ is the position and $v = D(2\sin t) = 2\cos t$ is the velocity of the cork at time t.

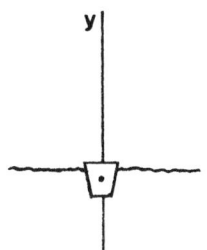

Summary:

t [sec]	0	$\pi/2$	π
v [cm/sec]	2	0	-2

Problem Set 3.5 The Chain Rule

3. $y = (5x^2+2x-8)^5.$

$D_x y = 5(5x^2+2x-8)^4\, D_x(5x^2+2x-8)$

$= 5(5x^2+2x-8)^4(10x+2) = 10(5x+1)(5x^2+2x-8)^4.$

Problem Set 3.5

6. $y = (2x^4-12x^2+11x-9)^{10}$

$D_x y = 10(2x^4-12x^2+11x-9)^9 \, D_x(2x^4-12x^2+11x-9)$

$ = 10(2x^4-12x^2+11x-9)^9(8x^3-24x+11)$

9. $y = (3x^4+x-8)^{-9}$

$D_x y = -9(3x^4+x-8)^{-10} \, D_x(3x^4+x-8)$

$ = -9(3x^4+x-8)^{-10}(12x^3+1)$

$ = \dfrac{-9(12x^3+1)}{(3x^4+x-8)^{10}}$

12. $y = \cos(4x^5-11x)$

$D_x y = [-\sin(4x^5-11x)] \, D_x(4x^5-11x)$

$ = [-\sin(4x^5-11x)](20x^4-11) = -(20x^4-11)\sin(4x^5-11x)$

15. $y = \left(\dfrac{x^2-1}{x+4}\right)^4$

$D_x y = 4\left(\dfrac{x^2-1}{x+4}\right)^3 D_x\left(\dfrac{x^2-1}{x+4}\right) = 4\left(\dfrac{x^2-1}{x+4}\right)^3 \dfrac{(x+4)(2x)-(x^2-1)(1)}{(x+4)^2}$

$ = 4\left(\dfrac{x^2-1}{x+4}\right)^3 \dfrac{2x^2+8x-x^2+1}{(x+4)^2} = \dfrac{4(x^2-1)^3(x^2+8x+1)}{(x+4)^5} = \dfrac{4(x-1)^3(x+1)^3(x^2+8x+1)}{(x+4)^5}$

18. $y = \cos\left(\dfrac{x^2-1}{x+4}\right)$

$D_x y = \left[-\sin\left(\dfrac{x^2-1}{x+4}\right)\right] D_x\left(\dfrac{x^2-1}{x+4}\right) \qquad \left[D_x\left(\dfrac{x^2-1}{x+4}\right) \text{ was obtained in Problem 15.}\right]$

$ = \left[-\sin\left(\dfrac{x^2-1}{x+4}\right)\right]\dfrac{x^2+8x+1}{(x+4)^2} = -\dfrac{x^2+8x+1}{(x+4)^2}\sin\left(\dfrac{x^2-1}{x+4}\right)$

21. $y = (2x-1)^3(x^2-3)^2$

$D_xy = (2x-1)^3 D_x[(x^2-3)^2] + (x^2-3)^2 D_x[(2x-1)^3]$ [Product Rule]

$= (2x-1)^3[2(x^2-3)D_x(x^2-3)] + (x^2-3)^2[3(2x-1)^2 D_x(2x-1)]$ [Chain Rule]

$= (2x-1)^3 \, 2(x^2-3)(2x) + (x^2-3)^2 \, 3(2x-1)^2(2)$

$= 2(2x-1)^2(x^2-3)[2x(2x-1) + 3(x^2-3)]$ [Factoring out common factor]

$= 2(2x-1)^2(x^2-3)(7x^2-2x-9) = 2(2x-1)^2(x^2-3)(7x-9)(x+1)$

24. $y = \dfrac{2x-3}{(x^2+4)^2}$

$D_xy = \dfrac{(x^2+4)^2 D_x(2x-3) - (2x-3) D_x[(x^2+4)^2]}{(x^2+4)^4}$ [Quotient Rule]

$= \dfrac{(x^2+4)^2(2) - (2x-3)[2(x^2+4) D_x(x^2+4)]}{(x^2+4)^4}$ [Chain Rule]

$= \dfrac{2(x^2+4)^2 - (2x-3) \, 2(x^2+4) \, (2x)}{(x^2+4)^4}$

$= \dfrac{2(x^2+4) - 4x(2x-3)}{(x^2+4)^3}$ [Reduce -- common factor of (x^2+4)]

$= \dfrac{-6x^2 + 12x + 8}{(x^2+4)^3} = \dfrac{-2(3x^2-6x-4)}{(x^2+4)^3}$

27. $D_t\left(\dfrac{3t-2}{t+5}\right)^3 = 3\left(\dfrac{3t-2}{t+5}\right)^2 D_t\left(\dfrac{3t-2}{t+5}\right) = 3\left(\dfrac{3t-2}{t+5}\right)^2 \dfrac{(t+5)(3)-(3t-2)(1)}{(t+5)^2}$

$= 3\left(\dfrac{3t-2}{t+5}\right)^2 \dfrac{3t+15-3t+2}{(t+5)^2} = 3\left(\dfrac{3t-2}{t+5}\right)^2 \dfrac{17}{(t+5)^2} = \dfrac{51(3t-2)^2}{(t+5)^4}$

30. $D_s\left(\dfrac{s^2-9}{s+4}\right)^3 = 3\left(\dfrac{s^2-9}{s+4}\right)^2 \dfrac{(s+4)(2s)-(s^2-9)(1)}{(s+4)^2} = 3\left(\dfrac{s^2-9}{s+4}\right)^2 \dfrac{2s^2+8s-s^2+9}{(s+4)^2}$

$= 3\left(\dfrac{s^2-9}{s+4}\right)^2 \dfrac{s^2+8s+9}{(s+4)^2} = \dfrac{3(s^2-9)^2(s^2+8s+9)}{(s+4)^4} = \dfrac{3(s+3)^2(s-3)^2(s^2+8s+9)}{(s+4)^4}$

Problem Set 3.5

33. $D_x\left(\dfrac{\sin x}{\cos 2x}\right)^3 = 3\left(\dfrac{\sin x}{\cos 2x}\right)^2 D_x\left(\dfrac{\sin x}{\cos 2x}\right)$

$= \dfrac{3\sin^2 x}{\cos^2 2x} \cdot \dfrac{(\cos 2x)(\cos x) - (\sin x)[(-\sin 2x)(2)]}{(\cos 2x)^2}$

$= \dfrac{3\sin^2 x(\cos 2x \cos x + 2 \sin 2x \sin x)}{\cos^4 2x}$

36. $G(t) = (t^2+9)^3(t^2-2)^4$

$G'(t) = (t^2+9)^3 D_t[(t^2-2)^4] + (t^2-2)^4 D_t(t^2+9)^3$

$= (t^2+9)^3[4(t^2-2)^3 D_t(t^2-2)] + (t^2-2)^4[3(t^2+9)^2 D_t(t^2+9)]$

$= (t^2+9)^3(4)(t^2-2)^3(2t) + (t^2-2)^4(3)(t^2+9)^2(2t)$

$G'(1) = (10)^3(4)(-1)^3(2) + (-1)^4(3)(10)^2(2) = -8000 + 600 = -7400$

[Note: We could have first simplified G'(t) to $2t(t^2+9)^2(t^2-2)^3(7t^2+30)$ which would be easier to use if we were going to evaluate G'(t) for several values of t. But since we only have to substitute t=1, the extra algebraic work is hardly warranted.]

39. $D_x[\sin^4(x^2+3x)] = D_x[\sin(x^2+3x)]^4$

$= 4[\sin(x^2+3x)]^3 D_x[\sin(x^2+3x)]$

$= 4[\sin(x^2+3x)]^3[\cos(x^2+3x)] D_x(x^2+3x)$

$= 4[\sin(x^2+3x)]^3[\cos(x^2+3x)](2x+3)$

$= 4(2x+3) \sin^3(x^2+3x) \cos(x^2+3x)$

[or think of it as] $y = u^4$, $u = \sin v$, $v = x^2+3x$

Then $D_x y = D_u y\, D_v u\, D_x v$

$= (4u^3)(\cos v)(2x+3)$

$= 4(\sin^3 v)(\cos v)(2x+3)$

$= 4(2x+3) \sin^3(x^2+3x) \cos(x^2+3x)$

42. $D_u\left[\cos^4\left(\frac{u+1}{u-1}\right)\right] = 4\left[\cos^3\left(\frac{u+1}{u-1}\right)\right]\left[-\sin\left(\frac{u+1}{u-1}\right)\right]\frac{(u-1)(1) - (u+1)(1)}{(u-1)^2}$

$= 4\left[\cos^3\left(\frac{u+1}{u-1}\right)\right]\left[-\sin\left(\frac{u+1}{u-1}\right)\right]\frac{-2}{(u-1)^2}$

$= \frac{8}{(u-1)^2}\sin\left(\frac{u+1}{u-1}\right)\cos^3\left(\frac{u+1}{u-1}\right)$

45. $D_x(\sin[\cos(\sin 2x)]) = \cos[\cos(\sin 2x)]\, D[\cos(\sin 2x)]$
$= \cos[\cos(\sin 2x)]\, [-\sin(\sin 2x)]\, D(\sin 2x)$
$= \cos[\cos(\sin 2x)]\, [-\sin(\sin 2x)]\, (\cos 2x)\, D(2x)$
$= \cos[\cos(\sin 2x)]\, [-\sin(\sin 2x)]\, (\cos 2x)\, (2)$
$= -2\cos[\cos(\sin 2x)]\, \sin(\sin 2x)\, (\cos 2x)$

48. (a) $x = 4\cos 2t$ and $y = 7\sin 2t$

Thus, $\left(\frac{x}{4}\right)^2 + \left(\frac{y}{7}\right)^2 = \left(\frac{4\cos 2t}{4}\right)^2 + \left(\frac{7\sin 2t}{7}\right)^2 = \cos^2 2t + \sin^2 2t = 1$

(b) $L = \sqrt{(x-0)^2 + (y-0)^2}$

$= \sqrt{(4\cos 2t)^2 + (7\sin 2t)^2} = \sqrt{16\cos^2 2t + 49\sin^2 2t}$

$= \sqrt{16(1-\sin^2 2t) + 49\sin^2 2t} = \sqrt{16 + 33\sin^2 2t}$

(c) D_tL is the rate of change with respect to time of the distance of P from the origin.

$D_tL = D_t\sqrt{16 + 33\sin^2 2t} = \frac{1}{2\sqrt{16 + 33\sin^2 2t}}\, D_t(16 + 33\sin^2 2t)$

$= \frac{(66\sin 2t)\, D_t(\sin 2t)}{2\sqrt{16 + 33\sin^2 2t}} = \frac{(66\sin 2t)(\cos 2t)\, D_t(2t)}{2\sqrt{16 + 33\sin^2 2t}}$

$= \frac{(66\sin 2t)(\cos 2t)(2)}{2\sqrt{16 + 33\sin^2 2t}} = \frac{66\sin 2t \cos 2t}{\sqrt{16 + 33\sin^2 2t}}$

At $t = \pi/8$: $D_tL = \frac{66(\sqrt{2}/2)(\sqrt{2}/2)}{\sqrt{16 + 33(\sqrt{2}/2)^2}} = \frac{33}{\sqrt{65/2}} \approx 5.7886$ [ft/sec]

Problem Set 3.5

51. (a) First convert minutes to seconds and revolutions to radians.

$$60 \text{ rev/min} = \frac{60 \text{ rev}}{1 \text{ min}} \times \frac{1 \text{ min}}{60 \text{ sec}} \times \frac{2\pi \text{ rad}}{1 \text{ rev}} = \frac{2\pi \text{ rad}}{1 \text{ sec}} = 2\pi \text{ rad/sec}$$

Let the angle of revolution be $\beta = (2\pi \text{ rad/sec})(t \text{ sec}) = 2\pi t$ rad.

Then $P = (\cos\beta, \sin\beta) = (\cos 2\pi t, \sin 2\pi t)$

(b) $y_Q = y_P + L$ [See figure at right.]

$\qquad = y_P + \sqrt{25 - x_P^2}$

$y_Q(t) = \sin 2\pi t + \sqrt{25 - \cos^2 2\pi t}$

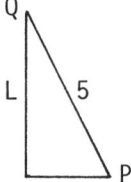

(c) The velocity of Q is then $v_Q(t)$

$= y_Q'(t) = (\cos 2\pi t)(2\pi) + \dfrac{1}{2\sqrt{25 - \cos^2 2\pi t}}(-2\cos 2\pi t)(-\sin 2\pi t)(2\pi)$

$= 2\pi \cos 2\pi t + \dfrac{2\pi \sin 2\pi t \cos 2\pi t}{\sqrt{25 - \cos^2 2\pi t}}$

Problem Set 3.6   Leibniz Notation

3. $y = \dfrac{3}{x+1}$

$y_1 = \dfrac{3}{3.34}$ and $y_2 = \dfrac{3}{3.31}$ so $\Delta y = y_2 - y_1 = \dfrac{3}{3.31} - \dfrac{3}{3.34} \approx 0.00814$

6. $y = x^3$

$\dfrac{\Delta y}{\Delta x} = \dfrac{(x+\Delta x)^3 - x^3}{\Delta x} = \dfrac{(x^3 + 3x^2\Delta x + 3x\Delta x^2 + \Delta x^3) - x^3}{\Delta x}$

$\qquad = \dfrac{\Delta x(3x^2 + 3x\Delta x + \Delta x^2)}{\Delta x} = 3x^2 + 3x\Delta x + \Delta x^2 \quad [\Delta x \neq 0]$

$\dfrac{dy}{dx} = \lim_{\Delta x \to 0}(3x^2 + 3x\Delta x + \Delta x^2) = 3x^2$

9. $y = u^3$ and $u = x^2 + 3x$

$\frac{dy}{dx} = \frac{dy}{du}\frac{du}{dx} = (3u^2)(2x+3) = 3(x^2+3x)^2(2x+3)$

12. Let $y = u^2$ and $u = \sin x$

$\frac{dy}{dx} = \frac{dy}{du}\frac{du}{dx} = (2u)(\cos x) = 2 \sin x \cos x = \sin 2x$ [Double-angle identity]

15. $y = \cos(x^2) \sin^2 x = (\cos x^2)(\sin x)^2$

$\frac{dy}{dx} = (\cos x^2) \frac{d}{dx}[(\sin x)^2] + (\sin x)^2 \frac{d}{dx}(\cos x^2)$

$= (\cos x^2) \, 2(\sin x) \frac{d}{dx}(\sin x) + (\sin x)^2 (-\sin x^2) \frac{d}{dx}(x^2)$

$= (\cos x^2) \, 2(\sin x)(\cos x) + (\sin x)^2 (-\sin x^2)(2x)$

$= 2 \sin x \, (\cos x^2 \cos x - x \sin x^2 \sin x)$

18. $y = \sin[(x^2+3)^4]$

$\frac{dy}{dx} = \cos[(x^2+3)^4] \frac{d}{dx}[(x^2+3)^4] = \cos[(x^2+3)^4] \, [4(x^2+3)^3] \frac{d}{dx}(x^2+3)$

$= \cos[(x^2+3)^4] \, [4(x^2+3)^3] \, (2x) = 8x(x^2+3)^3 \cos(x^2+3)^4$

21. $\frac{d}{dt}(\sin^3 t + \cos^3 t) = 3 \sin^2 t \cos t + 3 \cos^2 t \, (-\sin t)$

$= 3 \sin t \cos t \, (\sin t - \cos t)$

24. $D_t[u^3+3u] = D_u[u^3+3u] \, D_t u = (3u^2+3)(2t) = [3(t^2)^2+3](2t) = 6t(t^4+1)$

27. (a) $(f+g)'(3) = f'(3) + g'(3) = (-1) + (-4) = -5$

(b) $(f \cdot g)'(3) = f'(3)g(3) + f(3)g'(3) = (-1)(3) + (2)(-4) = -11$

Problem Set 3.6

27. (c) $(f/g)'(3) = \dfrac{g(3)f'(3) - f(3)g'(3)}{[g(3)]^2} = \dfrac{(3)(-1) - (2)(-4)}{(3)^2} = \dfrac{5}{9} \approx 0.5556$

(d) $(f \circ g)'(3) = f'[g(3)] \cdot g'(3) = f'(3) \cdot (-4) = (-1)(-4) = 4$

30. (a) The tangent to the graph of f where x = 2 seems to pass through (0,0) and (4,5) so its slope is 5/4 and hence f'(2) ≈ 5/4 = 1.25. The tangent to the graph of g where x = 2 seems to be horizontal so g'(2) ≈ 0. Also, f(2) ≈ 2.7, and g(2) ≈ 1.0.

$(f/g)'(2) = \dfrac{g(2)f'(2) - f(2)g'(2)}{[g(2)]^2} \approx \dfrac{(1)(1.25) - (2.7)(0)}{(1)^2} = 1.25$

(b) f(3) ≈ 4.1. The tangent to the graph of g where x = 4.1 seems to pass through the points (2,0) and (5,4) so its slope is 4/3 and hence g'(4.1) ≈ 4/3. The tangent to the graph of f where x=3 seems to pass through the points (1,0) and (4,6) so its slope is 2 and hence f'(3) ≈ 2.

$(g \circ f)'(3) = g'[f(3)]f'(3) \approx g'(4.1) \cdot (2) \approx (4/3)(2) = 8/3 \approx 2.7$

Problem Set 3.7 Higher Order Derivatives

3. $y = (2x+5)^4$

$\dfrac{dy}{dx} = 4(2x+5)^3(2) = 8(2x+5)^3$

$\dfrac{d^2y}{dx^2} = 24(2x+5)^2(2) = 48(2x+5)^2$

$\dfrac{d^3y}{dx^3} = 96(2x+5)^1(2) = 192(2x+5)$

6. $y = \cos(x^2)$

$\dfrac{dy}{dx} = [-\sin(x^2)](2x) = -2x \sin(x^2)$

$\dfrac{d^2y}{dx^2} = (-2x)[\cos(x^2)](2x) + [\sin(x^2)](-2) = -4x^2 \cos(x^2) - 2\sin(x^2)$

$\dfrac{d^3y}{dx^3} = (-4x^2)[-\sin(x^2)](2x) + [\cos(x^2)](-8x) - 2[\cos(x^2)](2x)$

6. (continued)

$$= 8x^3\sin(x^2) - 8x\cos(x^2) - 4x\cos(x^2)$$
$$= 8x^3\sin(x^2) - 12x\cos(x^2)$$
$$= 4x[2x^2\sin(x^2) - 3\cos(x^2)]$$

9. $f(x) = 2x^3 - 7 \Rightarrow f'(x) = 6x^2 \Rightarrow f''(x) = 12x \Rightarrow f''(2) = 12(2) = 24$

12. $f(u) = (2u-5)^{-1}$

$f'(u) = -(2u-5)^{-2}(2) = -2(2u-5)^{-2}$

$f''(u) = 4(2u-5)^{-3}(2) = \dfrac{8}{(2u-5)^3}$; therefore, $f''(2) = \dfrac{8}{(4-5)^3} = -8$

15. [Before starting this problem, review the double-angle identity for the sine function on page 54 of the text. It is a useful identity for simplifying the work of derivative-taking in this and other problems.]

$f(x) = \sin^2(\pi x) = (\sin\pi x)^2$

$f'(x) = 2(\sin\pi x)(\cos\pi x)(\pi) = \pi\sin 2\pi x$ [double-angle identity]

$f''(x) = \pi(\cos 2\pi x)(2\pi) = 2\pi^2\cos 2\pi x$; so, $f''(2) = 2\pi^2\cos 4\pi = 2\pi^2 \approx 19.7392$

18. $D_x^n(a_n x^n + a_{n-1} x^{n-1} + \cdots + a_1 x + a_0)$

$= a_n D_x^n(x^n) + a_{n-1} D_x^n(x^{n-1}) + \cdots + a_1 D_x^n(x) + D_x^n(a_0)$

$= a_n n! + a_{n-1}(0) + \cdots + a_1(0) + (0) = a_n n!$

21. $f(x) = x^3 + 3x^2 - 45x - 6$

$f'(x) = 3x^2 + 6x - 45 = 3(x^2 + 2x - 15) = 3(x+5)(x-3)$

$\quad f'(x) = 0$ iff $x = -5$ or $x = 3$

$f''(x) = 6x + 6$

$\quad f''(-5) = 6(-5) + 6 = -24$; $f''(3) = 6(3) + 6 = 24$

Problem Set 3.7

24. $s(t) = t^3 - 6t^2 = t^2(t-6)$

(a) $v(t) = s'(t) = 3t^2 - 12t = 3t(t-4)$

Auxiliary axis for $v(t)$:

```
          (+)    (0)    (-)    (0)    (+)
         ─────────●─────────────●──────────
                  0             4            t
```

$a(t) = v'(t) = 6t - 12 = 6(t-2)$

Auxiliary axis for $a(t)$:

```
               (-)           (0)          (+)
         ──────────────────────●────────────────
                               2                  t
```

(b) The object is moving to the right (values of s increasing as t increases) when v is positive. Read from the auxiliary axis for v that this is when t is less than 0 or greater than 4.

(c) It is moving left (values of s decreasing as t increases) when v is negative; i.e., when t is between 0 and 4.

(d) The acceleration is negative when t is less than 2.

(e)
```
           t=4 ⊂══════════════════⇒
        ⇒═══════════════⊃ t=0
    ──────●──────────────────●───────────────────
         -32                 0                     s
```

27. $s(t) = t^2 + \dfrac{16}{t} = t^2 + 16t^{-1}$, $t > 0$

(a) $v(t) = s'(t) = 2t - 16t^{-2} = 2t - \dfrac{16}{t^2} = \dfrac{2t^3 - 16}{t^2} = \dfrac{2(t-2)(t^2+2t+4)}{t^2}$

Auxiliary axis for $v(t)$:

```
                  (-)           (0)          (+)
         ●──────────────────────●────────────────
         0                      2                  t
```

$a(t) = v'(t) = 2 + 32t^{-3} = 2 + \dfrac{32}{t^3} = \dfrac{2t^3 + 32}{t^3} = \dfrac{2(t^3 + 16)}{t^3}$

Auxiliary axis for $a(t)$:

```
                                (+)
         ●──────────────────────────────────────
         0                                         t
```

(b) The object is moving to the right (values of s increasing as t increases) when v is positive; i.e., when t is greater than 2.

(c) It is moving left (values of s decreasing as t increases) when v is negative; i.e., when t is between 0 and 2.

(d) The acceleration is never negative.

(e)
```
              t=2 ⊂══════════════⇒
                          ⇐═══════
    ──────────────●─────────────────────────
                  12                           s
```

30. $s(t) = \frac{1}{10}(t^4-14t^3+60t^2)$

$v(t) = s'(t) = \frac{1}{10}(4t^3-42t^2+120t) = \frac{1}{5}(2t^3-21t^2+60t)$

$a(t) = v'(t) = \frac{1}{5}(6t^2-42t+60) = \frac{6}{5}(t^2-7t+10) = \frac{6}{5}(t-2)(t-5)$

$a(t) = 0$ iff $t = 2$ or $t = 5$

$v(2) = (0.2)[2(2)^3-21(2)^2+60(2)] = 10.4$

$v(5) = (0.2)[2(5)^3-21(5)^2+60(5)] = 5$

33. Equations of Motion: $s(t) = -16t^2+48t+256$, $t \geq 0$ [ft]

$v(t) = s'(t) = -32t+48$ [ft/sec]

(a) Initial velocity is v when t=0. $v(0) = -32(0)+48 = 48$ [ft/sec].

(b) Maximum height is reached when v=0. $v(t)=0$ iff $t= \frac{48}{32} = 1.5$ [sec].

(c) Maximum height is $s(1.5) = -16(1.5)^2+48(1.5)+256 = 292$ [ft].

(d) Hits ground when s=0. $s(t)=0$ iff $-16t^2+48t+256 = 0$.

Using the quadratic formula obtain $t = \frac{3 \pm \sqrt{73}}{2} \approx -2.77, 5.77$ [sec].

The former value is outside the domain. Therefore, the object hits the ground in about 5.77 seconds.

(e) $v\frac{3 + \sqrt{73}}{2} = -32\frac{3 + \sqrt{73}}{2} + 48 \approx -136.70$ [ft/sec]

Therefore, it hits the ground with a speed of about 136.70 ft/sec.

36. Equations of Motion: $s(t) = v_0 t + 16t^2$ and $v(t) = s'(t) = v_0 + 32t$.

Note: Downward is the direction of increasing values of s. We are given that $v(3) = 140$ [ft/sec] and are asked to find s(3).

$v(3) = 140 \Rightarrow v_0 + 32(3) = 140 \Rightarrow v_0 = 44$

$\Rightarrow s(t) = 44t+16t^2 \Rightarrow s(3) = 44(3)+16(3)^2 = 276$

Therefore, the cliff is 276 ft high.

Problem Set 3.7

39. In this problem we will use the fact that $f(x) = |x|$ => $f'(x) = \frac{|x|}{x}$.
    [See Problem 52 of Section 3.5.]

    $s(t) = t^3 - 3t^2 - 24t - 6$

    $v(t) = s'(t) = 3t^2 - 6t - 24 = 3(t-4)(t+2)$

    Auxiliary axis for v(t): $\underline{\quad (+) \quad\quad (0) \quad\quad (-) \quad\quad (0) \quad\quad (+) \quad\quad}$
    $ -2 4 t$

    $a(t) = v'(t) = 6t - 6 = 6(t-1)$

    Auxiliary axis for a(t): $\underline{\quad\quad (-) \quad\quad\quad (0) \quad\quad\quad (+) \quad\quad}$
    $ 1 t$

    Let $r(t) = |v(t)|$ denote the speed. The rate of change of the speed with respect to time is then

    $$r'(t) = \frac{|v(t)|}{v(t)} v'(t) = \frac{|v(t)|}{v(t)} a(t)$$

    Use the equation for r'(t) and the auxiliary axes for v(t) and a(t) to develop the table.

Interval	v(t)	a(t)	Hence r'(t)
$(-\infty, -2)$	+	−	−
$(-2, 1)$	−	−	+
$(1, 4)$	−	+	−
$(4, \infty)$	+	+	+

    The point is slowing down (speed is decreasing) when r'(t) is negative; i.e., when t is less than −2 or between 1 and 4.

Problem Set 3.8 Implicit Differentiation

[We will use D for D_x and D^n for D_x^n in this section.]

3. $xy = 4$ => $D(xy) = D(4)$ => $x(Dy) + y(1) = 0$

   Therefore, $Dy = \frac{-y}{x}$

6. $x^3 - 3x^2y + 19xy = 0$

$D(x^3 - 3x^2y + 19xy) = D(0)$

$3x^2 - [(3x^2)(Dy) + (y)(6x)] + [(19x)(Dy) + (y)(19)] = 0$

$(-3x^2 + 19x)Dy = -3x^2 + 6xy - 19y$

Therefore, $Dy = \dfrac{-3x^2 + 6xy - 19y}{(-3x^2 + 19x)}$

9. $6x - \sqrt{2xy} + xy^3 = y^2$

$D(6x - \sqrt{2xy} + xy^3) = D(y^2)$

$6 - \dfrac{1}{2\sqrt{2xy}} D(2xy) + [(x) D(y^3) + (y^3)(1)] = 2y\, Dy$

$6 - \dfrac{[(2x)(Dy) + (y)(2)]}{2\sqrt{2xy}} + (x)(3y^2 Dy) + y^3 = 2y\, Dy$

$12\sqrt{2xy} - 2x\, Dy - 2y + 6xy^2\sqrt{2xy}\, Dy + 2y^3\sqrt{2xy} = 4y\sqrt{2xy}\, Dy$

$(-2x + 6xy^2\sqrt{2xy} - 4y\sqrt{2xy})Dy = -12\sqrt{2xy} + 2y - 2y^3\sqrt{2xy}$

Therefore, $Dy = \dfrac{-6\sqrt{2xy} + y - y^3\sqrt{2xy}}{-x + 3xy^2\sqrt{2xy} - 2y\sqrt{2xy}}$

12. $\cos(xy) = y^2 + 2x$

$D[\cos(xy)] = D[y^2 + 2x]$
$[-\sin(xy)]\, D(xy) = 2y\, Dy + 2$
$[-\sin(xy)][x\, Dy + y(1)] = 2y\, Dy + 2$
$-x \sin(xy)\, Dy - y \sin(xy) = 2y\, Dy + 2$
$[-x \sin(xy) - 2y]\, Dy = 2 + y \sin(xy)$

Therefore, $Dy = \dfrac{2 + y \sin(xy)}{-2y - x \sin(xy)}$

15. $\sin(xy) = y$
$D[\sin(xy)] = D[y]$
$[\cos(xy)]\, D(xy) = Dy$
$[\cos(xy)][x\, Dy + y(1)] = Dy$
$[\cos(xy)][x\, Dy + y] = Dy$

Problem Set 3.8

15. (continued)

[We could (1) solve for Dy (as was done in Problems 1-12) and then (2) replace x by $\pi/2$ and y by 1 to find the value of Dy at the point $(\pi/2,1)$, but it will involve less work if we reverse those steps.]

At $(\pi/2,1)$: $[\cos(\pi/2)][(\pi/2) \text{ Dy} + (1)] = \text{Dy}$
$[0][(\pi/2) \text{ Dy} + 1] = \text{Dy}$
$\text{Dy} = 0$

Therefore, at $(\pi/2,1)$, the slope of the tangent line is zero, so the tangent line is horizontal. Its equation is then $y = 1$.

18. $\sqrt{y} + xy^2 = 5$

$D(\sqrt{y} + xy^2) = D(5)$

$\frac{1}{2\sqrt{y}} \text{Dy} + [(x) D(y^2) + (y^2)(1)] = 0$

$\frac{1}{2\sqrt{y}} \text{Dy} + x(2y \text{ Dy}) + y^2 = 0$

Slope at $(4,1)$: $\frac{1}{2\sqrt{1}} \text{Dy} + (4)(2)(1) \text{Dy} + (1)^2 = 0 \Rightarrow \text{Dy} = \frac{-2}{17}$

Equations of tangent at $(4,1)$: $y - 1 = \frac{-2}{17}(x - 4)$ [Point-slope form]

$y = \frac{-2}{17}x + \frac{25}{17}$ [Slope-intercept form]

21. $y = x^{1/3} + x^{-1/3}$

$\frac{dy}{dx} = (1/3)x^{-2/3} + (-1/3)x^{-4/3} = \frac{1}{3\sqrt[3]{x^2}} - \frac{1}{3\sqrt[3]{x^4}} = \frac{\sqrt[3]{x^2} - 1}{3\sqrt[3]{x^4}}$

24. $y = (x^3 - 2x)^{1/3}$

$\frac{dy}{dx} = \frac{1}{3}(x^3 - 2x)^{-2/3}(3x^2 - 2) = \frac{3x^2 - 2}{3\sqrt[3]{(x^3 - 2x)^2}}$

27. $y = \sqrt{x^2 + \sin x}$

$\frac{dy}{dx} = \frac{1}{2\sqrt{x^2 + \sin x}}(2x + \cos x) = \frac{2x + \cos x}{2\sqrt{x^2 + \sin x}}$

30. $y = (1 + \sin 5x)^{1/4}$

$\frac{dy}{dx} = \frac{1}{4}(1 + \sin 5x)^{-3/4}[0 + (\cos 5x)(5)] = \frac{5 \cos 5x}{4\sqrt[4]{(1+\sin 5x)^3}}$

33. $s^2 t + t^3 = 1$

$\frac{d}{dt}(s^2 t + t^3) = \frac{d}{dt}(1)$ $\frac{d}{ds}(s^2 t + t^3) = \frac{d}{dt}(1)$

$[(s^2)(1) + t(2s\frac{ds}{dt})] + 3t^2 = 0$ $[(s^2)\frac{dt}{ds} + (t)(2s)] + 3t^2 \frac{dt}{ds} = 0$

$2st\frac{ds}{dt} = -3t^2 - s^2$ $(s^2 + 3t^2)\frac{dt}{ds} = -2st$

Therefore, $\frac{ds}{dt} = \frac{-3t^2-s^2}{2st} = \frac{s^2+3t^2}{-2st}$ Therefore, $\frac{dt}{ds} = \frac{-2st}{s^2+3t^2}$

Notice that $\frac{ds}{dt} = \frac{1}{dt/ds}$

36. $8(x^2+y^2)^2 = 100(x^2-y^2)$

$D[8(x^2+y^2)^2] = D[100(x^2-y^2)]$

$16(x^2+y^2)(2x + 2y\, Dy) = 100(2x - 2y\, Dy)$

At (3,1): $16(9+1)(6 + 2\, Dy) = 100(6 - 2\, Dy)$
          $960 + 320\, Dy = 600 - 200\, Dy$
          $520\, Dy = -360$
          $Dy = -360/520 = -9/13$

Slope of the tangent at (3,1): $-9/13$.

Slope of the normal at (3,1): $13/9$.

Equations of normal at (3,1): $y-1 = (13/9)(x-3)$ [Point-slope form]
                               $y = (13/9)x - 10/3$ [Slope-intercept form]

39. $2x^2 y - 4y^3 = 4$

First Derivative: $D[2x^2 y - 4y^3] = D[4]$

$[(2x^2)(Dy) + (y)(4x)] - [(12y^2)(Dy)] = 0$

Problem Set 3.8   65

39. (continued)

$$2x^2 Dy + 4xy - 12y^2 Dy = 0 \quad (*)$$

Second derivative: $D[2x^2 Dy + 4xy - 12y^2 Dy] = D[0]$

$[(2x^2)(D^2 y)+(Dy)(4x)]+[(4x)(Dy)+(y)(4)]-[(12y^2)(D^2 y)+(Dy)(24y)(Dy)] = 0$

$2x^2 D^2 y + 4x\, Dy + 4x\, Dy + 4y - 12y^2 D^2 y - 24y(Dy)^2 = 0$

$(2x^2 - 12y^2) D^2 y + 8x\, Dy - 24y(Dy)^2 + 4y = 0 \quad (**)$

Now use (*) at (2,1): $2(4)Dy + 4(2)(1) - 12(1)Dy = 0 \Rightarrow Dy = 2$

Now use (**) at (2,1) and the fact that $Dy = 2$ at (2,1):

$[2(4)-12(1)]D^2 y + 8(2)(2) - 24(1)(4) + (4) = 0 \Rightarrow D^2 y = -15$

42. $xy = 1$ $x^2 - y^2 = 1$

$D(xy) = D(1)$ $D(x^2 - y^2) = D(1)$

$(x)(Dy) + (y)(1) = 0$ $2x - 2y\, Dy = 0$

$Dy = \dfrac{-y}{x}$ $Dy = \dfrac{x}{y}$

At points of intersection, the slopes of tangents are negative reciprocals so the tangents, and hence the curves, are perpendicular.

45. Determine the point of intersection in the first quadrant:

$\begin{bmatrix} x^2 - xy + 2y^2 = 28 \\ y = 2x \end{bmatrix} \Rightarrow x^2 - x(2x) + 2(2x)^2 = 28$

$\Rightarrow x^2 = 4$

$\Rightarrow x = 2$ (in first quadrant)

Therefore, the point of intersection is (2,4).

Next determine the derivatives:

(*) $y = 2x$ (**) $x^2 - xy + 2y^2 = 28$

       $Dy = 2$ $D[x^2 - xy + 2y^2] = D[28]$

                                     $2x - [x\, Dy + y(1)] + 4y\, Dy = 0$

45. (continued)

$$(-x+4y) \, Dy + (2x-y) = 0$$

$$Dy = \frac{2x-y}{-x+4y}$$

Therefore, the slopes at $(2,4)$ are 2 (*) and 0 (**).

Then use the formula in Problem 44 with $m_1 = 2$ and $m_2 = 0$:

$$\tan\theta = \frac{(0) - (2)}{1 + (0)(2)} = -2, \text{ so } \theta \approx 2.0344 \text{ radians } [\text{About } 116.6°]$$

Problem Set 3.9 Related Rates

3. Step 1: Let t denote time elapsed after the plane is overhead [hr]. Let x and y be as indicated in Figure 7 of the text [mi].

Step 2: We are given that $\frac{dx}{dt} = 240$ [mph], the plane's speed.

We are to find $\frac{dy}{dt}$ when $t = 1/120$ [hr]

Step 3: $1 + x^2 = y^2$ [Pythagorean Theorem]

Step 4: $\frac{d}{dt}(1 + x^2) = \frac{d}{dt}(y^2)$

$2x \frac{dx}{dt} = 2y \frac{dy}{dt}$ [or] $x \frac{dx}{dt} = y \frac{dy}{dt}$

Step 5: When $t = 1/120$, $x = 2$ (from hint) and
$y = \sqrt{1 + 4} = \sqrt{5}$ (from equation in Step 3).

$(2)(240) = (\sqrt{5}) \frac{dy}{dx}$ (substituting into equation in Step 4).

Therefore, $\frac{dy}{dt} = \frac{480}{\sqrt{5}} \approx 214.6625$

Conclusion: The distance of the plane from the observer is increasing at about 214.66 miles per hour 30 seconds after the plane goes overhead.

6. Step 1: Let t denote time elapsed [sec].
Let x and y be as indicated in the figure to the right [ft].

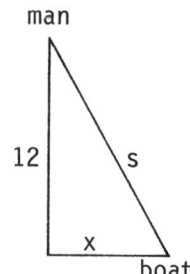

Step 2: We are given $\frac{ds}{dt}$ = -3 [ft/sec] since length of rope is decreasing at 3 ft/sec.

We are to find $\frac{dx}{dt}$ when s = 20.

Step 3: $144 + x^2 = s^2$ [Pythagorean Theorem]

Step 4: $\frac{d}{dt}(144 + x^2) = \frac{d}{dt}(s^2)$

$2x \frac{dx}{dt} = 2s \frac{ds}{dt}$ [or] $x \frac{dx}{dt} = s \frac{ds}{dt}$

Step 5: When s = 20, $x = \sqrt{(20)^2 - 144} = 16$

Therefore, $(16) \frac{dx}{dt} = (20)(-3)$, so $\frac{dx}{dt} = -3.75$

Conclusion: When 20 feet of rope is still out, the boat is approaching the dock (x is decreasing) at 3.75 ft/sec.

9. Step 1: Let t denote time elapsed [sec].
Let r and h be as indicated in Figure 8 [ft].
Let V denote the volume of the conical pile [ft^3]

Step 2: We are given $\frac{dV}{dt}$ = 16 [ft^3/sec] since the volume of the pile is increasing at the same rate as the sand pouring from the pipe.

We are to find $\frac{dh}{dt}$ when h = 4.

Step 3: $V = (1/3)\pi r^2 h$ and $h = (1/4)(2r) = (1/2)r$, so
$V = (1/3)\pi (2h)^2 h = (4\pi/3)h^3$

Step 4: $\frac{dV}{dt} = (4\pi/3)(3h^2)\frac{dh}{dt} = 4\pi h^2 \frac{dh}{dt}$

Step 5: When h = 4, $(16) = 4\pi(4)^2 \frac{dh}{dt}$, so $\frac{dh}{dt} = \frac{1}{4\pi} \approx 0.07958$

Conclusion: When the pile if 4 feet high, the altitude is increasing at about 0.07958 ft/sec [almost 1 in/sec].

12. Step 1: Let t denote time elapsed [sec].
    Let x and y be as indicated in the figure to the right [units].

    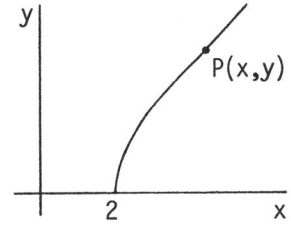

    Step 2: We are given $\frac{dx}{dt} = 5$ [units/sec]

    We are to find $\frac{dy}{dt}$ when x = 3.

    Step 3: $y = \sqrt{x^2 - 4}$, $x \geq 2$

    Step 4: $\frac{dy}{dt} = \frac{1}{2\sqrt{x^2-4}} \cdot 2x \frac{dx}{dt} = \frac{5x}{\sqrt{x^2-4}}$ since $\frac{dx}{dt} = 5$

    Step 5: When x = 3, $\frac{dy}{dt} = \frac{5(3)}{\sqrt{(9)-4}} = \frac{15}{\sqrt{5}} \approx 6.7082$

    Conclusion: The y-coordinate is increasing at about 6.7082 units per second, when x = 3.

15. Step 1: Let t denote time elapsed [min].
    Let x [km] and β [rad] be as indicated in the figure to the right.

    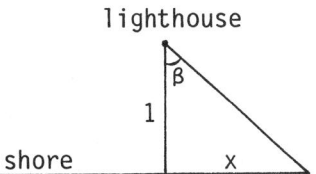

    Step 2: We are given $\frac{d\beta}{dt} = 4\pi$ [rad/min]

    We are to find $\frac{dx}{dt}$ when x = 1/2.

    Step 3: $\tan\beta = x/1$, so $x = \tan\beta$

    Step 4: $\frac{dx}{dt} = \sec^2\beta \frac{d\beta}{dt} = 4\pi\sec^2\beta$

    Step 5: When x = 1/2, $\sec\beta = \sqrt{1 + \tan^2\beta} = \sqrt{1 + x^2} = \sqrt{1 + (1/4)} = \frac{\sqrt{5}}{2}$

    and so $\frac{dx}{dt} = 4\pi(\sqrt{5}/2)^2 = 5\pi \approx 15.7080$.

    Conclusion: The beam is moving along the shoreline at about 15.7080 kilometers per minute when it is 1/2 kilometer past the point opposite the lighthouse.

Problem Set 3.9

18. Step 1: Let t denote time elapsed [min].
    Let β [rad] be as indicated in
       the figure to the right.
    Let A denote the area [sq. cm.]

    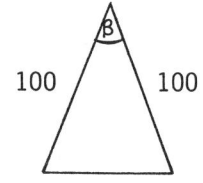

    Step 2: We are given $\frac{d\beta}{dt} = \frac{1}{10}$ [rad/min]

    We are to find $\frac{dA}{dt}$ when $\beta = \pi/6$.

    Step 3: $A = (1/2)(100)(100)\sin\beta = 5000 \sin\beta$ (from hint)

    Step 4: $\frac{dA}{dt} = 5000(\cos\beta)\frac{d\beta}{dt}$

    Step 5: When $\beta = \pi/6$, $\frac{dA}{dt} = 5000(\cos[\pi/6])(1/10) = 500(\sqrt{3}/2)$
    $= 250\sqrt{3} \approx 433.0127$.

    Conclusion: When the vertex angle opposite the base is π/6 radians,
    the area of the triangle is increasing at about 433.0127
    square centimeters per minute.

21. Step 1: Let t denote time elapsed [hr].
    Let h [ft] be as indicated in Figure 11 in the text.
    Let V denote the volume [cubic ft.]

    Step 2: We are given $\frac{dV}{dt} = -2$ [ft³/hr]   since the volume is decreasing at 2 cubic feet per hour.

    We are to find $\frac{dh}{dt}$ when $h = 3$.

    Step 3: $V = \pi h^2[8-(h/3)] = 8\pi h^2 - (1/3)\pi h^3$

    Step 4: $\frac{dV}{dt} = (16\pi h - \pi h^2)\frac{dh}{dt}$

    Step 5: When $h = 3$, $-2 = (16\pi[3]-\pi[9])\frac{dh}{dt}$,

    so $\frac{dh}{dt} = \frac{-2}{39\pi} \approx -0.01632$.

    Conclusion: When the water is 3 feet deep, the water level is
    decreasing at about 0.01632 feet per hour [almost 0.2
    inches per hour].

24. The only difference is that the formula for the volume of the water in the tank is $V = \pi h^2[r-(h/3)] = \pi h^2[20-(h/3)] = 20\pi h^2 - (1/3)\pi h^3$.

$\frac{dV}{dt} = (40\pi h - \pi h^2)\frac{dh}{dt}$.

At $t=7$, $\frac{dV}{dt} \approx (40\pi[15] - \pi[15]^2)(-3) = -1125\pi$.

Conclusion: The residents were using water at the rate of $2400 + 1125\pi$ (about 5934.3) cubic feet per hour at 7:00 a.m.

Problem Set 3.10 Differentials and Approximations

3. $y = (3+2x^3)^{-4}$.

$dy = -4(3+2x^3)^{-5}(0+6x^2)dx = -24x^2(3+2x^3)^{-5}dx$.

6. $y = (6x^8-11x^5+x^2)^{-2/3}$.

$dy = -\frac{2}{3}(6x^8-11x^5+x^2)^{-5/3}(48x^7-55x^4+2x)dx$.

9. $y = f(x) = x^3$.

$dy = 3x^2dx$.

(a) When $x = 0.5$ and $dx = 1$,
$dy = 3(0.25)(1) = 0.75$.

(b) When $x = -1$ and $dx = 0.75$,
$dy = 3(1)(0.75) = 2.25$.

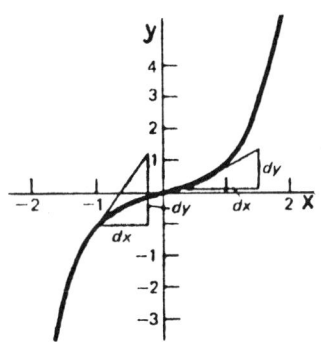

12. Let $y = f(x) = \frac{1}{x}$.

(a) For $x = 1$ and $\Delta x = dx = 0.5$:

$\Delta y = f(1+0.5) - f(1) = f(1.5) - f(1) = \frac{1}{1.5} - \frac{1}{1} = -\frac{1}{3} \approx -0.3333$

(b) For $x = -2$ and $\Delta x = dx = 0.75$:

$\Delta y = f(-2+0.75) - f(-2) = f(-1.25) - f(-2) = \frac{1}{-1.25} - \frac{1}{-2} = \frac{-3}{10} = -0.3$

15. Let $y = f(x) = \sqrt{x}$

$f(x+dx) \approx f(x) + dy = \sqrt{x} + \frac{1}{2\sqrt{x}}dx$

402 is near 400 and $\sqrt{400}$ is 20, so let $x = 400$ and $dx = 2$.

Then, $\sqrt{402} = f(402) = f(400 + 2) \approx \sqrt{400} + \frac{1}{2\sqrt{400}}(2) = 20 + 0.05 = 20.05$.

One calculator gives 20.04993766.

18. Let $y = f(x) = \sqrt[6]{x} = x^{1/6}$.

$f(x+dx) \approx f(x) + dy = \sqrt[6]{x} + (1/6)x^{-5/6}dx = \sqrt[6]{x} + \frac{1}{6\sqrt[6]{x^5}}dx$

64.05 is near 64 and $\sqrt[6]{64}$ is 2, so let $x = 64$ and $dx = 0.05$.

Then, $\sqrt[6]{64.05} = f(64.05) = f(64 + 0.05) \approx \sqrt[6]{64} + \frac{1}{6\sqrt[6]{(64)^5}}(0.05) = 2 + \frac{1}{3840}$

$= 2.000260416666\cdots$. One calculator gives 2.0002603319.

21. Let $V = f(r) = \frac{4}{3}\pi r^3$. Then $dV = 4\pi r^2 dr$.

When $r = 6$ ft and $dr = -0.3$ in $= \frac{-0.3 \text{ in}}{12 \text{ in/ft}} = \frac{-1}{40}$ ft,

$dV = 4\pi(6)^2\left(\frac{-1}{40}\right) = -3.6\pi$ ft^3.

Thus, volume of interior $\approx f(6) + dV = \frac{4}{3}\pi(6)^3 + (-3.6\pi) \approx 893.4690$ ft^3

24. $T(L) = 2\pi\sqrt{L/32} = 0.25\pi\sqrt{2L}$.

Let $t(L)$ denote the amount of time shown by the clock during one day when the length of the pendulum is L.

Note: (1) $86400/T(L)$ is the number of swings of the pendulum per day (since there are 86400 seconds in a day, and $T(L)$ is the number of seconds per swing).

(2) $T(4)$ is the number of seconds the clock advances with each swing of the pendulum (assuming the clock is accurate when $L = 4$).

Therefore, $t(L) = \dfrac{86400}{T(L)} T(4)$. [Note that $t(4) = 86400$ sec $= 1$ day]

$$= \dfrac{86400(0.25)\pi\sqrt{8}}{(0.25)\pi\sqrt{2L}} = 172800 \, L^{-1/2}$$

Therefore, $dt = -86400 \, L^{-3/2} dL = \dfrac{-86400}{\sqrt{L^3}} dL$

For L=4 and dL=-0.03: $dt = \dfrac{-86400}{\sqrt{64}}(-0.03) = 324$

That is, the clock has gained about 324 seconds (5 min, 24 sec) in 24 hours.

27. Let c denote the length of the third side, and β denote the angle between the sides of length 151 centimeters. The Law of Cosines gives

$$c = \sqrt{(151)^2 + (151)^2 - 2(151)(151)\cos\beta} = 151\sqrt{2}\sqrt{1-\cos\beta}$$

$$dc = 151\sqrt{2} \, \dfrac{1}{2\sqrt{1-\cos\beta}}(\sin\beta)d\beta = \dfrac{151\sqrt{2}\sin\beta}{2\sqrt{1-\cos\beta}} d\beta$$

For $\beta = 0.53$ radians, and $d\beta = \pm 0.005$ radians,

$$c = 151\sqrt{2}\sqrt{1-\cos(0.53)} \approx 79.097$$

$$dc = \dfrac{151\sqrt{2}\sin(0.53)}{2\sqrt{1-\cos(0.53)}}(\pm 0.005) \approx \pm 0.729$$

Therefore, $c = 79.097 \pm 0.729$ centimeters.

Problem Set 3.10

Problem Set 3.11 Chapter Review Problems

True-False Quiz

3. True. For example, if the velocity increases from -4 to -2, the speed would change from $|-4|$ to $|-2|$, which is a decrease.

   For a particular function, consider $s(t) = (t-4)^2$ when t changes from 2 to 3.

6. False. π^5 is a constant, so $D_x y = 0$.

9. False. c^{ex}: $D(x^2 \cdot x^3) = D(x^5) = 5x^4$,

   but $D(x^2) \cdot D(x^3) = (2x)(3x^2) = 6x^3$,

   and $5x^4 \neq 6x^3$ if, for example, x is 1.

12. False. Equations of lines are linear equations but the given equation is cubic in x. The derivative should have been evaluated at x=1 to obtain the slope of the tangent line at the point (1,1).

15. True. By the derivative rules stated in Theorems A-F of Section 3.3.

18. True. Leibniz notation form.

21. True. $(f \circ g)'(2) = f'[g(2)] \cdot g'(2) = f'[2] \cdot (2) = (2)(2) = 4$

24. True. $D(\sin x) = \cos x$

   $D^2(\sin x) = -\sin x$

   $D^3(\sin x) = -\cos x$ This pattern will continue in cycles of length four derivatives.

   $D^4(\sin x) = \sin x$

   $D^5(\sin x) = \cos x$

   For a formal proof, use the Principle of Mathematical Induction. What is above constitutes the "n=1" part of such a proof.

27. True. $V = \frac{4}{3}\pi r^3 \Rightarrow \frac{dV}{dt} = 4\pi r^2 \frac{dr}{dt}$

Set $\frac{dV}{dt} = 3$; obtain $\frac{dr}{dt} = \frac{3}{4\pi r^2} > 0$, so the radius is increasing.

However, $\frac{dr}{dt}$ is positive but approaching zero as r increases.

30. False. Ce^x: $dy = 5x^4 dx$.

At $x=1$ and $dx=-0.1$: $dy = 5(1)^4(-0.1) = -0.5 < 0$.

Miscellaneous Problems

3. (a) Derivative of $f(x) = 3x^2$ at $x = 2$.

(b) Derivative of $f(x) = \tan x$ at $x = \pi/4$.

(c) Derivative of $f(p) = 3/p$ at $p = x$.

6. $\frac{(x^2+1)(3) - (3x-5)(2x)}{(x^2+1)^2} = \frac{3x^2 + 3 - 6x^2 + 10x}{(x^2+1)^2} = \frac{3 + 10x - 3x^2}{(x^2+1)^2}$

9. $-\frac{1}{2}(x^2+4)^{-3/2}(2x) = \frac{-x}{\sqrt{(x^2+4)^3}}$

12. $2[\sin(\cos 4t)][\cos(\cos 4t)][-\sin 4t](4) = -8 \sin 4t \sin(\cos 4t) \cos(\cos 4t)$

15. $2x-y+2 = 0$, or equivalently, $y = 2x+2$ in slope-intercept form. The slope of this line is 2, so the slope of each line perpendicular to it is -1/2.

The equation of the given curve is $y = (x-2)^2$.
Its derivative is $Dy = 2(x-2)$.

Set $2(x-2) = -1/2$ and solve for x; obtain $x = 7/4$. Then substitute $x = 7/4$ into the equation of the curve; obtain $y = 1/16$.

The point is (7/4, 1/16).

Problem Set 3.11

18. Step 1: Let t denote the time elapsed [hr]
    Let r and V be the radius and volume of the sphere [m, m^3]

    Step 2: We are given that $\frac{dV}{dt} = 10$.

    We are to find $\frac{dr}{dt}$ when r=5.

    Step 3: $V = \frac{4}{3}\pi r^3$

    Step 4: $\frac{dV}{dt} = 4\pi r^2 \frac{dr}{dt}$, so $10 = 4\pi r^2 \frac{dr}{dt}$

    Step 5: When r=5, $10 = 4\pi(25)\frac{dr}{dt}$, so $\frac{dr}{dt} = \frac{1}{10\pi} \approx 0.03183$

    Conclusion: The radius is increasing at about 0.03183 m/hr.

21. $s(t) = t^3 - 6t^2 + 9t$

    $v(t) = 3t^2 - 12t + 9 = 3(t-1)(t-3)$

    Auxiliary axis for v(t): (+) (0) (-) (0) (+)
                                    1   3   t

    $a(t) = 6t - 12 = 6(t-2)$

    Auxiliary axis for a(t):   (-)   (0)   (+)
                                           2   t

    (a) The object is moving to the left when the velocity is negative; that is, when t is between 1 and 3.

    (b) The velocity is zero when t=1 and when t=3.
        $a(1) = 6(-1) = -6$ and $a(3) = 6(1) = 6$

    (c) The acceleration is positive when t is greater than 2.

24. $y^2 = 4x^3$ $2x^2 + 3y^2 = 14$

    $D(y^2) = D(4x^3)$ $D(2x^2 + 3y^2) = D(14)$

    $2y\, Dy = 12x^2$ $4x + 6y\, Dy = 0$

    $Dy = 6x^2/y$ $Dy = -2x/3y$

    Slopes at (1,2) are 3 and -1/3, respectively. Therefore, the lines tangent to the respective curves at (1,2) are perpendicular.

27. Step 1: Let t (sec) denote the time elapsed.
Let x and y (ft) be as indicated in the figure at the right.

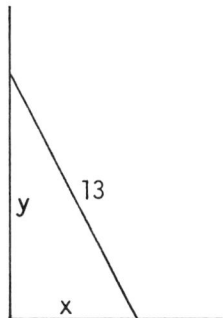

Step 2: We are given that $\frac{dx}{dt} = 2$.

We are to find $\frac{dy}{dt}$ when y=5.

Step 3: $x^2+y^2 = 13^2$. [Pythagorean Theorem]

Step 4: $\frac{d}{dt}(x^2+y^2) = \frac{d}{dt}(169)$.

$2x\frac{dx}{dt} + 2y\frac{dy}{dt} = 0$. (*)

Step 5: When y = 5, x = $\sqrt{13^2-5^2}$ = 12.

Substituting the values of x, y, and $\frac{dx}{dt}$ into (*),

$2(12)(2) + 2(5)\frac{dy}{dt} = 0$, so $\frac{dy}{dt} = \frac{-24}{5} = -4.8$.

Conclusion: The top of the ladder is moving down the wall at 4.8 feet per second when it is 5 feet above the ground.

Problem Set 3.11

CHAPTER 4 APPLICATIONS OF THE DERIVATIVE

Problem Set 4.1 Maxima and Minima

3. $G(x) = \frac{1}{5}(2x^3+3x^2-12x)$, $I = [-3,3]$.

   G is continuous on $[-3,3]$ so the Max-Min Existence Theorem applies.

   $G'(x) = \frac{1}{5}(6x^2+6x-12) = \frac{6}{5}(x^2+x-2) = \frac{6}{5}(x+2)(x-1)$.

   $G'(x) = 0$ iff $x = -2, 1$; $G'(x)$ is defined everywhere on $(-3,3)$.

   Critical Points: $-3, 3$ (end points); $-2, 1$ (stationary points).

x	G(x)
-3	1.8
-2	4
1	-1.4	⬅ Minimum value is -1.4; it occurs at x=1.
3	9	⬅ Maximum value is 9; it occurs at x=3.

6. $f(x) = x^3-3x+1$, $I = [-1.5, 3]$.

   f is continuous on $[-1.5,3]$ so the Max-Min Existence Theorem applies.

   $f'(x) = 3x^2-3 = 3(x^2-1) = 3(x+1)(x-1)$.

   $f'(x) = 0$ iff $x = -1, 1$; $f'(x)$ is defined everywhere on $(-1.5, 3)$.

   Critical Points: $-1.5, 3$ (end points); $-1, 1$ (stationary points).

x	f(x)
-1.5	2.125
-1	3
1	-1	⬅ Minimum value is -1; it occurs at x=1.
3	19	⬅ Maximum value is 19; it occurs at x=3.

9. $f(x) = \frac{x}{x^2+2}$, $I = [-1, 4]$.

   f is continuous on $[-1,4]$ so the Max-Min Existence Theorem applies.

   $f'(x) = \frac{(x^2+2)(1) - (x)(2x)}{(x^2+2)^2} = \frac{2-x^2}{(x^2+2)^2} = \frac{(\sqrt{2}+x)(\sqrt{2}-x)}{(x^2+2)^2}$.

   $f'(x) = 0$ iff $x = -\sqrt{2}$ [not in $(-1,4)$], $\sqrt{2}$; $f'(x)$ is defined everywhere on $(-1,4)$.

Problem Set 4.1   79

9. (continued)

Critical Points: -1, 4 (end points); $\sqrt{2}$ (stationary point).

x	f(x)	
-1	$-1/3 \approx -0.3333$	← Minimum value is -1/3; it occurs at x=-1.
$\sqrt{2}$	$\sqrt{2}/4 \approx 0.3536$	← Maximum value is $\sqrt{2}/4$; it occurs at x=$\sqrt{2}$.
4	$2/9 \approx 0.2222$	

12. $f(x) = |5-3x|$, $I = [0,3]$.

f is continuous on [0,3] so the Max-Min Existence Theorem applies.

$f'(x) = \frac{|5-3x|}{5-3x}(-3) = \frac{-3|5-3x|}{5-3x}$.

$f'(x) = 0$ nowhere; $f'(x)$ is undefined for $x = 5/3$.

Critical Points: 0, 3 (end points); 5/3 (singular point).

x	f(x)	
5/3	0	← Minimum value is 0; it occurs at x = 5/3.
0	5	← Maximum value is 5; it occurs at x = 0.
3	4	

15. $h(x) = x^{2/5}$, $I = (-1,32)$.

Note that I is not a closed interval so the Max-Min Existence Theorem does not apply. However, since h could be defined and would be continuous on [-1,32] we will consider h on that interval and then eliminate the end points from consideration in the final conclusion about the extreme values.

In Problem 13 it was found that, on $I = [-1,32]$, $g(x) = x^{2/5}$ has a maximum value of 4 which occurs at x=32 (an end point), and has a minimum value of 0 which occurs at x=0 (a singular point).

Therefore, h has no maximum value on (-1,32) [but gets arbitrarily close to 4 as x approaches 32], and has a minimum value of 0 at x=0.

18. Let x denote the number, and A denote the amount by which x exceeds x^2, so that $A = x-x^2$. If x is negative or greater than 1, x^2 is larger than x, so A is negative.

Thus, we wish to maximize the function $A(x) = x-x^2$ on $I = [0,1]$.

A is continuous on [0,1] so the Max-Min Existence Theorem applies.

$A'(x) = 1-2x$.

$A'(x) = 0$ iff $x = 1/2$; $A'(x)$ is defined everywhere on I.

Critical Points: 0, 1 (end points); 1/2 (stationary point).

x	A(x)
0	0
1/2	1/4
0	0

⬅ Maximum value is 1/4; it occurs at $x = 1/2$.

Conclusion: 1/2 is the number that exceeds its square by a maximum amount.

21. Let x denote the length (inches) of the edges of the squares to be cut out.
Let V denote the volume (cubic inches) of the resulting box.

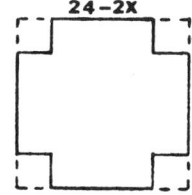

$V = (length)(width)(height) = (24-2x)(24-2x)x$
$= 4(x^3-24x^2+144x)$, where x is greater than 0 and less than 12.

That is, we wish to maximize the function
$V(x) = 4(x^3-24x^2+144x)$ on $I = (0,12)$. We will do so on [0,12].

V is continuous on [0,12] so the Max-Min Existence Theorem applies.

$V'(x) = 4(3x^2-48x+144) = 12(x^2-16x+48) = 12(x-4)(x-12)$.

$V'(x) = 0$ iff $x = 4, 12$ [not in (0,12)]; defined everywhere on (0,12).

Critical Points: 0, 12 (end points); 4 (stationary point).

x	V(x)
0	0
4	1024
12	0

⬅ Maximum value is 1024; it occurs at x=4.

Conclusion: The volume of the largest such box is 1024 cubic inches.

Problem Set 4.1

24. Let x denote the length (feet) of each section of fence perpendicular to the barn.
Let y denote the length (feet) of the section of fence parallel to the barn.
Let A denote the area (square feet) of the total enclosure.

Then (1) A = (width)(length) = xy,
(2) 4x+y = 80   or   y = 80-4x.

Therefore, A = x(80-4x) = $80x-4x^2$, where x is greater than zero and less than 20 (since there are 4 pieces of length x).

We will maximize the function A(x) = $80x-4x^2$ on I = [0,20].

A is continuous on I so the Max-Min Existence Theorem applies.

A'(x) = 80-8x = 8(10-x).

A'(x) = 0 iff x = 10; A'(x) is defined everywhere on (0,20).

Critical Points: 0, 20 (end points); 10 (stationary point).

x	A(x)
0	0
10	400
20	0

← Maximum value of A is 400; it occurs at x=10.

Conclusion: The dimensions for the total enclosure should be 10 feet (perpendicular to barn) by 40 feet (parallel to barn).

27. Let x and y (units) be as indicated in Figure 15 (in 1st quadrant).
Let A denote the area (square units) of the rectangle.

Then (1) y = $12-x^2$,
(2) A = (width)(height) = (2x)(y) = 2xy.

Therefore, A = $2x(12-x^2)$ = $24x-2x^3$, where x not negative or greater than $\sqrt{12}$.

We wish to maximize the function A(x) = $24x-2x^3$ on I = [0,$\sqrt{12}$].

A is continuous on [0,$\sqrt{12}$] so the Max-Min Existence Theorem applies.

A'(x) = $24-6x^2$ = $-6(x^2-4)$ = -6(x+2)(x-2).

A'(x) = 0 iff x = -2 [not in (0,$\sqrt{12}$)], 2; defined everywhere on I.

Critical Points: 0, $\sqrt{12}$ (end points); 2 (stationary point).

27. (continued)

x	A(x)
0	0
2	32	← Maximum value of A is 32; it occurs at x=2.
√12	0

Conclusion: The dimensions of the rectangle of maximum area are 4 units (wide) by 8 units (high).

30. Let r and h denote the radius and altitude (meters) of the resulting cone.
Let V denote the volume (cubic meters) of the cone.
Let c denote the circumference (meters) of the base of the cone.
Let s denote the arc length (meters) of the sector.

Cross Section of Cone

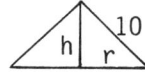

Then (1) $V = (1/3)\pi r^2 h$.

(2) $r^2 + h^2 = 100$. [Pythagorean Theorem]

Therefore, $V = (\pi/3)(100-h^2)h = (\pi/3)(100h-h^3)$, where h is positive and less than 10.

We will maximize the function $V(h) = (\pi/3)(100h-h^3)$ on $I = [0,10]$.

V is continuous on [0,10] so the Max-Min Existence Theorem applies.

$V'(h) = (\pi/3)(100-3h^2)$.

$V'(h) = 0$ iff $h = \frac{-10}{\sqrt{3}}$ [not in (0,10)], $\frac{10}{\sqrt{3}}$; defined everywhere on I.

Critical Points: 0, 10 (end points); $\frac{10}{\sqrt{3}}$ (stationary point).

h	V(h)
0	0
10/√3	2000π/9√3	← Maximum value of V is 2000π/9√3;
10	0	it occurs at h = 10/√3, r = (10/3)√6.

Conclusion: $\theta = s/10 = (20\pi-c)/10 = (20\pi-2\pi r)/10$
$= 2\pi(1 - \sqrt{6}/3) \approx 1.1530$ rad (66.06°).

33. $x^3 < x^2$ for all x in [-0.5, 0.5], except for x=0,

but $\lim_{x \to 0} x^3 < \lim_{x \to 0} x^2$ is not true, since each limit is zero.

Problem Set 4.1

Problem Set 4.2 Monotonicity and Concavity

3. $F(x) = x^3 - 1$ is continuous on R.

$F'(x) = 3x^2$.

Auxiliary axis for $F'(x)$:

```
        (+)      (0)      (+)
    ─────────────●─────────────── x
                 0
```

Therefore, F is increasing on $(\infty, 0]$ and on $[0, \infty)$; and so, on R.

6. $g(t) = \frac{t^4}{2} - \frac{4t^3}{3}$ is continuous on R.

$g'(t) = 2t^3 - 4t^2 = 2t^2(t-2)$.

Auxiliary axis for $g'(t)$:

```
        (-)     (0)     (-)     (0)     (+)
    ─────────────●───────────────●─────────── t
                 0               2
```

Therefore, g is decreasing on $(-\infty, 2]$; increasing on $[2, \infty)$.

9. $H(t) = \sin^2 2t$, t in $[0, \pi]$. H is continuous on $[0, \pi]$.

$H'(t) = 2(\sin 2t)(\cos 2t)(2) = 4 \sin 2t \cos 2t = 2 \sin 4t$. (Double-angle id.)

Set $H'(t) = 0$: $\sin 4t = 0$.
$4t = 0, \pm\pi, \pm 2\pi, \pm 3\pi, \pm 4\pi, \cdots$.

$t = 0, \pm\frac{\pi}{4}, \pm\frac{\pi}{2}, \pm\frac{3\pi}{4}, \pm\pi, \cdots$.

The only values in $(0, \pi)$ are $\frac{\pi}{4}, \frac{\pi}{2},$ and $\frac{3\pi}{4}$.

Auxiliary axis for $H'(t)$:

```
        (+)    (0)    (-)    (0)    (+)    (0)    (-)
    ─────●─────────────────────────────────────────── t
         0     π/4    π/2    3π/4    π
```

Therefore, H is increasing on $[0, \pi/4]$ and on $[\pi/2, 3\pi/4]$;
decreasing on $[\pi/4, \pi/2]$ and on $[3\pi/4, \pi]$.

12. $f(x) = 4 - x^2$.
$f'(x) = -2x$.
$f''(x) = -2$.

Therefore f is concave down on R. There are no inflection points.

15. $g(x) = 3x^2 - x^{-2}$.

$g'(x) = 6x + 2x^{-3}$.

$g''(x) = 6 - 6x^{-4} = \dfrac{6x^4 - 6}{x^4} = \dfrac{6(x^2+1)(x+1)(x-1)}{x^4}$.

Auxiliary axis for $g''(x)$:

```
      (+)   (0)   (-)   (u)   (-)   (0)   (+)
   ─────────┼─────────────┼─────────────┼────────── x
            -1            0             1
```

Therefore, g is concave up on $(-\infty,-1)$ and on $(1,\infty)$;
concave down on $(-1,0)$ and on $(0,1)$.

Inflection points occur at x=-1 since g is concave up to the left of -1 and concave down to the right of -1, and at x=1 since g is concave down to the left of 1 and concave up to the right of 1. The inflection points are (-1,2) and (1,2), since $g(-1) = 2$ and $g(1) = 2$.

18. $g(x) = 2x^2 + \cos^2 x$.

$g'(x) = 4x + 2(\cos x)(-\sin x)$
      $= 4x - 2\sin x \cos x = 4x - \sin 2x$. [Double-angle identity]

$g''(x) = 4 - (\cos 2x)(2) = 2(2 - \cos 2x)$ which is always positive since $\cos 2x$ is always less than 2.

Therefore, g is concave up on **R**. There are no inflection points.

21. $g(x) = 3x^4 - 4x^3 + 2$, $D_g = \mathbf{R}$.

$g'(x) = 12x^3 - 12x^2 = 12x^2(x-1)$.

```
 g'    (-)   (0)   (-)   (0)   (+)
   ──────────┼───────────┼────────── x
             0           1
```

$g''(x) = 36x^2 - 24x = 12x(3x-2)$.

```
 g''   (+)   (0)   (-)   (0)   (+)
   ──────────┼───────────┼────────── x
             0          2/3
```

Use the information shown on the two auxiliary axes to develop the following table summarizing the "activity" of g. Then plot at least x-intercepts, y-intercepts and points at which $g'(x)$ or $g''(x)$ is 0.

Inc = increasing; Dec = decreasing; CU = concave up; CD = concave down.

Intervals and Split Points	Activity of g
$(-\infty, 0)$	Dec and CU
at 0	Levels off
$(0, 2/3)$	Dec and CD
at 2/3	Inflection
$(2/3, 1)$	Dec and CU
at 1	Levels off
$(1, \infty)$	Inc and CU

x	g(x)
0	2
2/3	1.41
1	1
-1	9
2	18

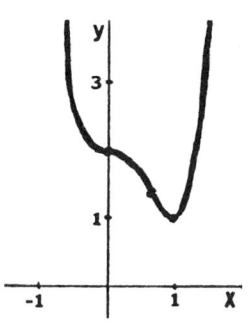

Problem Set 4.2

24. $H(x) = \dfrac{x^2}{x^2+1}$, $D_H = \mathbb{R}$ [an even function]

$H'(x) = \dfrac{(x^2+1)(2x) - (x^2)(2x)}{(x^2+1)^2} = \dfrac{2x}{(x^2+1)^2}$

$H''(x) = \dfrac{(x^2+1)^2(2) - (2x)[2(x^2+1)(2x)]}{(x^2+1)^4} = \dfrac{-2(\sqrt{3}x+1)(\sqrt{3}x-1)}{(x^2+1)^3}$

```
H'    (-)    (0)    (+)              H"   (-)    (0)    (+)    (0)    (-)
     ─────────┼──────────── x            ──────────┼──────────────┼──────── x
              0                                  -√3/3          √3/3
                                                 -0.58          0.58
```

Intervals and Split Points	Activity of H
$(-\infty, -0.58)$	Dec and CD
at -0.58	Inflection
$(-0.58, 0)$	Dec and CU
at 0	Levels off
$(0, 0.58)$	Inc and CU
at 0.58	Inflection
$(0.58, \infty)$	Inc and CD

x	g(x)
0	0
0.58	0.25
1	0.5

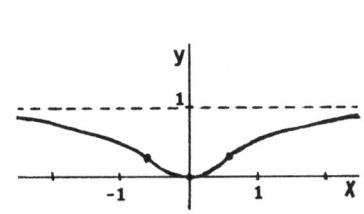

27. $f(x) = x^{2/3}(1-x) = x^{2/3} - x^{5/3}$, $D_f = \mathbb{R}$

$f'(x) = (2/3)x^{-1/3} - (5/3)x^{2/3} = \dfrac{5x-2}{-3\sqrt[3]{x}}$

$f''(x) = (-2/9)x^{-4/3} - (10/9)x^{-1/3} = \dfrac{2+10x}{-9\sqrt[3]{x^4}}$

```
f'    (-)    (u)    (+)    (0)    (-)         f"   (+)    (0)    (-)    (u)    (-)
     ──────────┼──────────────┼──────── x        ──────────┼──────────────┼──────── x
               0             0.4                         -0.2             0
```

Intervals and Split Points	Activity of f
$(-\infty, -0.2)$	Dec and CU
at -0.2	Inflection
$(-0.2, 0)$	Dec and CD
at 0	Cusp
$(0, 0.4)$	Inc and CD
at 0.4	Levels off
$(0.4, \infty)$	Dec and CD

x	g(x)
0	0
-0.2	0.41
0.4	0.33
1	0
-1	2
2	-1.59
-2	4.76
3	-4.16

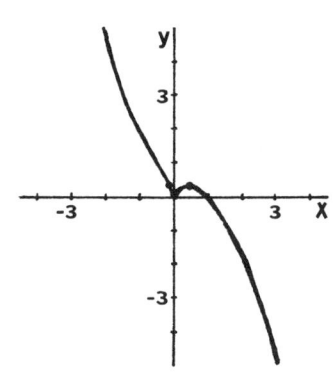

30. f' (-) (0) (-)
       ─┼────┼────┼──── x
        0   2   6

   f" (-) (+)   (-)
      ─┼──┼──┼────┼──── x
       0 1 2   6

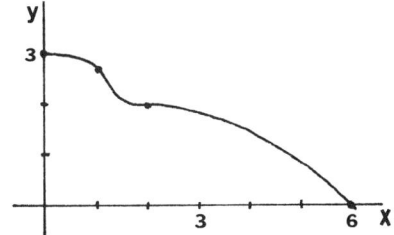

There are infinitely many graphs that satisfy all the conditions. One such graph is at the right.

33. $f(x) = ax^2 + bx + c$, $a \neq 0$

$f'(x) = 2ax + b$

$f''(x) = 2a$

$a > 0 \Rightarrow f$ is concave up on **R**; $a < 0 \Rightarrow f$ is concave down on **R**. Hence, no change of concavity, so no inflection point for any quadratic function.

36. $f'(x) = \dfrac{x^2 - x + 1}{x^2 + 1}$ exists everywhere since x^2+1 is never zero.

$f'(x)$ is never zero since the discriminant of x^2-x+1 is negative.

$f'(0) = 1$, so (by Problem 35) $f'(x)$ is positive everywhere.

Therefore, f is increasing on **R** (everywhere).

Problem Set 4.3 Local Maxima and Minima

3. $f(x) = \frac{1}{2}x - \sin x$ on $(0, 2\pi)$. f is continuous on $(0, 2\pi)$.

$f'(x) = \frac{1}{2} - \cos x$

   (-)   (0)   (+)   (0)   (-)
  ─┼─────┼─────┼─────┼─────┼── x
   0   π/3   5π/3   2π

$f''(x) = \sin x$

   Critical points: $\pi/3$, $5\pi/3$ (stationary points)

3. (continued)

   1st Derivative Test: Use the auxiliary axis for f' to conclude that $\pi/3$ gives a local minimum; $5\pi/3$ gives a local maximum.

   2nd Derivative Test: $f''(\pi/3) = \sin(\pi/3) = \sqrt{3}/2 > 0$ so $\pi/3$ gives a local minimum; $f''(5\pi/3) = \sin(5\pi/3) = -\sqrt{3}/2 < 0$ so $5\pi/3$ gives a local maximum.

6. $g(x) = 3x^4 - 4x^3$. g is continuous on **R**.

   $g'(x) = 12x^3 - 12x^2 = 12x^2(x-1)$

   ```
      (-)    (0)    (-)    (0)    (+)
   ─────────┼──────────────┼────────── x
            0              1
   ```

   Critical points: 0 and 1 (stationary points)

   1st Derivative Test: Use the auxiliary axis for g' to conclude that 1 gives a local minimum; 0 does _not_ give a local extremum.

   2nd Derivative Test: $g''(0) = 0$ so no conclusion about a local extremum at x=0. $g''(1) = 12 > 0$ so 1 gives a local minimum.

9. $h(x) = x^4 + 2x^3$. h is continuous on **R**.

   $h'(x) = 4x^3 + 6x^2 = 2x^2(2x+3)$

   ```
      (-)    (0)    (+)    (0)    (+)
   ─────────┼──────────────┼────────── x
           -3/2            0
   ```

   Therefore, critical points are 0, -3/2 (stationary). Use the auxiliary axis for h' to conclude by the First Derivative Test that -1.5 gives a local minimum value of $h(-3/2) = -27/16$; 0 does _not_ give a local extremum.

12. $h(t) = 2t + t^{2/3}$. h is continuous on **R**.

    $h'(t) = 2 + (2/3)t^{-1/3} = \dfrac{2(3\sqrt[3]{t} + 1)}{3\sqrt[3]{t}}$

    ```
       (+)    (0)    (-)    (u)    (+)
    ─────────┼──────────────┼────────── t
           -1/27             0
    ```

    Use the auxiliary axis for h' to conclude by the First Derivative Test that the stationary point -1/27 gives a local maximum value of $h(-1/27) = 1/27$; the singular point 0 gives a local minimum value of $h(0) = 0$.

15. $f(t) = \dfrac{\sin t}{2 + \cos t}$, $0 < t < 2\pi$. f is continuous on $(0, 2\pi)$.

$f'(t) = \dfrac{(2 + \cos t)(\cos t) - (\sin t)(-\sin t)}{(2 + \cos t)^2} = \dfrac{2\cos t + \cos^2 t + \sin^2 t}{(2 + \cos t)^2}$

$= \dfrac{2\cos t + 1}{(2 + \cos t)^2}$

```
          (+)       (0)      (-)      (0)     (+)
  <-------|---------|--------|--------|-------|------> t
          0       2π/3      4π/3             2π
```

Use the auxiliary axis for f' to conclude by the First Derivative Test that the stationary point $2\pi/3$ gives a local maximum value of $f(2\pi/3) = \sqrt{3}/3$; the stationary point $4\pi/3$ gives a local minimum value of $f(4\pi/3) = -\sqrt{3}/3$.

18. $F(x) = 6\sqrt{x} - 3x$ on $[0, \infty)$. F is continuous on $[0, \infty)$.

$F'(x) = 6 \cdot \dfrac{1}{2\sqrt{x}} - 3 = \dfrac{3(1-\sqrt{x})}{\sqrt{x}}$

```
          (+)       (0)      (-)
  <-------|---------|----------------> x
          0         1
```

Critical Points: 0 (end point); 1 (stationary point)

x	F(x)
0	0
1	3
9	-9

By observing the auxiliary axis for F' and the table of values, we can conclude: (a) The global maximum value is $F(1) = 3$,
(b) There is no global minimum value. [f(0) was a candidate but, for example, $f(9) < f(0)$.]

21. $g(x) = x^2 + \dfrac{16x^2}{(8-x)^2}$, $x > 8$. g is continuous on $(8, \infty)$.

$g'(x) = 2x + \dfrac{(8-x)^2(32x) - (16x^2)2(8-x)(-1)}{(8-x)^4}$

$= 2x + \dfrac{(8-x)(32x) - (16x^2)2(-1)}{(8-x)^3} = 2x + \dfrac{256x}{(8-x)^3}$

$= 2x\left[1 + \dfrac{128}{(8-x)^3}\right] = 0$ if $x = 0$ [not on $(8, \infty)$] or $x = 8 + \sqrt[3]{128}$
≈ 13.0397.

```
      (-)       (0)       (+)
  <---|---------|----------|-----> x
      8       13.04
```

Therefore, the global minimum value is $g(8 + \sqrt[3]{128}) \approx 277.1477$.

Problem Set 4.3

Problem Set 4.4 More Max-Min Problems

3. Step 1: Let d denote the distance between (x,y) and (10,0).

   Let $s = d^2$

   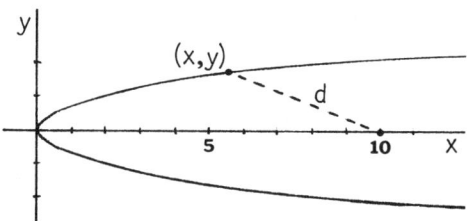

   Step 2: Note that if s is minimal, then d is minimal, since $d = \sqrt{s}$ and the square root function is an increasing function.

   $s = (x-10)^2 + (y-0)^2$

   Step 3: $x = 2y^2$ for points on the curve, so s can be expressed as a function of y only by $s(y) = (2y^2-10)^2 + y^2$

   Step 4: y can be any real number. However, s is an even function, so we can restrict our search to $[0,\infty)$ and then use the symmetry.

   Step 5: $s'(y) = 2(2y^2-10)(4y) + 2y = 16y^3 - 78y = 2y(8y^2-39)$

   $= 2y(\sqrt{8}y+\sqrt{39})(\sqrt{8}y-\sqrt{39})$

   ```
        (-)      (0)    (+)
   ←————————————————————————→ y
     0         √39/8
   ```

   Step 6: The global minimum of s is where $y = \sqrt{39/8} \approx 2.21$, and $x = 2(\sqrt{39/8})^2 = 9.75$.

   Conclusion: The points on $x = 2y^2$ that are closest to (10,0) are approximately (9.75, 2.21) and (9.75, -2.21).

6. Step 1: Let C denote the total cost of fencing [dollars]. Let x and y be as indicated in Figure 8 [feet].

   Step 2: Minimize $C = 2(4x+2y) + 1(y) = 8x+5y$.

   Step 3: $xy = 900$ [area of each pen]. Therefore, C can be expressed as a function of x only as $C(x) = 8x + 5(900/x) = 8x + 4500x^{-1}$.

   Step 4: $x > 0$

   Step 5: $C'(x) = 8 - 4500x^{-2} = \dfrac{8x^2-4500}{x^2} = \dfrac{8(x+\sqrt{562.5})(x-\sqrt{562.5})}{x^2}$

   ```
        (-)      (0)    (+)
   ←————————————————————————→ x
     0        √562.5
   ```

   Step 6: C is minimum at $x = \sqrt{562.5} \approx 23.72$; $y = 900/\sqrt{562.5} \approx 37.95$

   Conclusion: The fence will be least expensive if x is about 23.72 feet and y is about 37.95 feet [$x \approx 23'9''$; $y \approx 37'11''$]

9. Step 1: Let x and y (ft) be as indicated in the figure to the right.
Let k (a constant) be the cost per square foot of base.
Let C denote the total cost.

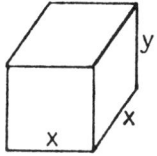

Step 2: Minimize C = (cost of base)+(cost of 4 sides)+(cost of top)
= $k(x^2) + k(4xy) + 2k(x^2) = 3kx^2+4kxy$.

Step 3: $12000 = x^2y$. Therefore, C can be expressed as a function of x by $C(x) = 3kx^2 + 4kx(12000/x^2) = 3kx^2 + \frac{48000k}{x}$.

Step 4: $x > 0$.

Step 5: $C'(x) = 6kx - 48000kx^{-2} = \frac{6kx^3-48000k}{x^2} = \frac{6k(x-20)(x^2+20x+400)}{x^2}$.

```
        (-)      (0)      (+)
  •—————————————•——————————————→
  0             20                x
```

Step 6: C is minimum when x = 20; y = $12000/(20)^2$ = 30.

Conclusion: The most economical dimensions for the cistern are a base that is 20'x20' and a depth of 30'.

12. Step 1: Let B_1 and B_2 denote the illumination from the sources.

Let $B = B_1 + B_2$, and let x be as indicated.

Step 2: Minimize B.

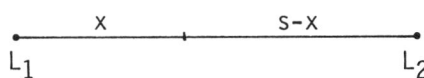

Step 3: $B = B1 + B2 = \frac{kI_1}{x^2} + \frac{kI_2}{(s-x)^2}$ for some k, the constant of proportionality.

Therefore, B can be expressed as a function of x by

$B(x) = kI_1x^{-2} + kI_2(s-x)^{-2}$.

Step 4: $0 < x < s$.

Step 5: $B'(x) = -2kI_1x^{-3} - 2kI_2(s-x)^{-3}(-1)$
$= \frac{-2k[I_1(s-x)^3 - I_2x^3]}{x^3(s-x)^3}$.

Problem Set 4.4

12. (continued)

$$B'(x) = 0 \text{ iff } I_1(s-x)^3 - I_2 x^3 = 0$$

$$\text{iff } \left(\frac{s-x}{x}\right)^3 = \frac{I_2}{I_1} \quad \text{iff } \frac{s}{x} - 1 = \sqrt[3]{I_2/I_1}$$

$$\text{iff } x = \frac{s}{1 + \sqrt[3]{I_2/I_1}}. \quad [\text{Call this number } B_0.]$$

```
     (-)    (0)    (+)
 ├─────────┼─────────────┤
 0         B_0           s
```

Step 6: S is minimum at B_0.

Conclusion: The sum of the illuminations will be minimum B_0 feet from light source L_1.

15. **Step 1:** Let x denote the distance (mi) indicated.
    Let T_b and T_w denote the times (hr) by boat and walking, respectively.
    Let $T = T_b + T_w$.

    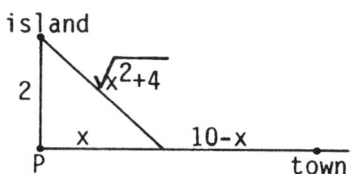

    Step 2: Minimize $T = T_b + T_w$.

    Step 3: $T(x) = (\sqrt{x^2+4})/20 + (10-x)/4$. $\left[\text{Recall: time} = \frac{\text{distance}}{\text{velocity}}\right]$

    Step 4: $0 \le x \le 10$.

    Step 5: $T'(x) = \dfrac{x - 5\sqrt{x^2+4}}{20\sqrt{x^2+4}}$ which is never 0 since $5\sqrt{x^2+4} > x$.

    Hence, there are no stationary points. There are no singular points since x^2+4 is never zero. Therefore, the minimum must occur at an end point. The correct end point is obviously at $x=10$ (boat all the way) rather than $x=0$.

    Conclusion: Take the boat right to the town.

18. **Step 1:** Let (X,Y) denote the coordinates of the point of tangency.
    Let m denote the slope of the tangent line.
    Let A denote the area of the triangle.

    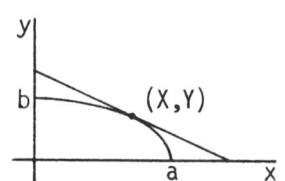

18. (continued)

Step 2: Minimize $A = \frac{1}{2}(\text{x-intercept of tangent})(\text{y-intercept of tangent})$

Step 3: $b^2x^2 + a^2y^2 = a^2b^2$ or $y = \frac{b}{a}\sqrt{a^2-x^2}$,

so $Y = \frac{b}{a}\sqrt{a^2-X^2}$, replacing (x,y) by (X,Y).

Therefore, $2b^2x + 2a^2y\, Dy = 0$ or $Dy = \frac{-b^2x}{a^2y}$,

so the slope at (X,Y) is $m = \frac{-b^2X}{a^2Y} = \frac{-bX}{a\sqrt{a^2-X^2}}$

Then, the equation of the tangent is

$$y - \frac{b}{a}\sqrt{a^2-X^2} = \frac{-bX}{a\sqrt{a^2-X^2}}(x - X) \quad \text{[point-slope form]}$$

$$y = \frac{-bX}{a\sqrt{a^2-X^2}}x + \frac{ab}{\sqrt{a^2-X^2}} \quad \text{[slope-intercept form]}$$

y-intercept of tangent line = $\frac{ab}{\sqrt{a^2-X^2}}$

x-intercept of tangent line = $\frac{a^2}{X}$

Therefore, $A(X) = \frac{1}{2}\left(\frac{a^2}{X}\right)\left(\frac{ab}{\sqrt{a^2-X^2}}\right) = \frac{a^3b}{2X\sqrt{a^2-X^2}}$

Step 4: $0 < X < a$

Step 5: We wish to minimize $A(X)$. Since $A(X) > 0$ and the numerator is a constant, this is equivalent to maximizing the denominator.

That is, maximize $f(X) = X\sqrt{a^2-X^2}$, $0 < X < a$.

$f'(X) = \frac{(a+\sqrt{2}X)(a-\sqrt{2}X)}{\sqrt{a^2-X^2}}$

```
         (+)      (0)     (-)
    ┼─────────────┼─────────────→ X
    0           a/√2            a
```

Step 6: Therefore f is maximum at $X = a/\sqrt{2}$; $Y = b/\sqrt{2}$; $m = -b/a$.

Conclusion: Equations of the tangent line are
$y - (b/\sqrt{2}) = (-b/a)(x - [a/\sqrt{2}])$ [Point-slope form]
$y = (-b/a)x + \sqrt{2}b$ [Slope-intercept form]
$bx + ay - \sqrt{2}ab = 0$ [General form]

21. Step 1: Let r denote the radius of the sphere (a constant)
Let x denote the radius of the base of the cylinder.
Let 2y denote the altitude of the cylinder.
Let S denote the curved surface area.

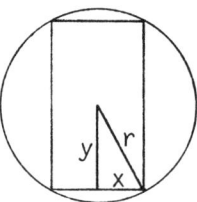

Step 2: Maximize S = 2π(radius of base)(altitude) = 2π(x)(2y) = 4πxy.

Step 3: $x^2 + y^2 = r^2$ so $y = \sqrt{r^2 - x^2}$

Then express S as a function of x only by $S(x) = 4\pi x\sqrt{r^2 - x^2}$

Step 4: 0 < x < r

Step 5: Maximizing S is equivalent to maximizing $f(x) = x\sqrt{r^2 - x^2}$ (since 2π is positive.) This is the same function we maximized in Problem 18, except that we have x and r (instead of X and a).

Step 6: Thus, (see Problem 18) S is maximum if $x = r/\sqrt{2}$;
$$y = \sqrt{r^2 - (r/\sqrt{2})^2} = r/\sqrt{2}.$$

Conclusion: The maximum surface area is obtained if the altitude of the cylinder is the same as the diameter of its base.

24. Step 1: Let x and y be as indicated [in].
Let V denote the volume of the box (a constant) [in³].
Let C denote the cost of the box.
Let k (a constant) denote the cost per square inch of the sides.

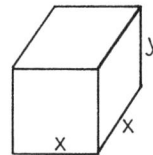

Step 2: Minimize C.

Step 3: $V = x^2y$, so $y = V/x^2$.
C = (cost of 4 sides) + (cost of bottom) + (cost of top)
$= k(4xy) + 1.2k(x^2) + 1.5k(x^2) = k(4xy + 2.7x^2)$

Therefore, C can be expressed as a function of x only by
$$C(x) = k(4x[V/x^2] + 2.7x^2) = k[4Vx^{-1} + 2.7x^2]$$

Step 4: 0 < x

Step 5: $C'(x) = k[-4Vx^{-2} + 5.4x] = \dfrac{k(-4V + 5.4x^3)}{x^2}$

which is zero iff $x = (4V/5.4)^{1/3} = (20V/27)^{1/3}$

24. (continued)

Step 6: $C''(x) = k[8Vx^{-3} + 5.4]$ which is positive when $C'(x) = 0$.

Therefore, using the Second Derivative Test, we conclude that C is minimum when $x = (20V/27)^{1/3}$.

The proportion is: $\dfrac{x}{y} = \dfrac{x}{V/x^2} = \dfrac{x^3}{V} = \dfrac{(20V/27)}{V} = \dfrac{20}{27} \approx 74\%$

Conclusion: The minimum cost is achieved if the base dimensions are about 74% of the height.

27. Step 1: Let s be as indicated.
Let P denote the perimeter.

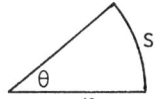

Step 2: Maximize $P = s + 2r$.

Step 3: $s = r\theta$ and $A = \dfrac{1}{2}\theta r^2$ [Formulas for arc length and area]

Then P can be expressed as a function of r only by
$$P(r) = r\theta + 2r = r(2A/r^2) + 2r = 2r + 2Ar^{-1}$$

Step 4: $0 < r$

Step 5: $P'(r) = 2 - 2Ar^{-2} = \dfrac{-2(A-r^2)}{r^2} = 0$ iff $r = \sqrt{A}$.

Step 6: $P''(r) = 4Ar^{-3} = \dfrac{4A}{r^3} > 0$ where $P'(r) = 0$.

Therefore, P is minimum where $r = \sqrt{A}$.

Conclusion: $r = \sqrt{A}$; $\theta = 2A/r^2 = 2$ radians (about 114.6°)

30. For a reflection, the light is traveling in the same medium.

Therefore, $c_1 = c_2$.

Therefore, $\sin\theta_1 = \sin\theta_2$.

Therefore, $\theta_1 = \theta_2$, since each angle is in $[0, \pi/2)$.

Problem Set 4.4

Problem Set 4.5 Economic Applications

3. $P(x) = R(x) - C(x) = [300x - (x^2/2)] - [8000 + 110x] = -8000 + 190x - (x^2/2)$

6. $C(x) = 1100 + \dfrac{x^2}{1200}$

   Average Cost $= \dfrac{C(x)}{x} = \dfrac{1100}{x} + \dfrac{x}{1200}$

   At $x = 900$: $\dfrac{C(900)}{900} = \dfrac{1100}{900} + \dfrac{900}{1200} \approx 1.97$

   Marginal Cost $= C'(x) = x/600$

   At $x = 900$: $C'(900) = 900/600 = 1.50$

9. (a) $R(x) = x \cdot p(x) = 20x + 4x^2 - x^3/3$ is the revenue function.

   $R'(x) = 20 + 8x - x^2$
   $= (10-x)(2+x)$ is the marginal revenue function.

   (b)
   ```
         (+)      (0)      (-)
   ├──────────────┼──────────────── x
   0             10
   ```
   R is increasing on $[0, 10]$.

   (c) $R''(x) = 8 - 2x = 2(4-x)$ [This is the 1st derivative of R'.]
   $R''(x) = 0$ iff $x = 4$
   $R^{(3)}(x) = -2$ [This is the 2nd derivative of R'.]

   Therefore $R^{(3)}(4)$ is negative. Make use of the Second Derivative Test and conclude that the marginal revenue is maximum when $x = 4$.

12. $p(x) = 12.00 - \dfrac{x-400}{10}(0.20) = 20 - 0.02x,\ \ x \geq 400$

    $R(x) = x \cdot p(x) = 20x - 0.02x^2$

    $R'(x) = 20 - 0.04x = -0.04(x-500)$

    ```
               (+)      (0)      (-)
    ├──────────┼─────────┼──────────── x
              400       500
    ```

    R is maximum for 500 passengers.

15. Fixed cost = 6000

Variable cost = cost of material + cost of labor

$$= \begin{cases} 1.00x + 0.40x, & \text{if } 0 \le x \le 4500 \\ 1.00x + 0.40(4500) + 0.60(x-4500), & \text{if } x > 4500 \end{cases}$$

(a) Total cost = Fixed cost + Variable cost

$$= \begin{cases} 6000 + 1.40x, & \text{if } 0 \le x \le 4500 \\ 5100 + 1.60x, & \text{if } x > 4500 \end{cases}$$

(b) $p(x) = \begin{cases} 7.00, & \text{if } 0 \le x \le 4000 \\ 7.00 - (0.10)\frac{x-4000}{100} = 11.00 - 0.001x, & \text{if } x > 4000 \end{cases}$

(c) $P(x) = R(x) - C(x) = xp(x) - C(x)$

$$= \begin{cases} 7.00x - [6000 + 1.40x], & \text{if } 0 \le x \le 4000 \\ x[11.00 - 0.001x] - [6000 + 1.40x], & \text{if } 4000 < x \le 4500 \\ x[11.00 - 0.001x] - [5100 + 1.60x], & \text{if } x > 4500 \end{cases}$$

$$= \begin{cases} 5.60x - 6000, & \text{if } 0 \le x \le 4000 \\ -6000 + 9.60x - 0.001x^2, & \text{if } 4000 < x \le 4500 \\ -5100 + 9.40x - 0.001x^2, & \text{if } x > 4500 \end{cases}$$

$$P'(x) = \begin{cases} 5.60, & \text{if } 0 < x < 4000 \\ 9.60 - 0.002x, & \text{if } 4000 < x < 4500 \\ 9.40 - 0.002x, & \text{if } x > 4500 \end{cases}$$

$P'(x)$ is undefined for $x = 0, 4000,$ and 4500.
$P'(x) = 0$ for $x = 4700$.
P is continuous on $[0,\infty)$, and the auxiliary axis for $P'(x)$ is

```
        (+)        (u)       (+)       (u)      (+)       (0)       (-)
  •──────────•───────────•──────────•──────────•─────────•──────────────→
  0         4000        4500                  4700                    x
```

P is maximum if 4700 units/month are produced.

18. $R(x) = x \cdot p(x)$

$\dfrac{dR}{dm} = \dfrac{dR}{dx}\dfrac{dx}{dm} = \left[x\dfrac{dp}{dx} + p(1)\right]\dfrac{dx}{dm} = \dfrac{dx}{dm}\left(p + x\dfrac{dp}{dx}\right)$

Problem Set 4.5

Problem Set 4.6 Limits at Infinity, Infinite Limits

3. $\lim\limits_{x\to-\infty} \dfrac{2x^2-x+5}{5x^2+6x-1} = \lim\limits_{x\to-\infty} \dfrac{2 - (1/x) + (5/x^2)}{5 + (6/x) - (1/x^2)} = \dfrac{2 - 0 + 0}{5 + 0 - 0} = \dfrac{2}{5}$

6. $\lim\limits_{x\to\infty} \dfrac{6x^2+8x-8}{2x^2+5x+2} = \lim\limits_{x\to\infty} \dfrac{6 + (8/x) - (8/x^2)}{2 + (5/x) + (2/x^2)} = \dfrac{6 + 0 - 0}{2 + 0 + 0} = 3$

9. $\lim\limits_{x\to\infty} \sqrt[3]{\dfrac{1+8x^2}{x^2+4}} = \lim\limits_{x\to\infty} \sqrt[3]{\dfrac{(1/x^2) + 8}{1 + (4/x^2)}} = \sqrt[3]{\dfrac{0 + 8}{1 + 0}} = 2$

12. $\lim\limits_{x\to\infty} \dfrac{\sqrt{2x+1}}{x+4} = \lim\limits_{x\to\infty} \dfrac{\sqrt{(2/x) + (1/x^2)}}{1 + (4/x)}$ [Divided numerator by $\sqrt{x^2}$, and denominator by x; these are equal if $x > 0$.]

$= \dfrac{\sqrt{0 + 0}}{1 + 0} = 0$

15. $\lim\limits_{y\to-\infty} \dfrac{9y^3+1}{y^2-2y+2} = \lim\limits_{y\to-\infty} \dfrac{9y + (1/y^2)}{1 - (2/y) + (2/y^2)} = -\infty$

18. $\lim\limits_{x\to 3^-} \dfrac{3+t}{3-t} = +\infty$ Since the numerator approaches 6 and the denominator approaches 0 from the positive side, the fraction is positive and gets larger. (To see this, think of t having a value close to and to the left of 3, like 2.99, for example.) This concept will be abbreviated:

$\left[\begin{array}{c} \to 6 \\ \to 0^+ \end{array}\right]$

21. $\lim\limits_{x\to 1.5^-} \dfrac{4x+1}{2x-3} = -\infty$ since $\left[\begin{array}{c} \to 7 \\ \to 0^- \end{array}\right]$

24. $\lim\limits_{x\to 3} \dfrac{2x}{x-3}$ doesn't exist because $\lim\limits_{x\to 3^-} \dfrac{2x}{x-3} = -\infty$ since $\left[\begin{array}{c} \to 6 \\ \to 0^- \end{array}\right]$

and $\lim\limits_{x\to 3^+} \dfrac{2x}{x-3} = +\infty$ since $\left[\begin{array}{c} \to 6 \\ \to 0^+ \end{array}\right]$

27. $\lim_{x \to 0^+} \frac{[\![x]\!]}{x} = \lim_{x \to 0^+} \frac{0}{x}$ since $[\![x]\!] = 0$ if x is near and on the right of 0.

$\qquad = \lim_{x \to 0^+} (0) = 0$

30. $\lim_{x \to 0^+} \frac{|x|}{x} = \lim_{x \to 0^+} \frac{x}{x}$ since $|x| = x$ if x is on the right of 0.

$\qquad = \lim_{x \to 0^+} (1) = 1$

33. $f(x) = \frac{3}{x+1}$

Vertical asymptotes: Denominator is 0 when x = -1 so we suspect that the line x=-1 is a vertical asymptote.

It is since $\lim_{x \to -1^-} \frac{3}{x+1} = -\infty \quad \left[\begin{array}{c} \to 3 \\ \to 0^- \end{array}\right]$

and $\lim_{x \to -1^+} \frac{3}{x+1} = +\infty \quad \left[\begin{array}{c} \to 3 \\ \to 0^+ \end{array}\right]$

Horizontal asymptotes: The horizontal line y=0 since

$\lim_{x \to -\infty} \frac{3}{x+1} = 0$ and $\lim_{x \to \infty} \frac{3}{x+1} = 0$

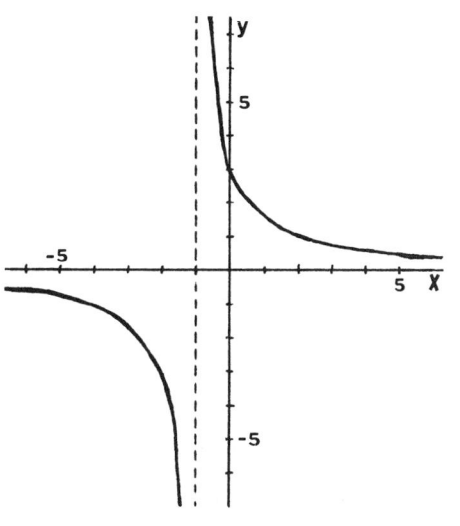

36. $F(x) = \frac{3}{9-x^2} = \frac{3}{(3+x)(3-x)}$

$\lim_{x \to -3^-} F(x) = -\infty$ since $\left[\begin{array}{c} \to 3 \\ \to 0^- \end{array}\right]$

$\lim_{x \to -3^+} F(x) = +\infty$ since $\left[\begin{array}{c} \to 3 \\ \to 0^+ \end{array}\right]$

x=-3 is a vertical asymptote.

$\lim_{x \to 3^-} F(x) = +\infty$ since $\left[\begin{array}{c} \to 3 \\ \to 0^+ \end{array}\right]$

$\lim_{x \to 3^+} F(x) = -\infty$ since $\left[\begin{array}{c} \to 3 \\ \to 0^- \end{array}\right]$

x=3 is a vertical asymptote.

$\lim_{x \to \pm\infty} \frac{3}{9-x^2} = \lim_{x \to \pm\infty} \frac{(3/x^2)}{(9/x^2) - 1} = \frac{0}{0 - 1} = 0$ Therefore, y=0 is a horizontal asymptote.

Problem Set 4.6

36. (continued)

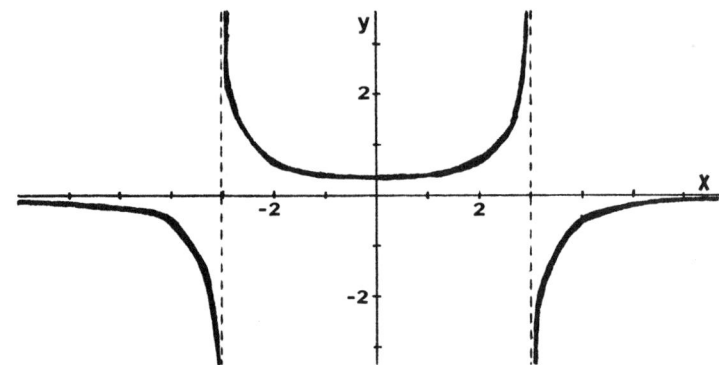

39. $\dfrac{2x^4+3x^3-2x-4}{x^3-1} = 2x+3 - \dfrac{1}{x^3-1}$, so $y = 2x+3$ is an oblique asymptote

since $\lim\limits_{x \to \pm\infty}\left[\dfrac{2x^4+3x^3-2x-4}{x^3-1} - (2x+3)\right] = \lim\limits_{x \to \pm\infty}\left[\left(2x+3 - \dfrac{1}{x^3-1}\right) - (2x+3)\right]$

$= \lim\limits_{x \to \pm\infty} \dfrac{-1}{x^3-1} = 0$

42. (a) Let f be defined on $[c,\infty)$ for some number c.

Then $\lim\limits_{x \to \infty} f(x) = \infty$ iff for each positive number M, there is a corresponding number N such that $x > c$ and $x > N \Rightarrow f(x) > M$

(b) Let f be defined on $(-\infty,c]$ for some number c.

Then $\lim\limits_{x \to -\infty} f(x) = \infty$ iff for each positive number M, there is a corresponding number N such that $x < c$ and $x < N \Rightarrow f(x) > M$

Problem Set 4.7 Sophisticated Graphing

3. $F(x) = \frac{1}{20}(x^4 - 18x^2 + 20)$

Domain: R

Symmetry: With respect to the y-axis [F is an even function.]

Intercepts: Set F(x) equal to zero and use the quadratic equation to solve for x^2, then obtain the square roots.

$$x^2 = \frac{18 \pm \sqrt{324-80}}{2} = 9 \pm \sqrt{61}, \text{ so } x = \pm\sqrt{9 \pm \sqrt{61}}$$

That is, x-intercepts are approximately ±4.10, ±1.09.
y-intercept is 1.

Monotonicity: $F'(x) = \frac{1}{20}(4x^3 - 36x) = \frac{1}{5}x(x+3)(x-3)$

```
        Dec           Inc           Dec           Inc
        (-)    (0)    (+)    (0)    (-)    (0)    (+)
       ─────────────────────────────────────────────── x
               -3            0             3
```

Concavity: $F''(x) = \frac{1}{20}(12x^2 - 36) = \frac{3}{5}(x+\sqrt{3})(x-\sqrt{3})$

```
         CU              CD              CU
        (+)    (0)      (-)     (0)     (+)
       ─────────────────────────────────────── x
              -√3              √3
              -1.73            1.73
```

Asymptotes: None

Summary	
At 0	Levels off
(0, 1.73)	Dec and CD
At 1.73	Inflection
(1.73, 3)	Dec and CU
At 3	Levels off
(3, ∞)	Inc and CU

x	F(x)	
0	1	(local max)
1.73	-1.25	(inflection point)
3	-3.05	(global min)
5	9.75	

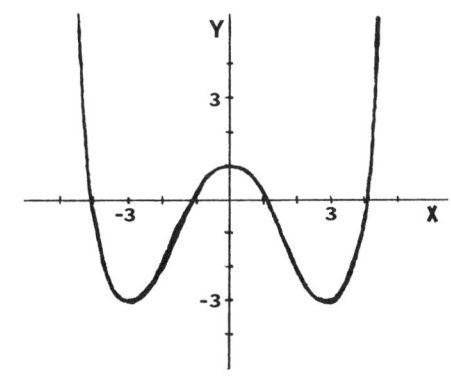

Use symmetry to obtain the graph for x < 0.

6. $g(x) = \frac{(x+2)^2}{x} = \frac{x^2+4x+4}{x} = x + 4 + 4x^{-1}$

Domain: $\{x : x \neq 0\}$

Symmetry: None of the basic three kinds, but notice (after you finish the graph) that it looks like there is symmetry with respect to the point $(0,4)$. [Try to prove that there is.]

Intercepts: x-intercept is -2. [Let $g(x) = 0$, and solve for x.]
No y-intercept [x can't be zero.]

Monotonicity: $g'(x) = 1 - 4x^{-2} = \frac{x^2-4}{x^2} = \frac{(x+2)(x-2)}{x^2}$

```
         Inc           Dec          Dec           Inc
         (+)    (0)    (-)    (u)   (-)    (0)    (+)
        ─────────●────────────●────────────●──────────── x
                -2            0            2
```

Concavity: $F''(x) = 8/x^3$

```
                        CD           CU
                        (-)    (u)   (+)
                       ──────────────●──────────── x
                                     0
```

Asymptotes: $x = 0$ (vertical) since $\lim\limits_{x \to 0^-} g(x) = -\infty$ $\begin{bmatrix} \to 4 \\ \to 0^- \end{bmatrix}$

and $\lim\limits_{x \to 0^+} g(x) = +\infty$ $\begin{bmatrix} \to 4 \\ \to 0^+ \end{bmatrix}$

$y = x+4$ (oblique) since $\lim\limits_{x \to \pm\infty} [g(x) - (x+4)] = \lim\limits_{x \to \pm\infty} \frac{4}{x} = 0$

Summary	
$(-\infty, -2)$	Inc and CD
At -2	Levels off
$(-2, 0)$	Dec and CD
At 0	Asymptote
$(0, 2)$	Dec and CU
At 2	Levels off
$(2, \infty)$	Inc and CU

x	g(x)
-2	0 (local max)
2	8 (local minimum)
4	9
-4	-1

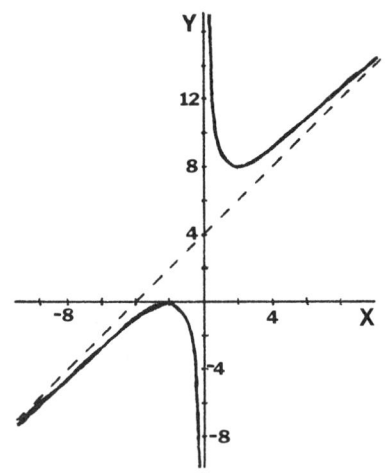

9. $f(x) = \dfrac{x^2}{x^2-4} = \dfrac{x^2}{(x+2)(x-2)}$

Domain: $\{x : x \neq -2, 2\}$

Symmetry: With respect to the y-axis [f is an even function.]

Intercepts: x-intercept is 0; y-intercept is 0.

Monotonicity: $f'(x) = \dfrac{-8x}{(x^2-4)^2} = \dfrac{-8x}{(x+2)^2(x-2)^2}$

```
         Inc         Inc         Dec         Dec
         (+)   (u)   (+)   (0)   (-)   (u)   (-)
        ─────●─────────────●─────────────●─────── x
             -2            0             2
```

Concavity: $f''(x) = \dfrac{8(3x^2+4)}{(x^2-4)^3} = \dfrac{8(3x^2+4)}{(x+2)^3(x-2)^3}$

```
         CU                CD                CU
         (+)   (u)         (-)         (u)   (+)
        ─────●─────────────────────────●─────── x
             -2                        2
```

Asymptotes: $x = -2$ (vertical) since $\lim\limits_{x \to -2^-} f(x) = \infty$ $\left[\begin{array}{c}\to 4 \\ \to 0^+\end{array}\right]$

and $\lim\limits_{x \to -2^+} f(x) = -\infty$ $\left[\begin{array}{c}\to 4 \\ \to 0^-\end{array}\right]$

$x = 2$ (vertical) since $\lim\limits_{x \to 2^-} f(x) = -\infty$ $\left[\begin{array}{c}\to 4 \\ \to 0^-\end{array}\right]$

and $\lim\limits_{x \to 2^+} f(x) = \infty$ $\left[\begin{array}{c}\to 4 \\ \to 0^+\end{array}\right]$

$y = 1$ (horizontal) since $\lim\limits_{x \to \pm\infty} f(x) = \lim\limits_{x \to \pm\infty} \dfrac{1}{1 - (4/x^2)} = 1$

Summary	
(0,2)	Dec and CD
At -2	Levels off
(2,∞)	Dec and CU

x	f(x)
0	0 (local max)
1	-1/3
3	1.8

Use symmetry to obtain the graph for $x < 0$.

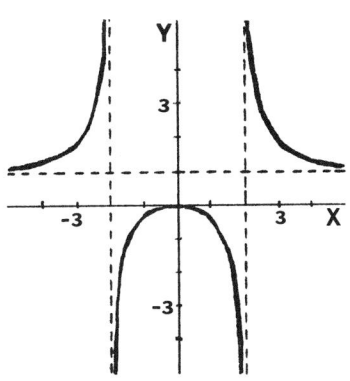

Problem Set 4.7

12. $g(x) = \sqrt{4-x} + 1$

   Domain: $(-\infty, 4]$ [The radicand, x-4, must be positive or zero.]

   Symmetry: None

   Intercepts: No x-intercept   [y is never less than 1.]
   y-intercept is 3.

   Monotonicity: $g'(x) = \dfrac{-1}{2\sqrt{4-x}}$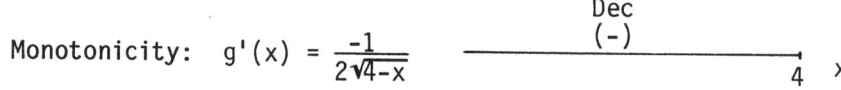

   Concavity: $g''(x) = \dfrac{-1}{4(4-x)^{3/2}}$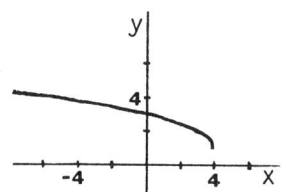

   Asymptotes: None

Summary	
$(-\infty, 4)$	Dec and CU

x	f(x)
-5	4
4	1

   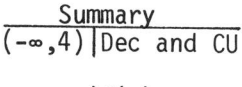

15. $f(x) = |\sin x|$. The graph is the same as that of $g(x) = \sin x$, except that all portions of the graph of g that are below the x-axis (where $y < 0$) are reflected with respect to the x-axis.

    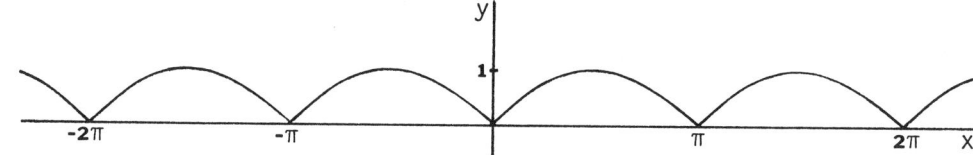

18.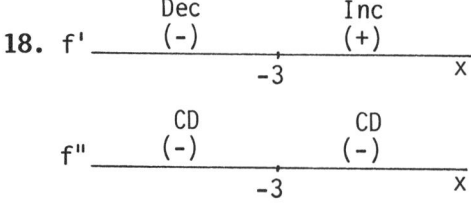

    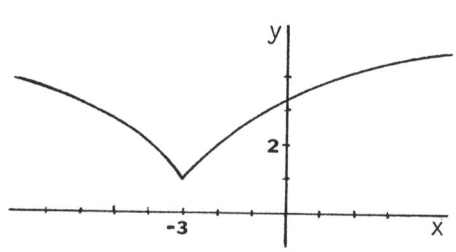

Problem Set 4.7

Problem Set 4.8 The Mean Value Theorem

3. The Mean Value Theorem applies since $g(x) = x^3/3$ is continuous on $[-2,2]$ and $g'(x) = x^2$ exists on $(-2,2)$. Therefore, for some c in $(-2,2)$

$$g'(c) = \frac{g(2)-g(-2)}{2-(-2)} = \frac{8/3-(-8/3)}{4} = \frac{4}{3}$$

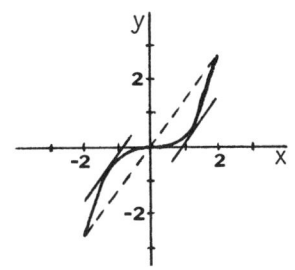

Therefore, $c^2 = 4/3$ or $c = \pm 2/\sqrt{3}$. That is, the tangent lines at $x = \pm 2/\sqrt{3} \approx \pm 1.15$ are parallel to the secant line connecting the points $(-2,-8/3)$ and $(2,8/3)$.

6. The Mean Value Theorem applies since $F(x) = (x+1)/(x-1)$ is continuous on $[1.5,5]$ and $F'(x) = -2/(x-1)^2$ exists on $(1.5,5)$. Therefore, for some c in $(1.5,5)$

$$F'(c) = \frac{F(5)-F(1.5)}{5-1.5} = \frac{(1.5)-(5)}{3.5} = -1$$

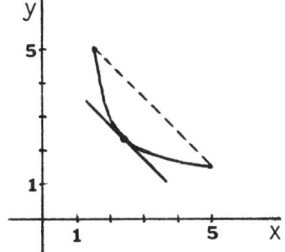

Therefore, $\frac{-2}{(c-1)^2} = -1$; $(c-1)^2 = 2$; $c-1 = \pm\sqrt{2}$; $c = 1 \pm \sqrt{2}$

$1-\sqrt{2}$ is not in $(1.5,5)$, so $c = 1+\sqrt{2} \approx 2.41$. That is, the tangent line at $x=1+\sqrt{2}$ is parallel to the secant line connecting the points $(1.5,5)$ and $(5,1.5)$.

9. The Mean Value Theorem does not apply since ϕ is discontinuous at 0 which is in the interval $[-1,0.5]$.

12.

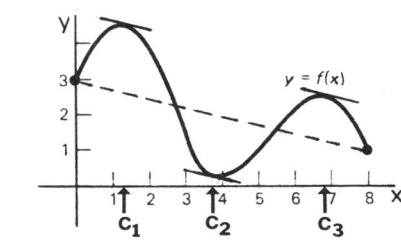

15. f is increasing on $(a, x_0]$ and on $[x_0, b)$ by the monotonicity theorem.

   Case I: If x_1 and x_2 are both in $(a, x_0]$ or both in $[x_0, b)$,
   then $f(x_1) < f(x_2)$ since f is increasing on those intervals.

   Case II: x_1 is in (a, x_0) and x_2 is in (x_0, b)
   $$\Rightarrow x_1 < x_0 \text{ and } x_0 < x_2$$
   $$\Rightarrow f(x_1) < f(x_0) \text{ and } f(x_0) < f(x_2)$$
   $$\Rightarrow f(x_1) < f(x_2)$$

   Therefore, for all x in (a,b), $x_1 < x_2 \Rightarrow f(x_1) < f(x_2)$, so f is increasing on (a,b).

18. Let $F(x) = \cos^2 x + \sin^2 x$

   Then $F'(x) = 2(\cos x)(-\sin x) + 2(\sin x)(\cos x) = 0$

   Therefore, $F(x) = \cos^2 x + \sin^2 x = C$ for some constant C [Problem 17]

   Therefore, $\cos^2(0) + \sin^2(0) = C$ [letting x=0]

   or $(1)^2 + (0)^2 = C$, so $C = 1$,

   and hence $\cos^2 x + \sin^2 x = 1$.

21. By the Intermediate Value Theorem, $f(x) = 0$ has <u>at least</u> one solution between a and b.

   Assume that $f(x) = 0$ has two distinct solutions, α and β, $\alpha < \beta$.

   Then f satisfies the conditions of Rolle's Theorem, so there must be a point c between α and β such that $f'(c) = 0$. But c is also between a and b so that contradicts that $f'(x) \neq 0$ on (a,b).

   Hence, the assumption that there are two distinct solutions must be incorrect.

24. Let $a \leq x_1 < x_2 < x_3 \leq b$, and $f(x_1) = f(x_2) = f(x_3) = 0$.

   Then (using Rolle's Theorem on $[x_1, x_2]$ and on $[x_2, x_3]$), there are numbers c_1 and c_2 such that c_1 is between x_1 and x_2, and c_2 is between x_2 and x_3, and such that $f'(c_1) = 0$ and $f'(c_2) = 0$.

24. (continued)

Now note that f' is a function that satisfies the conditions of Rolle's Theorem on the interval $[c_1, c_2]$, so there is at least one number c between c_1 and c_2, such that $f''(c) = 0$. [Remember that f" is the derivative of f'.]

27. (a) 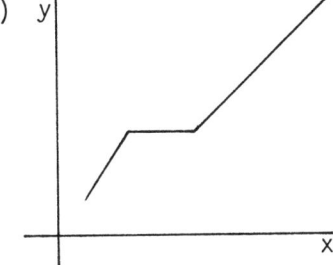 (b)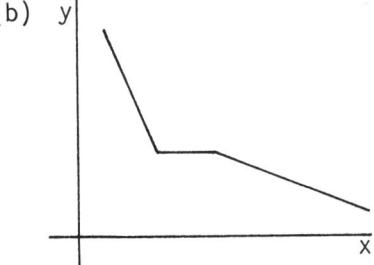

30. Let $f(x) = h(x) - g(x)$ on (a,b)
Then $f'(x) = h'(x) - g'(x) \geq 0$ on (a,b)

Therefore, $x_1 < x_2$ => $f(x_1) \leq f(x_2)$ [by Problem 28]
=> $h(x_1) - g(x_1) \leq h(x_2) - g(x_2)$
=> $g(x_2) - g(x_1) \leq f(x_2) - f(x_1)$

33. $F'(x) = 0 - f'(x) - [(b-x)f''(x) + f'(x)(-1)] - \frac{K}{2} 2(b-x)(-1)$
     $= -f'(x) - (b-x)f''(x) + f'(x) + (b-x)K$

Therefore, $F'(x) = -(b-x)f''(x) + (b-x)K$ (*)

Now note that (a) F is continuous on $[a,b]$, since f and f' are.
(b) F is differentiable on (a,b), since f and f' are.
(c) $F(a) = 0$ [Use the definition of K] and $f(b) = 0$

Therefore, there is a number c in (a,b) such that $F'(c) = 0$

That is, $F'(c) = -(b-c)f''(c) + (b-c)K = 0$ [from (*)], so $K = f''(c)$.

Then substitute K into (1) to get the desired result.

Problem Set 4.8

Problem Set 4.9 Chapter Review Problems

True-False Quiz

3. **True.** For example, the cosine function has a critical point at each multiple of π.

6. **False.** c^{ex}: $f(x) = x^3$ is increasing and differentiable on **R**, but $f'(0) = 0$.

9. **True.** See proof of Problem 33 of Section 4.2.

12. **False.** $\lim_{x \to 3} \frac{(x+2)(x-3)}{x-3} = \lim_{x \to 3}(x+2) = 5$ (instead of ∞ or $-\infty$)

15. **True.** f is continuous on $[0,2]$ and $f'(x) = \frac{1}{2\sqrt{x}}$ exists on $(0,2)$.

18. **False.** c^{ex}: If $f(x) = x^4$ and $c=0$, then $f'(0) = f''(0) = 0$, but $f(0)$ is the minimum value of f.

21. **True.** If a,b,c are the x-intercepts, with $a < b < c$, then by Rolle's Theorem, there are at least two points, one in (a,b) and one in (b,c), where there are horizontal tangent lines.

24. **True.** $f''(0) > 0$ on $[0,\infty)$ => f' is increasing on $[0,\infty)$
    Then $f'(x) > 0$ on $(0,\infty)$ [since $f'(0) = 0$].
    Hence, f is increasing on $[0,\infty)$.

27. **False.** c^{ex}: For the function graphed, f is concave up on **R**, but f has $y=0$ as a horizontal asymptote. Choose such a function where $f''(x) > 0$ everywhere.

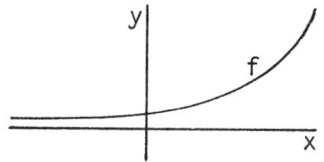

30. **True.** f is strictly monotonic on each open interval, so the minimum would occur at the left end point, but the left end point is not part of the open interval.

Miscellaneous Problems

3. $f(x) = 3x^4 - 4x^3$; $f'(x) = 12x^3 - 12x^2 = 12x^2(x-1)$

Critical Points: -2 and 3 (end points); 0 and 1 (stationary points)

x	f(x)	
-2	80	
0	0	
1	-1	← Minimum value is -1.
3	135	← Maximum value is 135.

6. $f(x) = (x-1)^3(x+2)^2$; $f'(x) = (x-1)^2(x+2)(5x+4)$

Critical Points: -2 and 2 (end points); -0.8 and 1 (stationary points)

x	f(x)	
-2	0	
-0.8	-8.40 (approx.)	← Minimum value is about -8.40.
1	0	
2	16	← Maximum value is 16.

9. $f(x) = x^4 - 4x^5$

$f'(x) = 4x^3 - 20x^4 = 4x^3(1-5x)$ (-)   (0)   (+)   (0)   (-)
                                                         0   0.2   x

Therefore, f is increasing on [0, 0.2].

$f''(x) = 12x^2 - 80x^3 = 4x^2(3-20x)$ (+)   (0)   (+)   (0)   (-)
                                                           0   0.15   x

Therefore, f is concave down on $(0.15, \infty)$.

12. $f(x) = \dfrac{4}{x^2+1} + 2 = 4(x^2+1)^{-1} + 2$

                                            Inc   Dec
                                            (+)   (0)   (-)

$f'(x) = -4(x^2+1)^{-2}(2x) = \dfrac{-8x}{(x^2+1)^2}$ 0   x

f has a maximum value of 6 at x=0; f has no minimum value but f gets arbitrarily close to

$$\lim_{x \to \pm\infty}\left(\dfrac{4}{x^2+1} + 2\right) = 2.$$

Problem Set 4.9

15. $f(x) = x\sqrt{x-3}$

   Domain: $[3,\infty)$ [since the radicand must be greater than or equal to 0]

   Symmetry: None

   Intercepts: x-intercept of 3; no y-intercept

   Monotonicity: $f'(x) = \dfrac{3(x-2)}{2\sqrt{x-3}}$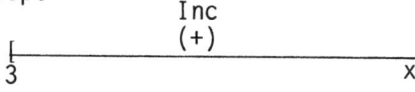

   Concavity: $f''(x) = \dfrac{3(x-4)}{4(x-3)^{3/2}}$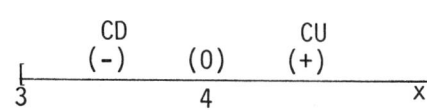

   Asymptotes: None

Summary	
(3,4)	Inc and CD
At 4	Inflection
(4,∞)	Inc and CU

x	f(x)	
3	0	(global minimum)
4	4	(inflection point)
5	7.07	

   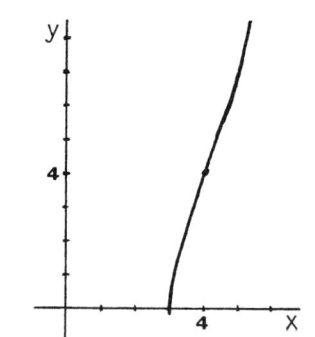

18. $f(x) = \dfrac{x^2-1}{x} = \dfrac{(x+1)(x-1)}{x}$ Also $\dfrac{x^2-1}{x} = x - \dfrac{1}{x}$

   Domain: $\{x : x \neq 0\}$

   Symmetry: With respect to the origin [since f is an odd function]

   Intercepts: x-intercepts of -1 and 1

   Monotonicity: $f'(x) = \dfrac{x^2+1}{x^2}$

   ```
              Inc           Inc
              (+)    (u)    (+)
   ─────────────────0──────────── x
   ```

   Concavity: $f''(x) = -2/x^3$

   ```
              CU            CD
              (+)    (u)    (-)
   ─────────────────0──────────── x
   ```

   Asymptotes: x=0 (vertical) since $\lim_{x \to 0^-} f(x) = \infty$ and $\lim_{x \to 0^+} f(x) = -\infty$

   y=x (oblique) since $\lim_{x \to \pm\infty} [f(x) - x] = \lim_{x \to \pm\infty} \dfrac{-1}{x} = 0$

110   Problem Set 4.9

18. (continued)

Summary	
$(-\infty,0)$	Inc and CU
$(0,\infty)$	Inc and CD

x	f(x)
0.5	-1.5
2	1.5
4	3.75

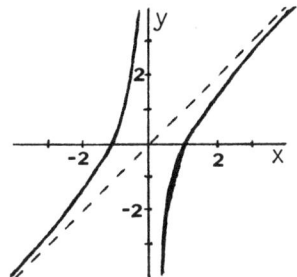

21. Step 1: Let y be as indicated [inches]
Let A be the cross-sectional area

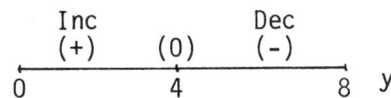

Step 2: Maximize A = (width)(height) = (16-2y)y

Step 3: $A(y) = (16-2y)y = 16y - 2y^2$

Step 4: y is in (0,8). We will use [0,8].

Step 5: $A'(y) = 16 - 4y = -4(y-4)$

```
         Inc         Dec
         (+)   (0)   (-)
    |_____|_____|
    0          4          8   y
```

Step 6: A will be maximum where y=4.

Conclusion: 4 inches should be turned up at each side.

24. Step 1: Let A denote the surface area of the trough.

Step 2: Minimize A = (area of the ends) + (area of curved section)
$= (\pi r^2) + (\pi r h)$

[The ends can form one disk; the curved section is half of a cylindrical surface.]

Step 3: The volume given is 128π so $128\pi = (1/2)\pi r^2 h$; $h = 256/r^{-2}$.

Therefore, A can be expressed as a function of r only as

$A(r) = \pi r^2 + \pi r(256/r^{-2}) = \pi r^2 + 256\pi r^{-1}$

Problem Set 4.9

24. (continued)

Step 4: $0 < r$

Step 5: $A'(r) = 2\pi r - 256\pi r^{-2} = \dfrac{2\pi(r^3-128)}{r^2} = 0$ iff $r = \sqrt[3]{128} \approx 5.04$

```
         Dec           Inc
         (-)    (0)    (+)
    |-----------•----------
    0          5.04              r
```

Step 6: A is minimum where $r = \sqrt[3]{128} \approx 5.04$; $h = \dfrac{256}{(\sqrt[3]{128})^2} \approx 10.08$.

Conclusion: The radius is about 5.04 feet; length is about 10.08 feet.

27. Let $y = f(x) = x^4 - 6x^3 + 12x^2 - 3x + 1$.

Then $f'(x) = 4x^3 - 18x^2 + 24x - 3$; $f''(x) = 12(x^2 - 3x + 2) = 12(x-1)(x-2)$

```
         CU           CD            CU
         (+)   (0)    (-)    (0)    (+)
    _____•_____•_____
                1            2               x
```

Thus, inflection points occur where $x = 1, 2$.

The corresponding y-values are $f(1) = 5$ and $f(2) = 11$, so the inflection points are $(1,5)$ and $(2,11)$.

The slopes of the respective tangent lines are $f'(1) = 7$ and $f'(2) = 5$.

Therefore, equations of the tangent lines are:

$y - 5 = 7(x-1)$ and $y - 11 = 5(x-2)$ [point-slope form]
$y = 7x - 2$ and $y = 5x + 1$ [slope-intercept form]

CHAPTER 5   THE INTEGRAL

Problem Set 5.1   Antiderivatives (Indefinite Integrals)

3. $\int (3x^2+\sqrt{2})dx = x^3+\sqrt{2}x + C$

6. $\int x^{-3/4}dx = \dfrac{x^{1/4}}{1/4} + C = 4x^{1/4} + C$

9. $\int (18x^8-25x^4+3x^2)dx = 18\dfrac{x^9}{9} - 25\dfrac{x^5}{5} + 3\dfrac{x^3}{3} + C = 2x^9-5x^5+x^3+C$

12. $\int (x^{-3}+6x^{-7})dx = \dfrac{x^{-2}}{-2} + \dfrac{6x^{-6}}{-6} + C = \dfrac{-1}{2x^2} - \dfrac{1}{x^6} + C$

15. $\int (x^3 + x^{1/2})dx = \dfrac{x^4}{4} + \dfrac{x^{3/2}}{3/2} + C = \dfrac{x^4}{4} + \dfrac{2x^{3/2}}{3} + C$

18. $\int (y^4-3y^2)dy = \dfrac{y^5}{5} - y^3 + C$

21. $\int (3\sin t - 2\cos t)dt = 3(-\cos t) - 2(\sin t) + C = -3\cos t - 2\sin t + C$

24. $\int (x^2-4)^3 2x\, dx = \int u^3 du$

    $= \dfrac{u^4}{4} + C = \dfrac{(x^2-4)^4}{4} + C$

    | Let $u = x^2-4$
    | Then $du = 2x\, dx$

27. $\int 3x^4(2x^5+9)^3 dx = \int (2x^5+9)^3 \dfrac{3}{10} 10x^4 dx = \int \dfrac{3}{10} u^3 du$

    $= \dfrac{3}{10} \dfrac{u^4}{4} + C = \dfrac{3(2x^5+9)^4}{40} + C$

    | Let $u = 2x^5+9$
    | Then $du = 10x^4 dx$

30. $\int (5x^2+1)\sqrt{5x^3+3x-2}\, dx$

    $= \int \dfrac{1}{3}(5x^3+3x-2)^{1/2}\, 3(5x^2+1)dx$

    | Let $u = 5x^3+3x-2$
    | Then $du = (15x^2+3)dx$
    | $= 3(5x^2+1)dx$

Problem Set 5.1

30. (continued)

$$= \int \frac{1}{3} u^{1/2} du = \frac{1}{3} \frac{u^{3/2}}{3/2} + C = \frac{2(5x^3+3x-2)^{3/2}}{9} + C$$

33. $\int \sin^4 x \cos x \, dx = \int u^4 du$ $\boxed{\text{Let } u = \sin x \\ \text{Then } du = \cos x \, dx}$

$$= \frac{u^5}{5} + C = \frac{\sin^5 x}{5} + C$$

36. $\int \cos(3x+1) \sin(3x+1) \, dx$ $\boxed{\text{Let } u = \sin(3x+1) \\ \text{Then } du = 3\cos(3x+1) \, dx}$

$$= \int \frac{1}{3} \sin(3x+1) \, 3\cos(3x+1) dx$$

$$= \int \frac{1}{3} u \, du = \frac{1}{3} \frac{u^2}{2} + C = \frac{\sin^2(3x+1)}{6} + C$$

39. $f''(x) = 3x + 1; \quad f'(x) = \frac{3x^2}{2} + x + C_1; \quad f(x) = \frac{3}{2} \frac{x^3}{3} + \frac{x^2}{2} + C_1 x + C_2$

$$= \frac{x^3}{2} + \frac{x^2}{2} + C_1 x + C_2$$

42. $f''(x) = x^{4/3}; \quad f'(x) = \frac{3}{7} x^{7/3} + C_1; \quad f(x) = \frac{3}{7} \frac{3}{10} x^{10/3} + C_1 x + C_2$

$$= \frac{9 x^{10/3}}{70} + C_1 x + C_2$$

45. To prove this we need to take the derivative with respect to x of the right-hand side of the equation and end up with the integrand of the left-hand side.

$$D[f(x)g(x) + C] = f(x)g'(x) + g(x)f'(x) + 0 = f(x)g'(x) + g(x)f'(x)$$

48. $\int \left[\frac{-x^3}{(2x+5)^{3/2}} + \frac{3x^2}{\sqrt{2x+5}} \right] dx = \int [x^3 \, D(2x+5)^{-1/2} + (2x+5)^{-1/2} \, D(x^3)] dx$

$$= x^3 (2x+5)^{-1/2} + C \quad [f(x) = x^3, \, g(x) = (2x+5)^{-1/2}]$$

51. $\int f''(x)dx = f'(x) + C = x\left(\dfrac{1}{2\sqrt{x^3+1}} \cdot 3x^2\right) + \sqrt{x^3+1} \ (1) + C$

$= \dfrac{3x^3 + 2(x^3+1)}{2\sqrt{x^3+1}} + C = \dfrac{5x^3+2}{2\sqrt{x^3+1}} + C.$

54. $\int \sin^3[(x^2+1)^4]\cos[(x^2+1)^4](x^2+1)^3 x\,dx$ $\boxed{\text{Let} \quad u = \sin[(x^2+1)^4] \\ \text{Then } du = \cos[(x^2+1)^4]4(x^2+1)^3 2x\,dx}$

$= \int \dfrac{1}{8} \sin^3[(x^2+1)^4]\cos[(x^2+1)^4]4(x^2+1)^3 2x\,dx$

$= \int \dfrac{1}{8} u^3\,du = \dfrac{1}{8}\dfrac{u^4}{4} + C = \dfrac{\sin^4[(x^2+1)^4]}{32} + C.$

Problem Set 5.2 Differential Equations

3. $y = C_1\sin x + C_2\cos x$; $\dfrac{dy}{dx} = C_1\cos x - C_2\sin x$; $\dfrac{d^2y}{dx^2} = -C_1\sin x - C_2\cos x.$

   Therefore, $\dfrac{d^2y}{dx^2} + y = (-C_1\sin x - C_2\cos x) + (C_1\sin x + C_2\cos x) = 0$

6. $y = \dfrac{x^{-1}}{-1} + x^2 + C = -x^{-1}+x^2+C.$ (general solution)

   $y=5, x=1 \Rightarrow C=5.$

   Therefore, $y = -x^{-1}+x^2+5$ is the particular solution.

9. $\dfrac{dy}{dt} = t^3 y^2$; $y=1$ at $t=2$.

   $y^{-2}dy = t^3 dt$; $\int y^{-2}dy = \int t^3 dt$; $-y^{-1} = (1/4)t^4 + C.$ (general solution)

   $y=1, t=2 \Rightarrow C=-5.$

   $-y^{-1} = \dfrac{t^4}{4} - 5$ or $y = \dfrac{-4}{t^4-20}.$ (particular solution)

12. $\frac{du}{dt} = u^2(t^2-3t)$; $u=4$ at $t=0$

$u^{-2}du = (t^2-3t)dt$; $\int u^{-2}du = \int(t^2-3t)dt$; $-u^{-1} = \frac{t^3}{3} - 3t^2 + C$ (gen. sol.)

$u=4$, $t=0$ => $C = -1/4$

$\frac{-1}{u} = \frac{t^3}{3} - \frac{3t^2}{2} - \frac{1}{4} = \frac{4t^3-18t^2-3}{12}$ or $u = \frac{-12}{4t^3-18t^2-3}$ (particular sol.)

15. Slope at (x,y) is $\frac{dy}{dx}$, and it is given that this equals $4x$.

$\frac{dy}{dx} = 4x$ => $y = 2x^2 + C$ (general solution)

$x=1$, $y=2$ => $C = 0$

Therefore, $y = 2x^2$ is the particular solution.

18. $a(t) = (1+t)^{-3}$

v is an antiderivative of a. $v(t) = \frac{(1+t)^{-2}}{-2} + C_1$; $t=0$, $v=4$ => $C_1 = \frac{9}{2}$

Then $v(t) = \frac{(1+t)^{-2}}{-2} + \frac{9}{2} = \frac{-1}{2(1+t)^2} + \frac{9}{2}$

s is an antiderivative of v. $s(t) = \frac{(1+t)^{-1}}{2} + \frac{9t}{2} + C_2$

$t=0$, $s=6$ => $C_2 = \frac{11}{2}$. Then $s(t) = \frac{1}{2(1+t)} + \frac{9t}{2} + \frac{11}{2}$

Finally, $s(2) = \frac{1}{6} + 9 + \frac{11}{2} = \frac{44}{3} \approx 14.6667$; $v(2) = \frac{-1}{18} + \frac{9}{2} = \frac{40}{9} \approx 4.4444$

21. $v_0 = 96$ and $s_0 = 0$ [That is, $v=96$ and $s=0$ when $t=0$]

Making use of the results of Example 3, the equations of motion are

$a(t) = -32$, $v(t) = -32t + 96$, $s(t) = -16t^2 + 96t + 0$

The maximum height is reached when $v = 0$.

$0 = -32t+96$ => $t=3$

$s(3) = -16(3)^2 + 96(3) = 144$. Maximum height reached is 144 feet.

24. $v(t) = -5.28t + 56$; $s(t) = -2.64t^2+56t+1000$ [from Problem 23]

$v=0 \Rightarrow t = \dfrac{56}{5.28}$

$s(56/5.28) = -2.64(56/5.28)^2 + 56(56/5.28) + 1000 \approx 1297.$

The maximum height reached is about 1297 feet.

27. The analysis in Example 5 is valid for escape velocity from any large body. Escape velocity is $\sqrt{2GR}$ where $-G$ is the gravitational acceleration and R is the radius of the large body.

$g \approx 32$ ft/sec^2 = 32/5280 mi/sec^2

Moon: $\sqrt{2GR} = \sqrt{2[(0.165)(32/5280)](1080)} \approx 1.47$ mi/sec

Venus: $\sqrt{2GR} = \sqrt{2[(0.85)(32/5280)](3800)} \approx 6.26$ mi/sec

Jupiter: $\sqrt{2GR} = \sqrt{2[(2.6)(32/5280)](43000)} \approx 36.8$ mi/sec

Sun: $\sqrt{2GR} = \sqrt{2[(28)(32/5280)](432000)} \approx 383$ mi/sec

30. $a(t) = 8$

$v(t) = 8t + v_0$

$s(t) = 4t^2 + v_0 t \quad [s_0 = 0]$

$s = 75, t = 3.75 \Rightarrow 75 = 4(3.75)^2 + v_0(3.75)$
$\Rightarrow v_0 = 5 \quad [\text{ft/sec}]$

Problem Set 5.3 Sums and Sigma Notation

3. $\dfrac{2}{3+1} + \dfrac{2}{4+1} + \dfrac{2}{5+1} = \dfrac{37}{30} \approx 1.2333$

6. $\dfrac{(-1)^2}{2(5)} + \dfrac{(-1)^3}{3(7)} + \dfrac{(-1)^4}{4(9)} = \dfrac{101}{1260} \approx 0.08016$

9. $\displaystyle\sum_{i=1}^{98} i$

Problem Set 5.3

12. $\displaystyle\sum_{j=1}^{50} \frac{(-1)^{j+1}}{j}$

15. $\displaystyle\sum_{k=1}^{n} f(c_k)$

18. $\sum(3a_i - 2b_i) = 3\sum a_i - 2\sum b_i = 3(40) - 2(50) = 20.$ Each summation is for i=1 to i=10.

21. $\left(\frac{1}{1} - \frac{1}{2}\right) + \left(\frac{1}{2} - \frac{1}{3}\right) + \left(\frac{1}{3} - \frac{1}{4}\right) + \cdots + \left(\frac{1}{40} - \frac{1}{41}\right) = \frac{1}{1} - \frac{1}{41} = \frac{40}{41} \approx 0.9756$

24. $(a_3-a_2) + (a_4-a_3) + (a_5-a_4) + \cdots + (a_{m+1}-a_m) = a_{m+1}-a_2$

27. $\displaystyle\sum_{k=1}^{10}(k^3-k^2) = \sum_{k=1}^{10} k^3 - \sum_{k=1}^{10} k^2 = \left[\frac{10(10+1)}{2}\right]^2 - \frac{10(10+1)(20+1)}{6} = 2640$

30. $\displaystyle\sum_{i=1}^{n}(2i-3)^2 = \sum_{i=1}^{n}(4i^2-12i+9) = 4\sum_{i=1}^{n} i^2 - 12\sum_{i=1}^{n} i + \sum_{i=1}^{n} 9$

$= 4\,\frac{n(n+1)(2n+1)}{6} - 12\,\frac{n(n+1)}{2} + 9n = \frac{n(4n^2-12n+11)}{3}$

33. $\displaystyle\sum_{k=0}^{10} \frac{k}{k+1} = \sum_{i=1}^{11} \frac{i-1}{i}$

```
Let i = k+1
Then k = i-1
     k=10  =>  i=11
     k=0   =>  i=1
```

36. Let P_n be the statement $a + ar + ar^2 + \cdots + ar^n = \dfrac{a - ar^{n+1}}{1-r}$ $(r \neq 1)$

P_1: $a+ar = a(1+r) = \dfrac{a(1+r)(1-r)}{1-r} = \dfrac{a(1-r^2)}{1-r} = \dfrac{a - ar^2}{1-r}$ is true.

Assume P_i: $a + ar + ar^2 + \cdots + ar^i = \dfrac{a - ar^{i+1}}{1-r}$ to be true.

36. (continued)

Show P_{i+1}: $a + ar + ar^2 + \ldots + ar^i + ar^{i+1} = \dfrac{a - ar^{i+2}}{1-r}$ is then also true.

$$a + ar + ar^2 + \cdots + ar^i + ar^{i+1} = (a + ar + ar^2 + \cdots + ar^i) + ar^{i+1}$$

$$= \dfrac{a - ar^{i+1}}{1-r} + ar^{i+1} \quad \text{[assumed } P_i\text{]}$$

$$= \dfrac{a - ar^{i+1} + ar^{i+1}(1-r)}{1-r}$$

$$= \dfrac{a - ar^{i+1} + ar^{i+1} - ar^{i+2}}{1-r}$$

$$= \dfrac{a - ar^{i+2}}{1-r}$$

We have now proved: (i) P_1 is true, and (ii) $P_i \Rightarrow P_{i+1}$. Therefore, P_n is true for each positive integer n.

39. $\displaystyle\sum_{i=1}^{100} i = \dfrac{100(100+1)}{2} = 5050$ [Special Sum (1) on page 217 of the text]

42. (a) $\sum(x_i - \bar{x}) = \sum x_i - \sum \bar{x} = n\bar{x} - n\bar{x} = 0$, summations from i=1 to i=n.

Note: $\sum x_i = n\bar{x}$ since $\bar{x} = \dfrac{\sum x_i}{n}$; $\sum \bar{x} = n\bar{x}$ since $\bar{x}$ is a constant.

(b) $s^2 = \dfrac{1}{n}\sum(x_i - \bar{x})^2 = \dfrac{1}{n}\sum(x_i^2 - 2\bar{x}x_i + \bar{x}^2) = \dfrac{1}{n}\left(\sum x_i^2 - 2\bar{x}\sum x_i + \sum \bar{x}^2\right)$

$$= \dfrac{1}{n}\left(\sum x_i^2 - 2\bar{x}\, n\bar{x} + n\bar{x}^2\right) = \dfrac{1}{n}\left(\sum x_i^2 - n\bar{x}^2\right) = \dfrac{1}{n}\sum x_i^2 - \bar{x}^2$$

Problem Set 5.3

Problem Set 5.4 Introduction to Area

3. Let $f(x) = x+1$. $n = 8$. $\Delta x = 2/8 = 0.25$ for each subinterval.

$A(R_8) = [f(0)+f(0.25)+f(0.5)+f(0.75)+f(1)+f(1.25)+f(1.5)+f(1.75)](0.25)$
$= [1 + 1.25 + 1.5 + 1.75 + 2 + 2.25 + 2.5 + 2.75](0.25) = 3.75$

6. Let $g(x) = 0.5x^2+1$. $n = 4$. $\Delta x = 2/4 = 0.5$ for each subinterval.

$A(S_4) = [g(0.5) + g(1) + g(1.5) + g(2)](0.5)$
$= [1.125 + 1.5 + 2.125 + 3](0.5) = 3.875$

9. $f(x) = x^2+2$ on $[0,2]$. $n = 6$. $\Delta x = \frac{2-0}{6} = \frac{1}{3}$.

$A(S_6) = [f(1/3)+f(2/3)+f(1)+f(4/3)+f(5/3)+f(2)](1/3)$
$= [(19/9)+(22/9)+(3)+(34/9)+(43/9)+(6)](1/3)$
$= (199/9)(1/3) \approx 7.3704$

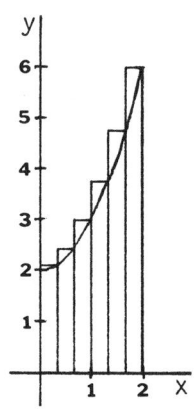

12. Let $f(x) = 0.5x^2+1$ on $[0,2]$. Since f is increasing on $[0,2]$, the circumscribed polygon is determined by the union of rectangles whose widths are $\Delta x = (2-0)/n = 2/n$, and whose heights are $f(x_i)$ where x_i is the right end point of the ith subinterval.

$0 = \frac{0}{n} \quad \frac{2}{n} \quad \frac{4}{n} \quad \frac{6}{n} \quad \cdots \quad \frac{2n-2}{n} \quad \frac{2n}{n} = 2$ (See graph in Problem 6.)

[In the following, all summations are from $i=1$ to $i=n$.]

$A(S_n) = \sum f\left(\frac{2i}{n}\right) \frac{2}{n} = \sum \left[\frac{1}{2}\left(\frac{2i}{n}\right)^2 + 1\right]\frac{2}{n} = \sum \left[\frac{2i^2}{n^2} + 1\right]\frac{2}{n} = \sum \left(\frac{4i^2}{n^3} + \frac{2}{n}\right)$

$= \frac{4}{n^3}\sum i^2 + \frac{2}{n}\sum 1 = \frac{4}{n^3} \cdot \frac{n(n+1)(2n+1)}{6} + \frac{2}{n} n = \frac{2}{3}(1)(1+\frac{1}{n})(2+\frac{1}{n}) + 2$

Area $= \lim_{n \to \infty} A(S_n) = \frac{2}{3}(1)(1)(2) + 2 = \frac{10}{3}$

15. Let $f(x) = x^3$ on $[0,1]$. Since f is increasing on $[0,1]$, the circumscribed polygon is determined by the union of rectangles whose widths are $\Delta x = (1-0)/n = 1/n$, and whose heights are $f(x_i)$ where x_i is the right end point of the ith subinterval.

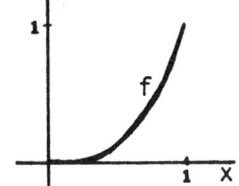

$0 = \frac{0}{n} \quad \frac{1}{n} \quad \frac{2}{n} \quad \frac{3}{n} \cdots \frac{n-1}{n} \quad \frac{n}{n} = 1$

[In the following, all summations are from i=1 to i=n.]

$A(S_n) = \sum f(\frac{i}{n})\frac{1}{n} = \sum (\frac{i}{n})^3 \frac{1}{n} = \sum \frac{i^3}{n^4} = \frac{1}{n^4} \sum i^3 = \frac{1}{n^4}\left(\frac{n(n+1)}{2}\right)^2 = \frac{1}{4}(1)^2(1+\frac{1}{n})^2.$

Area $= \lim_{n \to \infty}(S_n) = \frac{1}{4}(1)^2(1)^2 = \frac{1}{4}.$

18. Area over $[a,b]$ = Area over $[0,b]$ - Area over $[0,a]$

$= \frac{b^3}{3} - \frac{a^3}{3}.$ [from Problem 17]

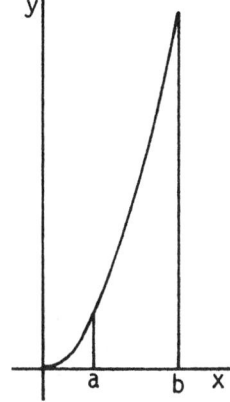

21. The distance the object travels is the area of the region under the velocity curve, $v(t) = t^2$, over the interval $[3,5]$. According to the result obtained in Problem 18, this area is

$\frac{5^3}{3} - \frac{3^3}{3} = \frac{98}{3} \approx 32.6667.$

Hence, the object travels about 32.6667 meters between t=3 and t=5.

Problem Set 5.4

Problem Set 5.5 The Definite Integral

3. $\sum_{i=1}^{5} f(\bar{x}_i)\Delta x_i = f(3)(3.75-3) + f(4)(4.25-3.75) + f(4.75)(5.5-4.25) + f(6)(6-5.5) + f(6.5)(7-6)$

$= (2)(0.75) + (3)(0.5) + (3.75)(1.25) + (5)(0.5) + (5.5)(1)$
$= 15.6875.$

6. $\sum_{i=1}^{8} f(\bar{x}_i)\Delta x_i = [f(0.25) + f(0.5) + f(0.75) + f(1) + f(1.25) + f(1.5) + f(1.75) + f(2)](0.25)$

$= (0.25)\left[\frac{(0.25)^3}{3} + \frac{(0.5)^3}{3} + \frac{(0.75)^3}{3} + \frac{(1)^3}{3} + \frac{(1.25)^3}{3} + \frac{(1.5)^3}{3} + \frac{(1.75)^3}{3} + \frac{(2)^3}{3} + 8\right]$

$= (0.25)\frac{(0.25)^3}{3}[1^3 + 2^3 + 3^3 + 4^3 + 5^3 + 6^3 + 7^3 + 8^3] + 2$

$= \frac{(0.25)^4}{3}\left(\frac{8(8+1)}{2}\right)^2 + 2 = 3.6875.$

9. $\int_0^3 \frac{x}{1+x}\, dx.$

12. The integral exists since $f(x) = x^2 + 2$ is continuous on $[0,4]$. Let P be a regular partition of $[0,4]$ of n subintervals. In each subinterval let the right end point be the sample point for that subinterval.

$0 = \frac{0}{n} \quad \frac{4}{n} \quad \frac{8}{n} \quad \frac{12}{n} \quad \cdots \quad \frac{4n-4}{n} \quad \frac{4n}{n} = 4 \qquad \Delta x_i = \frac{4-0}{n} = \frac{4}{n}.$

[In the following, each summation is from i=1 to i=n; $\bar{x}_i = \frac{4i}{n}$.]

Then $\int_0^4 (x^2+2)\,dx = \lim_{|P|\to 0} \sum f(\bar{x}_i)\Delta x_i = \lim_{n\to\infty}\left(\sum \frac{16i^2}{n^2} + 2\right)\frac{4}{n}$

$= \lim_{n\to\infty}\left(\frac{64}{n^3}\sum i^2 + \sum \frac{8}{n}\right) = \lim_{n\to\infty}\left(\frac{64}{n^3}\cdot\frac{n(n+1)(2n+1)}{6} + \frac{8}{n}n\right)$

$= \lim_{n\to\infty}\left(\frac{32}{3}(1)(1+\frac{1}{n})(2+\frac{1}{n}) + 8\right) = \frac{32}{3}(1)(1)(2) + 8 = \frac{88}{3} \approx 29.3333.$

15. The integral exists since $f(x) = x^2-2x$ is continuous on $[0,4]$. Let P be a regular partition of $[0,4]$ of n subintervals. In each subinterval let the right end point be the sample point for that subinterval.

$$0 = \frac{0}{n} \quad \frac{4}{n} \quad \frac{8}{n} \quad \frac{12}{n} \cdots \frac{4n-4}{n} \quad \frac{4n}{n} = 4 \qquad \Delta x_i = \frac{4-0}{n} = \frac{4}{n}.$$

[In the following, each summation is from $i=1$ to $i=n$; $\bar{x}_i = \frac{4i}{n}$].

Then
$$\int_0^4 (x^2-2x)dx = \lim_{|P| \to 0} \sum f(\bar{x}_i)\Delta x_i = \lim_{n \to \infty} \sum \left(\frac{16i^2}{n^2} - \frac{8i}{n}\right)\frac{4}{n}$$

$$= \lim_{n \to \infty} \left(\frac{64}{n^3}\sum i^2 - \frac{32}{n^2}\sum i\right) = \lim_{n \to \infty} \left(\frac{64}{n^3}\frac{n(n+1)(2n+1)}{6} - \frac{32}{n^2}\frac{n(n+1)}{2}\right)$$

$$= \lim_{n \to \infty} \left(\frac{32}{3}(1)(1+\tfrac{1}{n})(2+\tfrac{1}{n}) - 16(1)(1+\tfrac{1}{n})\right) = \frac{32}{3}(1)(1)(2) - 16(1)(1) = \frac{16}{3}.$$

18. Recall that the area bounded by a trapezoid is $\frac{h}{2}(b_1 + b_2)$ where b_1 and b_2 are the lengths of the parallel sides and h is the distance between the parallel sides.

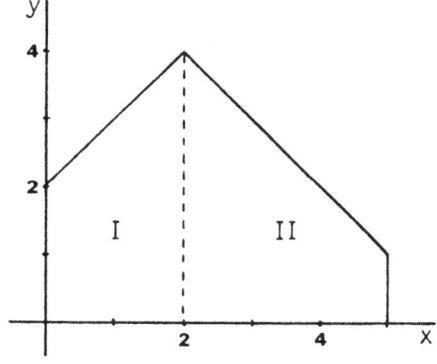

$$\int_0^5 f(x)\,dx = \text{Area(I)} + \text{Area(II)}$$

$$= \tfrac{2}{2}(2+4) + \tfrac{3}{2}(4+1) = 13.5.$$

21. (a) Is, since f is continuous on $[-2,2]$.

(b) Is, since f is continuous on $[-2,2]$.

(c) Is not, since f is not bounded on $[-2,2]$.

(d) Is not, since f is not even defined everywhere on $[-2,2]$; namely, it is not defined at $\pm\pi/2$. f is also not bounded on $[-2,2]$.

(e) Is, since f is bounded on $[-2,2]$ and f has only one discontinuity.

(f) Is, since f is bounded on $[-2,2]$ and f has only one discontinuity.

Problem Set 5.5

Problem Set 5.6 The Fundamental Theorem of Calculus

3. $\int_{-1}^{2} (3x^2-2x+3)dx = \left[x^3-x^2+3x\right]_{-1}^{2} = (8-4+6) - (-1-1-3) = 15$

6. $\int_{1}^{3} 2t^{-3}dt = \left[\frac{2t^{-2}}{-2}\right]_{1}^{3} = \left[\frac{-1}{t^2}\right]_{1}^{3} = \frac{-1}{9} - \frac{-1}{1} = \frac{8}{9} \approx 0.8889$

9. $\int_{-4}^{-2} (y^2+y^{-3})dy = \left[\frac{y^3}{3} + \frac{y^{-2}}{-2}\right]_{-4}^{-2} = \left(\frac{-8}{3} - \frac{1}{8}\right) - \left(\frac{-64}{3} - \frac{1}{32}\right) = \frac{1783}{96} \approx 18.5729$

12. $\int_{\pi/6}^{\pi/2} 2 \sin t \, dt = \left[-2 \cos t\right]_{\pi/6}^{\pi/2} = -2 \cos(\pi/2) - [-2 \cos(\pi/6)] = 0 + \sqrt{3} \approx 1.7321$

15. $\int (x^2+1)^{10}(2x)dx = \int u^{10}du$ Let $u = x^2+1$
       Then $du = 2x \, dx$

$= \frac{u^{11}}{11} + C = \frac{(x^2+1)^{11}}{11} + C$

Then $\int_{0}^{1} (x^2+1)dx = \left[\frac{(x^2+1)^{11}}{11}\right]_{0}^{1} = \frac{2048}{11} - \frac{1}{11} = \frac{2047}{11} \approx 186.0909$

18. $\int (y-1)^{1/2}dy = \int u^{1/2}du$ Let $u = y-1$
       Then $du = dy$

$= \frac{2u^{3/2}}{3} + C = \frac{2(y-1)^{3/2}}{3} + C$

Then $\int_{2}^{10} \sqrt{y-1}\,dy = \left[\frac{2(y-1)^{3/2}}{3}\right]_{2}^{10} = \frac{54}{3} - \frac{2}{3} = \frac{52}{3} \approx 17.3333$

21. $\int \sqrt{7+2t^2}\,(8t)dt = \int 2(7+2t^2)^{1/2}\,4t\,dt$ Let $u = 7+2t^2$
       Then $du = 4t\,dt$

$= \int 2u^{1/2}du = \frac{4u^{3/2}}{3} + C = \frac{4(7+2t^2)^{3/2}}{3} + C$

Then $\int_{-3}^{3} \sqrt{7+2t^2}(8t)dt = \left[\frac{4(7+2t^2)^{3/2}}{3}\right]_{-3}^{3} = \frac{500}{3} - \frac{500}{3} = 0$

24. $\int \sin^2 3x \cos 3x \, dx = \int \frac{1}{3}(\sin^2 3x)(3\cos 3x)dx$ $\boxed{\text{Let } u = \sin 3x \\ \text{Then } du = 3\cos 3x \, dx}$

$= \int \frac{1}{3}u^2 du = \frac{u^3}{9} + C = \frac{\sin^3 3x}{9} + C.$

Then $\int_0^{\pi/2} \sin^2 3x \cos 3x \, dx = \left[\frac{\sin^3 3x}{9}\right]_0^{\pi/2} = \frac{-1}{9} - 0 = \frac{-1}{9} \approx -0.1111.$

27. $\int (2x+1)^{1/2} dx = \int \frac{1}{2}(2x+1)^{1/2} \, 2dx$ $\boxed{\text{Let } u = 2x+1 \\ \text{Then } du = 2dx}$

$= \int \frac{1}{2}u^{1/2} du = \frac{1}{2} \frac{2u^{3/2}}{3} + C = \frac{(2x+1)^{3/2}}{3} + C.$ (We will use this for the second term.)

Then $\int_0^4 [x^{1/2} + (2x+1)^{1/2}]dx = \int_0^4 x^{1/2} dx + \int_0^4 (2x+1)^{1/2} dx$

$= \left[\frac{2x^{3/2}}{3}\right]_0^4 + \left[\frac{(2x+1)^{3/2}}{3}\right]_0^4 = (\frac{16}{3} - 0) + (9 - \frac{1}{3}) = 14.$

30. $\int_a^{8a} (a^{1/3} - x^{1/3})^3 dx = \int_a^{8a} (a - 3a^{2/3}x^{1/3} + 3a^{1/3}x^{2/3} - x)dx$

$= \left[ax - \frac{3a^{2/3} \, 3x^{4/3}}{4} + \frac{3a^{1/3} \, 3x^{5/3}}{5} - \frac{x^2}{2}\right]_a^{8a}$

$= \left(8a^2 - 36a^2 + \frac{288a^2}{5} - 32a^2\right) - \left(a^2 - \frac{9a^2}{4} + \frac{9a^2}{5} - \frac{a^2}{2}\right) = \frac{-49a^2}{20}.$

33. $\int_0^\pi \sin x \, dx = \left[-\cos x\right]_0^\pi = 1 - (-1) = 2.$

36. $\int_0^1 3f(x)dx = 3\int_0^1 f(x)dx = 3(4) = 12.$

39. $\int_0^1 [2g(x) - 3f(x)]dx = 2\int_0^1 g(x)dx - 3\int_0^1 f(x)dx = 2(-2) - 3(4) = -16.$

Problem Set 5.6

Problem Set 5.7 More Properties of the Definite Integral

3. $\int_1^2 g(x)dx$ [Think of going from x=1 to x=2 by going first from x=1 to x=0, and then going from x=0 to x=2.]

   $= \int_1^0 g(x)dx + \int_0^2 g(x)dx = -\int_0^1 g(x)dx + (4) = -(-1) + (4) = 5.$

6. $\int_1^1 g(x)dx = 0.$

9. $\int_1^2 f(x)dx + \int_2^0 f(x)dx = \int_1^0 f(x)dx = -\int_0^1 f(x)dx = -2.$

12. $G'(x) = x^2+x.$ [Theorem D with $f(t) = t^2+t$]

15. $G(x) = -\int_{\pi/4}^x u \tan u \, du,$ so $G'(x) = -x \tan x.$

18. $G'(x) = \sqrt{2 + \sin(x^2+1)} \, D_x(x^2+1) = \sqrt{2 + \sin(x^2+1)}(2x) = 2x\sqrt{2+\sin(x^2+1)}.$

21. $f'(x) = \dfrac{x}{\sqrt{a^2+x^2}}.$ [Theorem D]

   $f''(x) = \dfrac{\sqrt{a^2+x^2}\,(1) - (x)\dfrac{1}{2\sqrt{a^2+x^2}}(2x)}{a^2+x^2} = \dfrac{(a^2+x^2) - x^2}{(a^2+x^2)^{3/2}} = \dfrac{a^2}{(a^2+x^2)^{3/2}} > 0.$

   Therefore, f is concave up on R.

24. $\int_0^4 f(x)dx = \int_0^1 1\,dx + \int_1^2 x\,dx + \int_2^4 (4-x)dx = \left[x\right]_0^1 + \left[\dfrac{x^2}{2}\right]_1^2 + \left[4x - \dfrac{x^2}{2}\right]_2^4$

   $= (1 - 0) + (2 - \tfrac{1}{2}) + ([16-8] - [8-2]) = \dfrac{9}{2} = 4.5.$

27. $\int_0^1 1\,dx = \left[x\right]_0^1 = 1-0 = 1$ and $\int_0^1 (1+x^4)\,dx = \left[x + \frac{x^5}{5}\right]_0^1 = (1+\frac{1}{5}) - 0 = \frac{6}{5}$

Therefore, $1 \leq \int_0^1 \sqrt{(1+x^4)}\,dx \leq \frac{6}{5}$.

30. $\int \frac{x}{\sqrt{x^2+16}}\,dx = \int (x^2+16)^{-1/2} x\,dx$

   | Let $u = x^2+16$ |
   | Then $du = 2x\,dx$ |

   $= \int \frac{1}{2}(x^2+16)^{-1/2} 2x\,dx = \int \frac{1}{2} u^{-1/2}\,du = u^{1/2} + C = \sqrt{x^2+16} + C$

Then, $\int_0^3 \frac{x}{\sqrt{x^2+16}}\,dx = \left[\sqrt{x^2+16}\right]_0^3 = 5 - 4 = 1$

33. True. By the Comparison Property, since $\int_a^b (0)\,dx = 0$.

    That is, $f(x) \geq 0 \Rightarrow \int_a^b f(x)\,dx \geq \int_a^b (0)\,dx = 0$

36. False. c^{ex}: Let $f(x) = \begin{cases} 1, & \text{if } x = 0 \\ 0, & \text{if } 0 < x \leq 1 \end{cases}$. Then $\int_0^1 f(x)\,dx = 0$.

39. Since f is continuous on [a,c], it is continuous on [a,b] and on [b,c], so f is integrable on all three intervals.

To prove that $\int_a^c = \int_a^b + \int_b^c$, we will show that $\int_a^c - \int_a^b - \int_b^c = 0$.

$|\int_a^c - \int_a^b - \int_b^c| = |\int_a^c - \int_a^b - \int_b^c - \Sigma_1^n + \Sigma_1^m + \Sigma_{m+1}^n|$

$= |(\int_a^c - \Sigma_1^n) - (\int_a^b - \Sigma_1^m) - (\int_b^c - \Sigma_{m+1}^n)|$

$\leq |(\int_a^c - \Sigma_1^n)| + |(\int_a^b - \Sigma_1^m)| + |(\int_b^c - \Sigma_{m+1}^n)| \leq \frac{\varepsilon}{3} + \frac{\varepsilon}{3} + \frac{\varepsilon}{3} = \varepsilon$

if $|P|$ is sufficiently small. Thus, the constant $|\int_a^c - \int_a^b - \int_b^c|$ is less than every positive number ε, so it must be 0.

Problem Set 5.7

Problem Set 5.8 Aids in Evaluating Definite Integrals

3. $\int \cos(3x+2)\, dx = \int \frac{1}{3} \cos(3x+2)\, 3dx$

$\qquad$ Let $u = 3x+2$
$\qquad$ Then $du = 3dx$

$= \int \frac{1}{3} \cos u\, du = \int \frac{1}{3} \sin u + C = \frac{1}{3} \sin(3x+2) + C$

6. $\int x^2 (x^3+5)^9 dx = \int \frac{1}{3}(x^3+5)^9\, 3x^2 dx$

$\qquad$ Let $u = x^3+5$
$\qquad$ Then $du = 3x^2 dx$

$= \int \frac{1}{3} u^9 du = \frac{1}{3} \frac{u^{10}}{10} + C = \frac{1}{30}(x^3+5)^{10} + C$

9. $\int \frac{x\, \sin\sqrt{x^2+4}}{\sqrt{x^2+4}}\, dx$

$\qquad$ Let $u = \sqrt{x^2+4}$
$\qquad$ Then $du = \frac{1}{2\sqrt{x^2+4}} 2xdx$

$= \int \sin\sqrt{x^2+4}\, \frac{x}{\sqrt{x^2+4}}\, dx$

$\qquad = \frac{x}{\sqrt{x^2+4}}\, dx$

$= \int \sin u\, du = -\cos u + C = -\cos\sqrt{x^2+4} + C$

12. $\int x^2 \sin(x^3+5) \cos^9(x^3+5)\, dx$

$\qquad$ Let $u = \cos(x^3+5)$
$\qquad$ Then $du = -\sin(x^3+5)\, 3x^2 dx$
$\qquad\qquad = -3x^2 \sin(x^3+5)dx$

$= \int \frac{-1}{3} \cos^9(x^3+5)\, [-3x^2 \sin(x^3+5)]\, dx$

$= \int \frac{-1}{3} u^9 du = \frac{-1}{3} \frac{u^{10}}{10} + C = \frac{-1}{30} \cos^{10}(x^3+5) + C$

15. $\int_0^1 (3x+1)^3 dx$

$\qquad$ Let $u = 3x+1$
$\qquad$ Then $du = 3dx$
$\qquad$ $x=1 \Rightarrow u=4$
$\qquad$ $x=0 \Rightarrow u=1$

$= \int_0^1 \frac{1}{3}(3x+1)^3\, 3dx = \int_1^4 \frac{1}{3} u^3 du = \left[\frac{1}{3} \frac{u^4}{4}\right]_1^4 = \frac{64}{3} - \frac{1}{12} = 21.25$

18. $\int_0^{\sqrt{5}} (9-x^2)^{1/2} x\, dx$

$\qquad$ Let $u = 9-x^2$
$\qquad$ Then $du = -2xdx$
$\qquad$ $x=\sqrt{5} \Rightarrow u=4$
$\qquad$ $x=0 \Rightarrow u=9$

$= \int_0^{\sqrt{5}} \frac{-1}{2} (9-x^2)^{1/2}(-2x)dx$

18. (continued)

$$= \int_9^4 \frac{-1}{2} u^{1/2} \, du = \left[\frac{-1}{2} \frac{2u^{3/2}}{3}\right]_9^4 = \frac{-8}{3} - (-9) = \frac{19}{3} \approx 6.3333$$

21. $\int_0^{\pi/6} \sin^3\theta \cos\theta \, d\theta$

| Let $u = \sin\theta$ |
| Then $du = \cos\theta \, d\theta$ |
| $\theta = \pi/6 \Rightarrow u = \sin(\pi/6) = 1/2$ |
| $\theta = 0 \Rightarrow u = \sin(0) = 0$ |

$$= \int_0^{1/2} u^3 \, du = \left[\frac{u^4}{4}\right]_0^{1/2} = \frac{1}{64} = 0.015625$$

24. $\int_0^{1/2} \sin(2\pi x) \, dx$

| Let $u = 2\pi x$ |
| Then $du = 2\pi \, dx$ |
| $x = 1/2 \Rightarrow u = \pi$ |
| $x = 0 \Rightarrow u = 0$ |

$$= \int_0^{1/2} \frac{1}{2\pi} \sin(2\pi x) \, 2\pi \, dx$$

$$= \int_0^{\pi} \frac{1}{2\pi} \sin u \, du = \left[\frac{-1}{2\pi} \cos u\right]_0^{\pi} = \frac{-1}{2\pi}(-1) - \frac{-1}{2\pi}(1) = \frac{1}{\pi} \approx 0.3183$$

27. $\int_0^{\pi/2} \sin x \sin(\cos x) \, dx$

| Let $u = \cos x$ |
| Then $du = -\sin x \, dx$ |
| $x = \pi/2 \Rightarrow u = \cos(\pi/2) = 0$ |
| $x = 0 \Rightarrow u = \cos(0) = 1$ |

$$= \int_0^{\pi/2} -\sin(\cos x)(-\sin x) \, dx$$

$$= \int_1^0 -\sin u \, du = \left[\cos u\right]_1^0 = \cos(0) - \cos(1) = 1 - \cos(1) \approx 0.4597$$

30. $\int_1^2 \left(1 + \frac{1}{t}\right)^2 \left(\frac{1}{t^2}\right) dt$

| Let $u = 1 + \frac{1}{t}$ |
| Then $du = \frac{-1}{t^2} dt$ |
| $t = 2 \Rightarrow u = 3/2$ |
| $t = 1 \Rightarrow u = 2$ |

$$= \int_1^2 -\left(1 + \frac{1}{t}\right)^2 \left(\frac{-1}{t^2}\right) dt$$

$$= \int_2^{3/2} -u^2 \, du = \left[\frac{-u^3}{3}\right]_2^{3/2} = \frac{-9}{8} - \frac{-8}{3} = \frac{37}{24} \approx 1.5417$$

Problem Set 5.8

33. $\int_{-\pi/2}^{\pi/2} \frac{\sin x}{1 + \cos x} dx = 0$ since $f(x) = \frac{\sin x}{1 + \cos x}$ is an odd function.

36. $\int_{-1}^{1} (1+x+x^2+x^3) dx = \int_{-1}^{1} (1+x^2) dx + \int_{-1}^{1} (x+x^3) dx = 2\int_{0}^{1} (1+x^2) dx + 0$

$= 2\left[x + \frac{x^3}{3}\right]_0^1 = 2[(1 + \frac{1}{3}) - 0] = \frac{8}{3} \approx 2.6667$

39. f is even: $\int_{-b}^{-a} f(x) dx = \int_{-b}^{-a} f(-x) dx$

| Let $u = -x$ |
| Then $du = -dx$ |
| $x=-a \Rightarrow u=a$ |
| $x=-b \Rightarrow u=b$ |

$= \int_{b}^{a} -f(u) du$

$= \int_{a}^{b} f(u) du = \int_{a}^{b} f(x) dx$ [changing the dummy variable]

f is odd: $\int_{-b}^{-a} f(x) dx = \int_{-b}^{-a} -f(-x) dx$

$= \int_{b}^{a} f(u) du$

Same substitution as above for f even

$= -\int_{a}^{b} f(u) du = -\int_{a}^{b} f(x) dx$ [changing the dummy variable]

42. $\int_{0}^{4\pi} |\sin 2x| dx = \int_{0}^{4\pi} \frac{1}{2} |\sin 2x| 2dx$

| Let $u = 2x$ |
| Then $du = 2dx$ |
| $x=4\pi \Rightarrow u=8\pi$ |
| $x=0 \Rightarrow u=0$ |

$= \int_{0}^{8\pi} \frac{1}{2} |\sin u| du$

[Note: The period of $|\sin u|$ is π, and $|\sin u| = \sin u$ on $[0,\pi]$.]

$= 8\int_{0}^{\pi} \frac{1}{2} \sin u \, du = 4\left[-\cos u\right]_0^{\pi} = 4\left(-\cos(\pi) - [-\cos(0)]\right) = 4(2) = 8$

Problem Set 5.9 Chapter Review Problems

True-False Quiz

3. True. $\frac{dy}{dx} = -\sin x$, and $(-\sin x)^2 = 1 - (\cos x)^2$ or $\sin^2 x = 1 - \cos^2 x$

6. True. $(a_1+a_0) + (a_2+a_1) + (a_3+a_2) + \cdots + (a_{n-1}+a_{n-2}) + (a_n+a_{n-1})$
$= a_0 + a_n + 2(a_1 + a_2 + a_3 + \cdots + a_{n-2} + a_{n-1})$

9. False. See Problem 22 of Section 5.5.

12. False. See the write-up of Problem 36 of Section 5.7.

15. True. By the Interval Additive Property (Theorem A of Section 5.7) since $f(x) = \sin^2 x$ is integrable on $[1,7]$.

18. True. Each integral equals 4. Can show the equality without actually evaluating the integrals by using the substitution $u = x + \pi/2$.

21. True. By the Fundamental Theorem of Calculus. F and G are continuous on $[a,b]$, and both F and G are antiderivatives of $F' = G'$.

24. True. Since $ax^3 + cx$ is an odd function, and bx^2 is an even function.

27. True. This is a generalization of the Triangle Inequality. See property 3 of absolute values on Page 14. Could prove it using the Principle of Mathematical Induction.

30. False. c^{ex}: Define partitions P of the interval $[0,2]$ into n subintervals by subdividing $[0,1]$ into $(n-1)$ subintervals of equal length, and letting $[1,2]$ be the nth subinterval. Then $|P| = 1$, but the number of subintervals tends to ∞ as n tends to ∞.

More generally, $|P| \to 0$ implies that the number of subintervals tends to ∞, but the number of subintervals can tend to ∞ without $|P| \to 0$ being true.

Miscellaneous Problems

3. $\int y\sqrt{y^2-4}\, dy = \int y(y^2-4)^{1/2} dy = \int \frac{1}{2}(y^2-4)^{1/2}\, 2y\, dy$

| Let $u = y^2-4$ |
| Then $du = 2y\, dy$ |

$= \int \frac{1}{2} u^{1/2} du = \frac{1}{2} \frac{2u^{3/2}}{3} + C = \frac{(y^2-4)^{3/2}}{3} + C$

6. $\int \frac{y^2-1}{(y^3-3y)^2} dy = \int (y^3-3y)^{-2} (y^2-1)\, dy$

| Let $u = y^3-3y$ |
| Then $du = (3y^2-3)\, dy$ |
| $= 3(y^2-1)\, dy$ |

$= \int \frac{1}{3}(y^3-3y)^{-2}\, 3(y^2-1)\, dy = \int \frac{1}{3} u^{-2}\, du = \frac{1}{3}\frac{-1}{u} + C = \frac{-1}{3(y^3-3y)} + C$

9. $y^{-4} dy = t^2 dt$; $\int y^{-4} dy = \int t^2 dt$; $\frac{y^{-3}}{-3} = \frac{t^3}{3} + C$ (general solution)

$y=1, t=1 \Rightarrow C = -2/3$

$\frac{y^{-3}}{-3} = \frac{t^3}{3} - \frac{2}{3}$ or $y = (2-t^3)^{-1/3}$ (particular solution)

12. $a(t) = 15\sqrt{t} + 8 = 15t^{1/2} + 8$

$v(t) = 10t^{3/2} + 8t + v_0 = 10t^{3/2} + 8t - 6$

$s(t) = 4t^{5/2} + 4t^2 - 6t + s_0 = 4t^{5/2} + 4t^2 - 6t - 44$

$s(4) = 4(32) + 4(16) - 6(4) - 44 = 124$, so the particle is 124 feet to the right of the origin.

15. $\sum_{i=1}^{4} f(x_i)\Delta x_i = [f(0.5) + f(1) + f(1.5) + f(2)](0.5)$

$= [(-0.75) + (0) + (1.25) + (3)](0.5)$

$= 1.75$

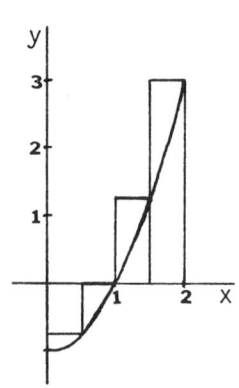

Problem Set 5.9

18. $\int_{2}^{5} 3x^2\sqrt{x^3-4}\,dx = \int_{2}^{5} 3x^2(x^3-4)^{1/2}dx$

> Let $u = x^3-4$
> Then $du = 3x^2 dx$
> $x=5 \Rightarrow u=121$
> $x=2 \Rightarrow u=4$

$= \int_{4}^{121} u^{1/2}du = \left[\frac{2u^{3/2}}{3}\right]_{4}^{121} = \frac{2(11)^3}{3} - \frac{16}{3} = 882.$

Then the average value of $f(x)$ on $[2,5]$ is $\frac{882}{5-2} = 294$.

21. $\sum_{i=1}^{10}(6i^2-8i) = 6\sum_{i=1}^{10}i^2 - 8\sum_{i=1}^{10}i = 6\frac{(10)(10+1)(20+1)}{6} - 8\frac{(10)(10+1)}{2} = 1870.$

24.

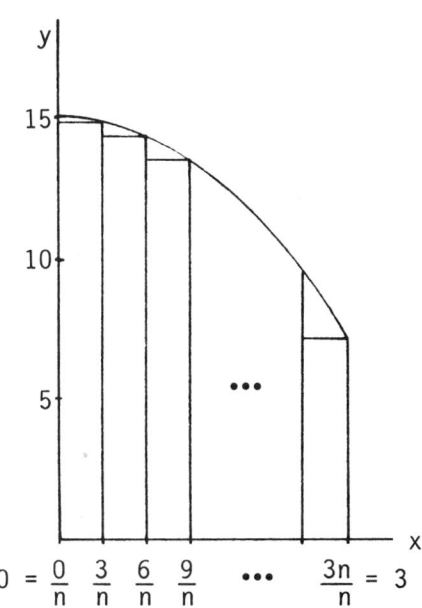

Area in polygon is $A(R_n) = \sum f(\frac{3i}{n})\frac{3}{n} = \sum\left(16 - (\frac{3i}{n})^2\right)\frac{3}{n} = \sum\left(16 - \frac{9i^2}{n^2}\right)\frac{3}{n}$

$= \sum\left(\frac{48}{n} - \frac{27i^2}{n^3}\right) = \sum\frac{48}{n} - \frac{27}{n^3}\sum i^2 = \frac{48}{n}n - \frac{27}{n^3}\frac{n(n+1)(2n+1)}{6}.$

$\lim_{n\to\infty} A(R_n) = \lim_{n\to\infty}\left(48 - \frac{27}{6}(1)(1+\frac{1}{n})(2+\frac{1}{n})\right) = 48 - \frac{9}{2}(1)(1)(2) = 39.$

Problem Set 5.9

27. (a) $\int_{-2}^{2} f(x)dx = 2\int_{0}^{2} f(x)dx$ [since f is an even function]

$= 2(-4) = -8$

(b) $\int_{-2}^{2} |f(x)|dx = \int_{-2}^{2} -f(x)dx$ [since $f(x) \leq 0$]

$= -(-8) = 8$ [from (a)]

(c) $\int_{-2}^{2} g(x)dx = 0$ [since g is an odd function]

(d) $\int_{-2}^{2} [f(x) + f(-x)]dx = \int_{-2}^{2} [f(x) + f(x)]dx = \int_{-2}^{2} 2f(x)dx$

$= 2\int_{-2}^{2} f(x)dx = 2(-8)$ [from (a)]

$= -16$

(e) $\int_{0}^{2} [2g(x) + 3f(x)]dx = 2\int_{0}^{2} g(x)dx + 3\int_{0}^{2} f(x)dx = 2(5) + 3(-4) = -2$

(f) $\int_{-2}^{0} g(x)dx = \int_{-2}^{0} -g(x)(-1)dx$

$= \int_{2}^{0} -g(-u)du = \int_{0}^{2} g(-u)du = \int_{0}^{2} -g(u)du = -5$

> Let $u = -x$
> Then $du = -dx$
> $x=0 \Rightarrow u=0$
> $x=-2 \Rightarrow u=2$

30. (a) $G'(x) = \dfrac{1}{x^2+1}$

(b) $G'(x) = \dfrac{1}{(x^2)^2+1} D_x(x^2) = \dfrac{2x}{x^4+1}$

(c) $G'(x) = \dfrac{1}{(x^3)^2+1} D_x(x^3) - \dfrac{1}{x^2+1} D_x(x) = \dfrac{3x^2}{x^6+1} - \dfrac{1}{x^2+1}$

CHAPTER 6 APPLICATIONS OF THE INTEGRAL

Problem Set 6.1 The Area of a Plane Region

3. Slice:

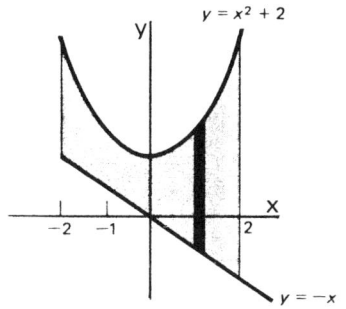

Approximate: $\Delta A \approx [(x^2+2) - (-x)]\Delta x$

Integrate:

$$A = \int_{-2}^{2} [(x^2+2) - (-x)]dx$$

$$= \int_{-2}^{2} (x^2+x+2)dx$$

6. Slice:

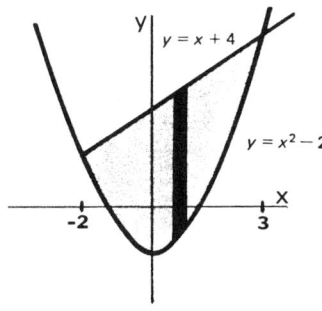

Approximate: $\Delta A \approx [(x+4) - (x^2-2)]\Delta x$

Integrate:

The curves intersect where $x = -2, 3$.

$$A = \int_{-2}^{3} [(x+4) - (x^2-2)]dx$$

$$= \int_{-2}^{3} (-x^2+x+6)dx$$

9. Slice: (on next page)

  Approximate: $\Delta A \approx [(\text{x-value at right}) - (\text{x-value at left})](\text{width})$

  $\qquad\qquad\quad = [(3-y^2) - (y+1)]\Delta y$

Problem Set 6-1

9. (continued)

Integrate:

$$A = \int_{-2}^{1} [(3-y^2) - (y+1)] dy$$

$$= \int_{-2}^{1} (-y^2 - y + 2) dy$$

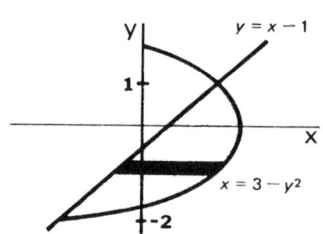

12. $\Delta A \approx (4x - x^2) \Delta x$

$$A = \int_{1}^{3} (4x - x^2) dx = \left[2x^2 - \frac{x^3}{3} \right]_{1}^{3}$$

$$= (18-9) - (2 - \tfrac{1}{3}) = \frac{22}{3}$$

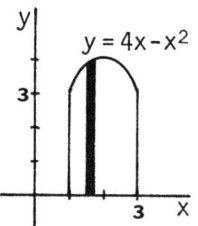

15. $\Delta A \approx (0 - x^3) \Delta x$, if x is on $[-1, 0]$
$\Delta A \approx x^3 \Delta x$, if x is on $[0, 2]$

$$A = \int_{-1}^{0} -x^3 dx + \int_{0}^{2} x^3 dx$$

$$= \left[\frac{-x^4}{4} \right]_{-1}^{0} + \left[\frac{x^4}{4} \right]_{0}^{2} = (0 - \tfrac{-1}{4}) + (4 - 0) = 4.25$$

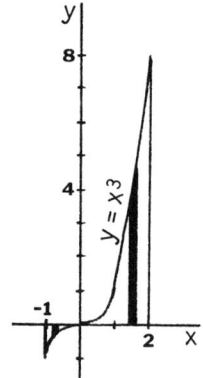

18. Solve $y = x^2 - 4x + 3$ and $y = x - 1$ simultaneously to find that the curves intersect where $x=1$ and $x=4$.

$\Delta A \approx [(x-1) - (x^2 - 4x + 3)] \Delta x$

$$A = \int_{1}^{4} [(x-1) - (x^2 - 4x + 3)] dx$$

$$= \int_{1}^{4} (-x^2 + 5x - 4) dx = \left[\frac{-x^3}{3} + \frac{5x^2}{2} - 4x \right]_{1}^{4} = (\tfrac{-64}{3} + 40 - 16) - (\tfrac{-1}{3} + \tfrac{5}{2} - 4) = \frac{9}{2}$$

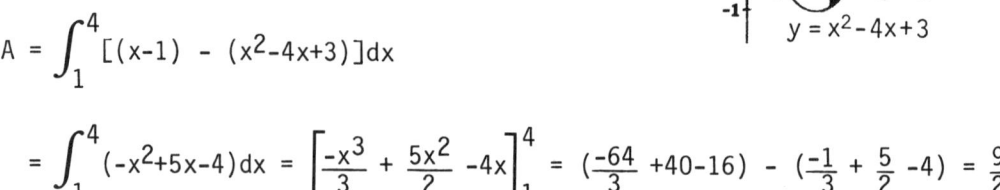

21. Solve $y = x^2 - 4x$ and $y = -x^2$ simultaneously to find that the curves intersect where $x=0$ and $x=2$.

$\Delta A \approx [(-x^2) - (x^2-4x)]\Delta x$

$A = \int_0^2 [(-x^2) - (x^2-4x)]dx$

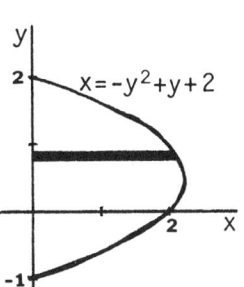

$= \int_0^2 (-2x^2+4x)dx = \left[\frac{-2x^3}{3} + 2x^2\right]_0^2 = (\frac{-16}{3} + 8) - 0 = \frac{8}{3}$

24. $\Delta A \approx (-y^2+y+2)\Delta x$

$A = \int_{-1}^2 (-y^2+y+2)dy = \left[\frac{-y^3}{3} + \frac{y^2}{2} + 2y\right]_{-1}^2$

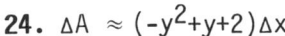

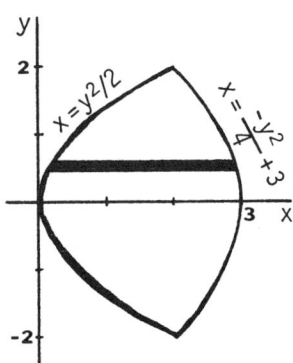

$= (\frac{-8}{3} + 2 + 4) - (\frac{1}{3} + \frac{1}{2} - 2) = \frac{9}{2} = 4.5$

27. Solve $y^2 - 2x = 0$ and $y^2 + 4x - 12 = 0$ simultaneously to find that the curves intersect where $y=-2$ and $y=2$.

$\Delta A \approx \left[\left(\frac{-y^2}{4} + 3\right) - \left(\frac{y^2}{2}\right)\right]\Delta y$

$A = \int_{-2}^2 \left(\frac{-3y^2}{4} + 3\right)dy$

$= 2\int_0^2 \left(\frac{-3y^2}{4} + 3\right)dy$ since the integrand is an even function

$= 2\left[\frac{-y^3}{4} + 3y\right]_0^2 = 2[(-2+6) - 0] = 8$

Problem Set 6.1

30.

Points on line	Equation of line
$(-1,4)$ and $(5,1)$	$y = (-1/2)x + (7/2)$
$(-1,4)$ and $(2,-2)$	$y = -2x + 2$
$(2,-2)$ and $(5,1)$	$y = x - 4$

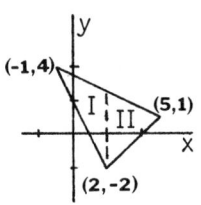

Area of triangle = A(I) + A(II)

$$= \int_{-1}^{2}[(-\tfrac{1}{2}x + \tfrac{7}{2}) - (-2x+2)]dx + \int_{2}^{5}[(-\tfrac{1}{2}x + \tfrac{7}{2}) - (x-4)]dx$$

$$= \int_{-1}^{2}(\tfrac{3x}{2} + \tfrac{3}{2})dx + \int_{2}^{5}(\tfrac{-3x}{2} + \tfrac{15}{2})dx = \left[\tfrac{3x^2}{4} + \tfrac{3x}{2}\right]_{-1}^{2} + \left[\tfrac{-3x^2}{4} + \tfrac{15x}{2}\right]_{2}^{5}$$

$$= [(3+3) - (\tfrac{3}{4} - \tfrac{3}{2})] + [(\tfrac{-75}{4} + \tfrac{75}{2}) - (-3+15)] = \tfrac{27}{2} = 13.5$$

Problem Set 6.2 Volumes of Solids: Slabs, Disks, Washers

3. (a) $\Delta V \approx \pi(4-x^2)^2 \Delta x$

$$V = \int_{0}^{2}\pi(4-x^2)^2 dx = \pi\int_{0}^{2}(16-8x^2+x^4)dx$$

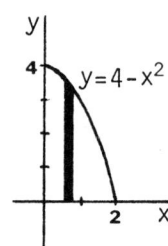

$$= \pi\left[16x - \tfrac{8x^3}{3} + \tfrac{x^5}{5}\right]_{0}^{2} = \tfrac{256\pi}{15} \approx 53.6165$$

(b) $\Delta V \approx \pi(\sqrt{4-y})^2 \Delta y = \pi(4-y)\Delta y$

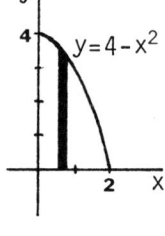

$$V = \int_{0}^{4}\pi(4-y)dy = \pi\left[4y - \tfrac{y^2}{2}\right]_{0}^{4}$$

$$= 8\pi \approx 25.1327$$

6. $\Delta V \approx \pi(x^3)^2 \Delta x = \pi x^6 \Delta x$

$$V = \pi\int_{0}^{2}x^6 dx = \pi\left[\tfrac{x^7}{7}\right]_{0}^{2}$$

$$= \tfrac{128\pi}{7} \approx 57.4463$$

9. $\Delta V \approx \pi(\sqrt{4-x^2})^2 \Delta x = \pi(4-x^2)\Delta x.$

$V = \pi\int_{-1}^{2}(4-x^2)dx = \pi\left[4x - \frac{x^3}{3}\right]_{-1}^{2}$

$= 9\pi \approx 28.2743.$

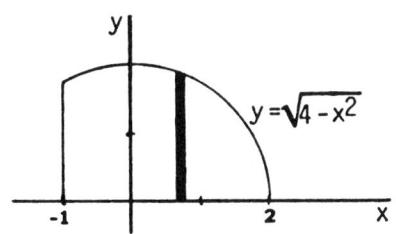

12. $\Delta V \approx \pi\left(\frac{2}{y}\right)^2 \Delta y = 4\pi y^{-2}\Delta y.$

$V = 4\pi\int_{1}^{6} y^{-2}dy = 4\pi\left[\frac{y^{-1}}{-1}\right]_{1}^{6} = 4\pi\left[\frac{-1}{y}\right]_{1}^{6}$

$= 4\pi(5/6) = 10\pi/3 \approx 10.4720.$

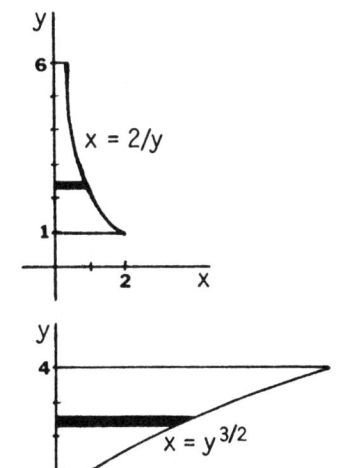

15. $\Delta V \approx \pi(y^{3/2})^2 \Delta y = \pi y^3 \Delta y.$

$V = \pi\int_{0}^{4} y^3 dy = \pi\left[\frac{y^4}{4}\right]_{0}^{4}$

$= 64\pi \approx 201.0619.$

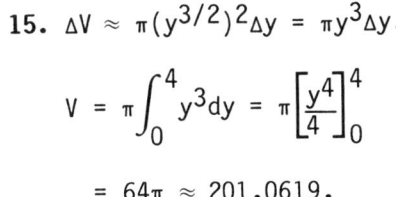

18. $\Delta V \approx \pi[(4x)^2 - (4x^2)^2]\Delta x = 16\pi(x^2-x^4)\Delta x.$

$V = 16\pi\int_{0}^{1}(x^2-x^4)dx = 16\pi\left[\frac{x^3}{3} - \frac{x^5}{5}\right]_{0}^{1}$

$= 16\pi(2/15) = 32\pi/15 \approx 6.7021.$

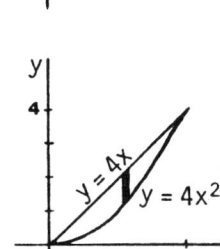

21. $\Delta V \approx \pi\left[\left(\frac{\sqrt{y}}{2}\right)^2 - \left(\frac{y}{4}\right)^2\right]\Delta y = \pi\left(\frac{y}{4} - \frac{y^2}{16}\right)\Delta y = \frac{\pi}{16}(4y-y^2)\Delta y.$

$V = \frac{\pi}{16}\int_{0}^{4}(4y-y^2)dy = \frac{\pi}{16}\left[2y^2 - \frac{y^3}{3}\right]_{0}^{4}$

$= \frac{\pi}{16}(32/3) = 2\pi/3 \approx 2.0944.$

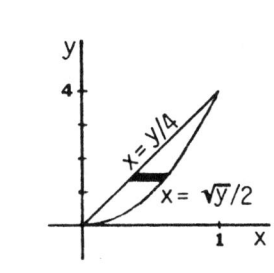
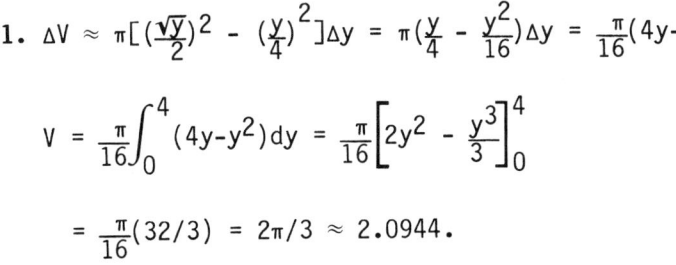

Problem Set 6.2

24. $\Delta V \approx$ (Area of triangle)(thickness of triangular piece)

$= \left(\frac{1}{2}(2\sqrt{4-x^2})(4)\right)\Delta x = 4\sqrt{4-x^2}\,\Delta x$

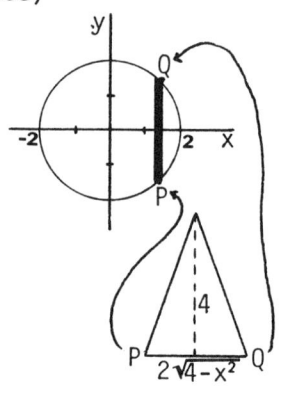

$V = 4\int_{-2}^{2}\sqrt{4-x^2}\,dx = 4\left(\frac{1}{2}\pi(2)^2\right)$

[since the integral is the area between the graph of $f(x) = \sqrt{4-x^2}$ and the x-axis (a semicircular region with radius 2).]

$= 8\pi \approx 25.1327$

27. See figure to the right for location of axes and cylinders.

$\Delta V \approx$ (area of square)(thickness of square piece)

$= (\sqrt{1-y^2})^2 \Delta y = (1-y^2)\Delta y$

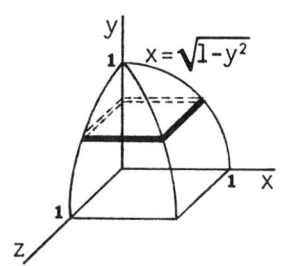

$V = \int_0^1 (1-y^2)\,dy = \left[y - \frac{y^3}{3}\right]_0^1 = 2/3$

30. (a) $\Delta V \approx \pi[(\text{outside radius})^2 - (\text{inside radius})^2]\Delta y$

$= \pi[(4)^2 - (4-y^{2/3})^2]\Delta y$

$= \pi[8y^{2/3} - y^{4/3}]\Delta y$

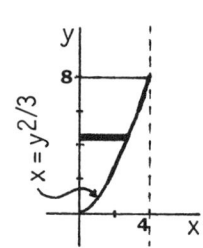

$V = \pi\int_0^8 (8y^{2/3} - y^{4/3})\,dy = \pi\left[\frac{24y^{5/3}}{5} - \frac{3y^{7/3}}{7}\right]_0^8$

$= \pi\left(\frac{768}{5} - \frac{384}{7}\right) = \frac{3456\pi}{35} \approx 310.2098$

(b) $\Delta V \approx \pi[8-x^{3/2}]^2 \Delta x$

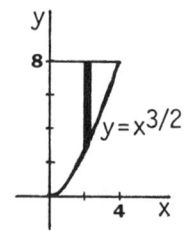

$V = \pi\int_0^4 (8-x^{3/2})^2\,dx = \pi\int_0^4 (64-16x^{3/2}+x^3)\,dx$

$= \pi\left[64x - \frac{32x^{5/2}}{5} + \frac{x^4}{4}\right]_0^4 = \frac{576\pi}{5} \approx 361.9115$

Problem Set 6.3 Volumes of Solids of Revolution: Shells

3. $\Delta V \approx 2\pi x \sqrt{x} \Delta x = 2\pi x^{3/2} \Delta x$

$V = 2\pi \int_0^4 x^{3/2} dx = 2\pi \left[\dfrac{2x^{5/2}}{5} \right]_0^4$

$= 2\pi (\dfrac{64}{5} - 0) = \dfrac{128\pi}{5} \approx 80.4248$

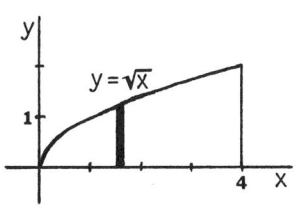

6. $\Delta V \approx 2\pi(2-x)(4-x^2)\Delta x = 2\pi(8-4x-2x^2+x^3)\Delta x$

$V = 2\pi \int_0^2 (8-4x-2x^2+x^3) dx$

$= 2\pi \left[8x - 2x^2 - \dfrac{2x^3}{3} + \dfrac{x^4}{4} \right]_0^2 = \dfrac{40\pi}{3} \approx 41.8879$

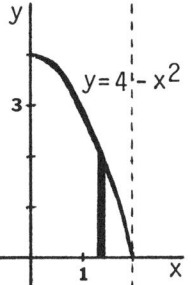

9. $\Delta V \approx 2\pi(y)(y^2)\Delta y = 2\pi y^3 \Delta y$

$V = 2\pi \int_0^2 y^3 dy = 2\pi \left[\dfrac{y^4}{4} \right]_0^2$

$= 8\pi \approx 25.1327$

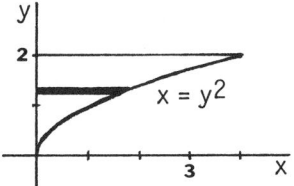

12. $\Delta V \approx 2\pi(3-y)(\sqrt{2y}+1)\Delta y = 2\pi(3\sqrt{2}y^{1/2}+3-\sqrt{2}y^{3/2}-y)\Delta y$

$V = 2\pi \int_0^2 (3+3\sqrt{2}y^{1/2}-y-\sqrt{2}y^{3/2}) dy$

$= 2\pi \left[3y + 2\sqrt{2}y^{3/2} - \dfrac{y^2}{2} - \dfrac{2\sqrt{2}y^{5/2}}{5} \right]_0^2$

$= 2\pi[(6 + 8 - 2 - \dfrac{16}{5}) - 0] = \dfrac{88\pi}{5} \approx 55.2920$

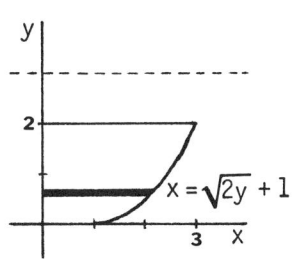

15. (a) $A = \int_1^3 x^{-3} dx$

(b) $V = \int_1^3 2\pi x (x^{-3}) dx$ (Shells)

(c) $V = \int_1^3 \pi\left((x^{-3}-[-1])^2 - (0-[-1])^2\right) dx$ (Washers)

(d) $V = \int_1^3 2\pi(4-x)(x^{-3}) dx$ (Shells)

18. $\Delta V \approx 2\pi(4-y)\left[\sqrt{y} - \dfrac{y^3}{32}\right]\Delta y$

$= 2\pi\left[4y^{1/2} - \dfrac{y^3}{8} - y^{3/2} + \dfrac{y^4}{32}\right]\Delta y$

$V = 2\pi\int_0^4 \left[4y^{1/2} - y^{3/2} - \dfrac{y^3}{8} + \dfrac{y^4}{32}\right] dy$

$= 2\pi\left[\dfrac{8y^{3/2}}{3} - \dfrac{2y^{5/2}}{5} - \dfrac{y^4}{32} + \dfrac{y^5}{160}\right]_0^4$

$= 2\pi[(\dfrac{64}{3} - \dfrac{64}{5} - 8 + \dfrac{32}{5}) - 0] = \dfrac{208\pi}{15} \approx 43.5634$

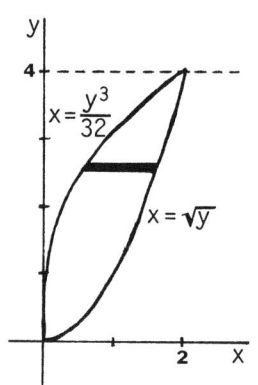

Problem Set 6.4 Length of a Plane Curve

3. $\dfrac{dy}{dx} = 3x^{1/2}$

$L = \int_{1/3}^7 \sqrt{1 + (3x^{1/2})^2}\, dx = \int_{1/3}^7 (1+9x)^{1/2} dx$

$= \int_{1/3}^7 \dfrac{1}{9}(1+9x)^{1/2}\, 9dx = \int_4^{64} \dfrac{1}{9} u^{1/2} du$

Let $u = 1+9x$
Then $du = 9dx$
$x=7 \Rightarrow u=64$
$x=1/3 \Rightarrow u=4$

3. (continued)

$$= \left[\frac{2u^{3/2}}{27}\right]_4^{64} = \frac{1008}{27} \approx 37.3333$$

6. $\frac{dy}{dx} = \frac{(6x)(4x^3) - (x^4+3)(6)}{36x^2} = \frac{18x^4-18}{36x^2} = \frac{x^4-1}{2x^2}$

$$L = \int_1^4 \sqrt{1 + \left[\frac{x^4-1}{2x^2}\right]^2} \, dx = \int_1^4 \sqrt{\frac{x^8+2x^4+1}{4x^4}} \, dx = \int_1^4 \sqrt{\frac{(x^4+1)^2}{4x^4}} \, dx = \int_1^4 \frac{x^4+1}{2x^2} \, dx$$

$$= \int_1^4 (\tfrac{1}{2}x^2 + \tfrac{1}{2}x^{-2}) \, dx = \left[\frac{x^3}{6} - \frac{1}{2x}\right]_1^4 = [(\tfrac{32}{3} - \tfrac{1}{8}) - (\tfrac{1}{6} - \tfrac{1}{2})] = \frac{87}{8} \approx 10.875$$

9. $x = t^3$, $y = t^2$, t on $[0,4]$

t	x	y
0	0	0
1	1	1
2	8	4
3	27	9
4	64	16

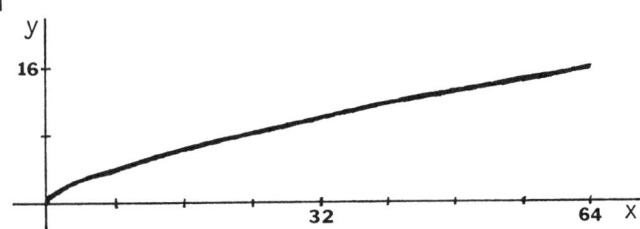

$\frac{dx}{dt} = 3t^2$, $\frac{dy}{dt} = 2t$

$$L = \int_0^4 \sqrt{(3t^2)^2 + (2t)^2} \, dt = \int_0^4 \sqrt{9t^4+4t^2} \, dt$$

$$= \int_0^4 \sqrt{9t^2+4} \, t \, dt = \int_0^4 \frac{1}{18}(9t^2+4)^{1/2} \, 18t \, dt$$

Let $u = 9t^2+4$
Then $du = 18t \, dt$
$t=4 \Rightarrow u=148$
$t=0 \Rightarrow u=4$

$$= \int_4^{148} \frac{1}{18} u^{1/2} \, du = \left[\frac{u^{3/2}}{27}\right]_4^{148} = \frac{296\sqrt{37} - 8}{27} \approx 66.3888$$

12. $\begin{bmatrix} x = 4\cos t + 5 \\ y = 4\sin t - 1 \end{bmatrix} \Rightarrow \begin{bmatrix} x - 5 = 4\cos t \\ y + 1 = 4\sin t \end{bmatrix} \Rightarrow (x-5)^2 + (y+1)^2 = 16$

[since $\cos^2 t + \sin^2 t = 1$]

The graph is a circle with center at $(5,-1)$ and radius 4.

Problem Set 6.4

12. (continued)

For $0 \leq t \leq 2\pi$, the length is the circumference of the circle, which is $2\pi(4) = 8\pi$ using the formula $c = 2\pi r$. We are now in position to varify that that formula (which has been accepted on faith until now) is correct.

$$L = \int_0^{2\pi} \sqrt{16\sin^2 t + 16\cos^2 t}\, dt = \int_0^{2\pi} 4\, dt = \left[4t\right]_0^{2\pi} = 8\pi - 0 = 8\pi \approx 25.1327$$

15. $\left(\dfrac{dx}{d\theta}\right)^2 + \left(\dfrac{dy}{d\theta}\right)^2 = [a(1-\cos\theta)]^2 + [a\sin\theta]^2 = a^2(1 - 2\cos\theta + \cos^2\theta) + a^2\sin^2\theta$

$= a^2(1 - 2\cos\theta + \cos^2\theta + \sin^2\theta) = a^2(1 - 2\cos\theta + 1)$

$= 2a^2(1 - \cos\theta) = 2a^2(2\sin^2\dfrac{\theta}{2})$ [Half-angle identity]

$= 4a^2\sin^2\dfrac{\theta}{2}$

Then $L = \int_0^{2\pi} \sqrt{4a^2\sin^2\dfrac{\theta}{2}}\, d\theta = \int_0^{2\pi} 2a\sin\dfrac{\theta}{2}\, d\theta = \left[-4a\cos\dfrac{\theta}{2}\right]_0^{2\pi} = 8a$

Problem Set 6.5 Area of a Surface of Revolution

3. $\dfrac{dy}{dx} = \dfrac{-x}{\sqrt{25-x^2}}$

$A = \int_{-2}^{3} 2\pi\sqrt{25-x^2}\sqrt{1 + \dfrac{x^2}{25-x^2}}\, dx$

$= 2\pi \int_{-2}^{3} \sqrt{(25-x^2) + x^2}\, dx = 2\pi \int_{-2}^{3} 5\, dx = 2\pi\left[5x\right]_{-2}^{3} = 50\pi \approx 157.0796$

6. $x = \sqrt{y}$, so $\frac{dx}{dy} = \frac{1}{2\sqrt{y}}$, y in $[0,12]$

$$A = \int_0^{12} 2\pi\sqrt{y}\sqrt{1 + \frac{1}{4y}}\, dy = 2\pi\int_0^{12}\sqrt{y + \frac{1}{4}}\, dy = \pi\int_0^{12}(4y+1)^{1/2}\, dy$$

[Use substitution of $u = 4y+1$ if needed to see the result.]

$$= \pi\left[\frac{(4y+1)^{3/2}}{6}\right]_0^{12} = \pi\left(\frac{343}{6} - \frac{1}{6}\right) = 57\pi \approx 179.0708$$

9. $\frac{dx}{dt} = 1$, $\frac{dy}{dt} = 3t^2$

Then $A = \int_0^1 2\pi(t^3)\sqrt{(1)^2 + (3t^2)^2}\, dt = \int_0^1 2\pi t^3(1+9t^4)^{1/2}\, dt$

[Use substitution of $u = 1+9t^4$ if needed to see the result.]

$$= \left[\frac{\pi(1+9t^4)^{3/2}}{27}\right]_0^1 = \frac{\pi(10\sqrt{10} - 1)}{27} \approx 3.5631$$

12. $\frac{dy}{dx} = 2x+2$

$$A = \int_1^3 2\pi([x^2+2x] - [-2])\sqrt{1 + (2x+2)^2}\, dx = 2\pi\int_1^3 (x^2+2x+2)\sqrt{4x^2+8x+5}\, dx$$

15. $\frac{dy}{dx} = \frac{2x+1}{2\sqrt{x^2+x}}$; $A = \int_2^4 2\pi(x-2)\sqrt{1 + \frac{(2x+1)^2}{4(x^2+x)}}\, dx$

18. $\frac{dx}{dt} = -a\sin t$, $\frac{dy}{dt} = a\cos t$ [assume $a > 0$]

$$A = \int_0^{\pi/2} 2\pi(a\cos t - [-a])\sqrt{a^2\sin^2 t + a^2\cos^2 t}\, dt$$

$$= 2\pi a\int_0^{\pi/2}(1 + \cos t)\sqrt{a^2}\, dt = 2\pi a^2\int_0^{\pi/2}(1 + \cos t)\, dt$$

Problem Set 6.5

Problem Set 6.6 Work

3. F = 200 dynes when x = 2 cm, so 200 = k(2) for some k (Hooke's Law). Thus, k = 100, so for this spring F(x) = 100x.

Therefore, the work done is $W = \int_0^4 100x \, dx = \left[50x^2\right]_0^4 = 800$ [ergs]

6. F = 1 pound when s = 8 inches, so $1 = k(8)^{4/3}$. Then, k = 1/16, and the relationship is $F(s) = \frac{1}{16} s^{4/3}$.

$W = \int_0^{27} \frac{1}{16} s^{4/3} ds = \left[\frac{1}{16} \frac{3s^{7/3}}{7}\right]_0^{27} = \frac{3(2187)}{112} = \frac{6561}{112} \approx 58.5804$ [in-lbs]

9. Approx. volume of a horizontal slab is (length)(width)(thickness), or

$\Delta V \approx (10)(\frac{-4}{5} y) \Delta y = -8y\Delta y$,

so approx. weight is $62.4(-8y\Delta y)$, or $-499.2y\Delta y$. The approximate distance to lift it is (5-y).

$\Delta W \approx (-499.2 y \Delta y)(5-y) = -499.2(5y - y^2)\Delta y$

$W = -499.2 \int_{-5}^0 (5y - y^2) dy = -499.2 \left[\frac{5y^2}{2} - \frac{y^3}{3}\right]_{-5}^0 = 52,000$ [ft-lbs]

12. $\Delta W \approx (62.4)[(10)(2\sqrt{9-y^2})\Delta y](5-y)$

$= 1248\sqrt{9-y^2}(5-y)\Delta y$

$W = 1248 \int_{-3}^0 \sqrt{9-y^2}(5-y) dy$

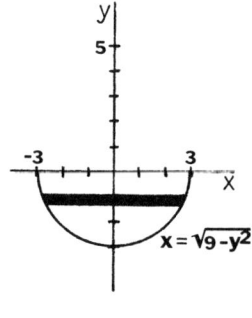

$= 1248 \int_{-3}^0 5\sqrt{9-y^2} dy - 1248 \int_{-3}^0 (9-y^2)^{1/2} y \, dy$

$= 6240(\frac{1}{4}\pi[3]^2) - 1248 \left[\frac{-(9-y^2)^{3/2}}{3}\right]_{-3}^0$

[The first integral is 5 times area of quarter circle; use $u = 9-y^2$ substitution for the second integral, if necessary to see result.]

$= 14040\pi - 1248(-9) \approx 55339.96$ [ft-lbs]

15. $F(x) = Af(x)$, so $\Delta W \approx Af(x)\Delta x$, and then $W = \int_{x_2}^{x_1} Af(x)dx$.

18. 80 lbs/in^2 is 11520 lbs/ft^2. $V = 1$ cubic foot and $p = 11520$ lbs/ft^2, so $11520(1)^{1.4} = C$ implies $C = 11520$.
Therefore, $p = 11520v^{-1.4}$, and so $\Delta W \approx p\Delta V = 11520v^{-1.4}\Delta V$

$$W = 11520\int_1^4 v^{-1.4}dV = 11520\left[\frac{v^{-0.4}}{-0.4}\right]_1^4 \approx 12258.74 \text{ [ft-lbs]}$$

21. Lifting force is $f(x) = kx^{-2}$. $5000 = k(4000)^{-2}$ implies $k = 8(10)^{10}$.
Therefore, $f(x) = 8(10)^{10}x^{-2}$.

$$W = 8(10)^{10}\int_{4000}^{4200} x^{-2}dx = 8(10)^{10}\left[\frac{-1}{x}\right]_{4000}^{4200} \approx 952,400 \text{ [mile-lbs]}$$
[or $5,029,000,000$ ft-lb]

Problem Set 6.7 Fluid Force

3. $\Delta F \approx$ (density of H_2O)(area of horizontal strip)(depth of strip)

$= (62.4)(\frac{2\sqrt{3}}{3} y \Delta y)(3\sqrt{3} - y)$

$= 41.6\sqrt{3}(3\sqrt{3}y - y^2)\Delta y$

$F = 41.6\sqrt{3}\int_0^{3\sqrt{3}} (3\sqrt{3}y - y^2)dy$

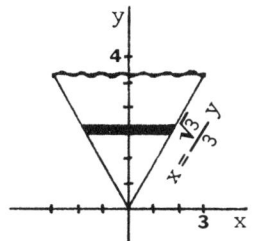

$= 41.6\sqrt{3}\left[\frac{3\sqrt{3}y^2}{2} - \frac{y^3}{3}\right]_0^{3\sqrt{3}} = 1684.8$ [lbs]

6. $\Delta F \approx (62.4)(2y^2\Delta y)(3-y) = 124.8(3y^2-y^3)\Delta y$

$F = 124.8\int_0^3 (3y^2-y^3)dy = 124.8\left[y^3 - \frac{y^4}{4}\right]_0^3$

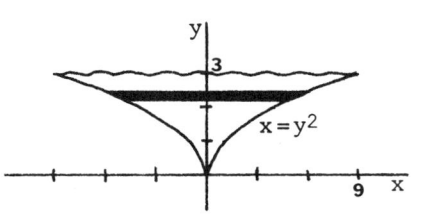

$= 842.4$ [lbs]

9.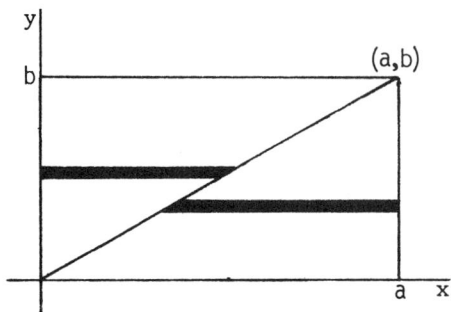

Equation of diagonal: $y = \frac{b}{a} x$ or $x = \frac{a}{b} y$

Upper Half: $\Delta F \approx \delta(\frac{a}{b} y \, \Delta y)(b-y) = \frac{\delta a}{b}(by-y^2)\Delta y$

$$F = \frac{\delta a}{b}\int_0^b (by-y^2)dy = \frac{\delta a}{b}\left[\frac{by^2}{2} - \frac{y^3}{3}\right]_0^b = \frac{\delta a}{b}\left(\frac{b^3}{2} - \frac{b^3}{3}\right) = \frac{\delta ab^2}{6}$$

Lower Half: $\Delta F \approx \delta(a - \frac{a}{b} y) \Delta y (b-y) = \frac{\delta a}{b}(b-y)^2 \Delta y$

$$F = \frac{\delta a}{b}\int_0^b (b^2-2by+y^2)dy = \frac{\delta a}{b}\left[b^2 y - by^2 + \frac{y^3}{3}\right]_0^b = \frac{\delta ab^2}{3}$$

Thus, (force on lower part) = 2(force on upper part)

12. $\Delta F \approx (50)(2\pi[5]\Delta y)(6-y) = 500\pi(6-y)\Delta y$

$$F = 500\pi \int_0^6 (6-y)dy = 500\pi\left[6y - \frac{y^2}{2}\right]_0^6$$

$= 500\pi(36 - 18) = 9000\pi \approx 28274.33$ [lbs]

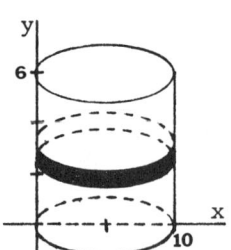

Problem Set 6.8 Moments, Center of Mass

3. $\bar{x} = \dfrac{M \text{ (moment)}}{m \text{ (mass)}}$

$\Delta M \approx [x][\delta(x)\Delta x] = x\sqrt{x}\Delta x = x^{3/2}\Delta x \qquad \Delta m = \delta(x)\Delta x = \sqrt{x}\Delta x$

$M = \displaystyle\int_0^9 x^{3/2}\,dx = \left[\dfrac{2x^{5/2}}{5}\right]_0^9 = 97.2 \qquad m = \displaystyle\int_0^9 x^{1/2}\,dx = \left[\dfrac{2x^{3/2}}{3}\right]_0^9 = 18$

Therefore, $\bar{x} = \dfrac{97.2}{18} = 5.4$ units from that end to the center of mass.

6. Moments: $M_y = (-3)(3) + (-2)(6) + (3)(2) + (4)(5) + (7)(1) = 12$
$\phantom{\text{Moments: }}M_x = (2)(3) + (-2)(6) + (5)(2) + (3)(5) + (-1)(1) = 18$

Total mass: $3 + 6 + 2 + 5 + 1 = 17$

Center of mass: $(\bar{x},\bar{y}) = \left(\dfrac{M_y}{m}, \dfrac{M_x}{m}\right) = \left(\dfrac{12}{17}, \dfrac{18}{17}\right)$

9. For a homogeneous rectangular lamina, the centroid is at the intersection of the diagonals of the rectangle.

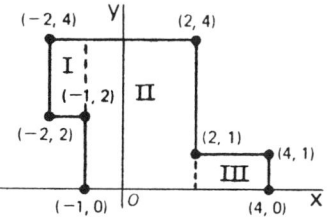

Rectangle I: The centroid is $(-3/2, 3)$ and the area is 2, so the moment about the y-axis is $(-3/2)(2) = -3$, and the moment about the x-axis is $(3)(2) = 6$.

Rectangle II: The centroid is $(1/2, 2)$ and the area is 12, so moment about the y-axis is $(1/2)(12) = 6$, and the moment about the x-axis is $(2)(12) = 24$.

Rectangle III: The centroid is $(3, 1/2)$ and the area is 2, so moment about the y-axis is $(3)(2) = 6$, and the moment about the x-axis is $(1/2)(2) = 1$

Entire Region: Moment about y-axis = $(-3) + (6) + (6) = 9$
$\phantom{\text{Entire Region: }}$Moment about x-axis = $(6) + (24) + (1) = 31$
$\phantom{\text{Entire Region: }}$Area = $(2) + (12) + (2) = 16$

$\bar{x} = \dfrac{9}{16},\ \bar{y} = \dfrac{31}{16}$ Centroid: $(9/16, 31/16)$

12. $\int_0^4 x(\frac{1}{2}x^2 - 0)dx = \int_0^4 \frac{1}{2}x^3 dx = \left[\frac{x^4}{8}\right]_0^4 = 32$

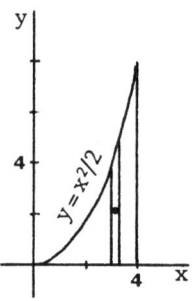

$\frac{1}{2}\int_0^4 [(\frac{1}{2}x^2)^2 - (0)^2]dx = \frac{1}{2}\left[\frac{x^5}{20}\right]_0^4 = \frac{128}{5}$

$\int_0^4 [\frac{1}{2}x^2 - 0]dx = \left[\frac{x^3}{6}\right]_0^4 = \frac{32}{3}$ (Area)

Then, $\bar{x} = \frac{32}{32/3} = 3$; $\bar{y} = \frac{128/5}{32/3} = 2.4$, so the centroid is (3,2.4).

15. $y = 2x-4$ and $y = 2\sqrt{x}$ intersect where x is 4.

$\int_0^4 x[2\sqrt{x} - (2x-4)]dx = \int_0^4 (2x^{3/2} - 2x^2 + 4x)dx$

$= \left[\frac{4x^{5/2}}{5} - \frac{2x^3}{3} + 2x^2\right]_0^4 = \frac{224}{15}$

$\frac{1}{2}\int_0^4 [(2\sqrt{x})^2 - (2x-4)^2]dx = \int_0^4 (-2x^2 + 10x - 8)dx$

$= \left[\frac{-2x^3}{3} + 5x^2 - 8x\right]_0^4 = \frac{16}{3}$

$\int_0^4 [2\sqrt{x} - (2x-4)]dx = \int_0^4 (2x^{1/2} - 2x + 4)dx$

$= \left[\frac{4x^{3/2}}{3} - x^2 + 4x\right]_0^4 = \frac{32}{3}$ (Area)

Then, $\bar{x} = \frac{224/15}{32/3} = 1.4$; $\bar{y} = \frac{16/3}{32/3} = 0.5$, so the centroid is (1.4,0.5).

18. The points of intersection are (-4,3) and (0,-1).

$\int_{-1}^3 y[(-y-1) - (y^2-3y-4)]dy = \int_{-1}^3 (-y^3 + 2y^2 + 3y)dy$

18. (continued)

$$= \left[\frac{-y^4}{4} + \frac{2y^3}{3} + \frac{3y^2}{2}\right]_{-1}^{3} = \frac{32}{3}$$

$$\frac{1}{2}\int_{-1}^{3}[(-y-1)^2 - (y^2-3y-4)^2]dy$$

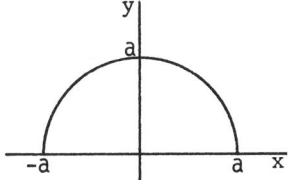

$$= \frac{1}{2}\int_{-1}^{3}(-y^4+6y^3-22y-15)dy$$

$$= \frac{1}{2}\left[\frac{-y^5}{5} + \frac{3y^4}{2} - 11y^2 - 15y\right]_{-1}^{3} = \frac{-192}{5}$$

$$\int_{-1}^{3}[(-y-1) - (y^2-3y-4)]dy = \int_{-1}^{3}(-y^2+2y+3)dy = \left[\frac{-y^3}{3} + y^2 + 3y\right]_{-1}^{3} = \frac{32}{3}$$

Then, $\bar{x} = \frac{-192/5}{32/3} = -3.6$; $\bar{y} = \frac{32/3}{32/3} = 1$, so the centroid is $(-3.6, 1)$.

21. Centroid is at $(0, \bar{y})$

In revolving the shown region about the the x-axis, the centroid travels $2\pi\bar{y}$.

The area of the region is $(1/2)\pi a^2$.

The volume of the sphere is $(4/3)\pi a^3$.

By Pappus: $(4/3)\pi a^3 = [(1/2)\pi a^2][2\pi\bar{y}]$, so $\bar{y} = \frac{4a}{3\pi} \approx 0.4244a$

The centroid is $(0, 4a/3\pi)$.

Problem Set 6.8

Problem Set 6.9 Chapter Review Problems

True-False Quiz

3. **False.** c^{ex}: For $f(x) = \sin x$, $g(x) = \cos x$, $a=0$, $b=\pi/2$, the integral equals zero, but the area of the region is not zero. The correct integral is $\int_a^b |f(x)-g(x)|\,dx$.

6. **False.** The volumes are $(1/3)\pi r^2 h$ and $(1/3)\pi(2r)^2(h/2) = 2(1/3)\pi r^2 h$, so the volume is doubled.

9. **False.** (Intuitive) Place the unit circle with center at the origin. Then define a curve that starts at (1,0), smoothly moves up and down touching the circle at the top and bottom in each cycle as x approaches zero. Have the curve complete one cycle as x goes from 2^{-n} to 2^{-n-1} (for $n = 0,1,2,3,\ldots$). Thus as x approaches zero the lengths of the cycles approach 4 (up 1, down 2, up 1), and the number of cycles approaches infinity.

   A particular counterexample (which doesn't satisfy all the conditions in the intuitive approach) is $f(x) = (1-x)\sin(1/x)$, for x in (0,1).

12. **False.** (Intuitive) There is a greater proportion of water toward the top of the cone, so for the cone a greater proportion of water has to be lifted less. (In fact, the ratio is 1/2.)

15. **True.** (Intuitive) Can be seen from symmetry.

18. **True.** (Intuitive) The density is the same going each way from the midpoint, so the wire is balanced at the midpoint.

Miscellaneous Problems

3. $V = \int_0^1 2\pi x(x-x^2)\,dx = 2\pi \int_0^1 (x^2-x^3)\,dx = 2\pi \left[\dfrac{x^3}{3} - \dfrac{x^4}{4}\right]_0^1 = \dfrac{\pi}{6} \approx 0.5236$

6. $\bar{x} = 0.5$ (by symmetry).

$$\frac{1}{2}\int_0^1 (x-x^2)^2 dx = \frac{1}{2}\int_0^1 (x^2 - 2x^3 + x^4) dx = \frac{1}{60} \quad \text{[Make use of Problem 2 work.]}$$

$$\int_0^1 (x-x^2) dx = 1/6 \quad \text{[from Problem 1]}. \quad \text{Therefore, } \bar{y} = \frac{1/60}{1/6} = 0.1$$

9. $W = \int_0^8 (62.4)\pi(5)^2(10-y)\,dy$

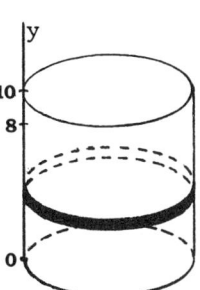

$= 1560\pi \left[10y - \frac{y^2}{2}\right]_0^8 = 74880\pi \approx 235242$ ft-lbs

12. $\int_0^3 x(3x-x^2)\,dx = \left[x^3 - \frac{x^4}{4}\right]_0^3 = 6.75$

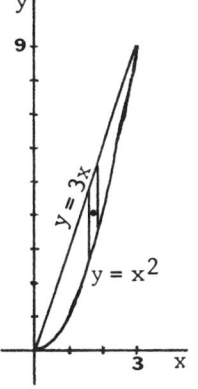

$\frac{1}{2}\int_0^3 [(3x)^2 - (x^2)^2]\,dx = \frac{1}{2}\left[3x^3 - \frac{x^5}{5}\right]_0^3 = 16.2$

From Problem 11, the area of the region is 4.5.

Therefore, $\bar{x} = \frac{6.75}{4.5} = 1.5$, $\bar{y} = \frac{16.2}{4.5} = 3.6$,

so the centroid is (1.5, 3.6)

15. $\Delta F \approx (62.4)(\frac{5}{4} y \Delta y)(8-y) = 78(8y - y^2)\Delta y$

$F = 78 \int_0^8 (8y - y^2)\,dy$

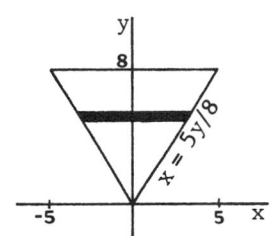

$= 78\left[4y^2 - \frac{y^3}{3}\right]_0^8 = 6656$ [pounds]

Problem Set 6.9

18. $\frac{dy}{dx} = \frac{x^2}{2} - \frac{x^{-2}}{2}$.

$A = \int_1^3 2\pi(\frac{x^3}{6} + \frac{x^{-1}}{2})\sqrt{1 + (\frac{x^2}{2} - \frac{x^{-2}}{2})^2}\, dx = \frac{\pi}{3}\int_1^3 (x^3+3x^{-1})\sqrt{(\frac{x^2}{2} + \frac{x^{-2}}{2})^2}\, dx$

$= \frac{\pi}{6}\int_1^3 (x^3+3x^{-1})(x^2+x^{-2})\, dx = \frac{\pi}{6}\int_1^3 (x^5+4x+3x^{-3})\, dx = \frac{208\pi}{9} \approx 72.6057$.

21. $\Delta V \approx (\sqrt{4-x^2})^2 \Delta x = (4-x^2)\Delta x$.

$V = \int_{-2}^{2} (4-x^2)\, dx = 2\int_0^2 (4-x^2)\, dx$

$= 2\left[4x - \frac{x^3}{3}\right]_0^2 = \frac{32}{3}$.

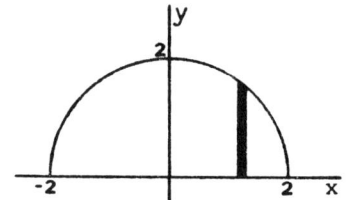

24. $\int_a^b 2\pi(x-a)[f(x)-g(x)]\, dx$. (Shells)

27. Surface area = (left surface area) + (right surface area) +

            + (outside surface area) + (inside surface area)

       = $\pi[f^2(a) - g^2(a)] + \pi[f^2(b) - g^2(b)] +$

          + $\int_a^b 2\pi f(x)\sqrt{1+[f'(x)]^2}\, dx + \int_a^b 2\pi g(x)\sqrt{1+[g'(x)]^2}\, dx$.

CHAPTER 7 TRANSCENDENTAL FUNCTIONS

Problem Set 7.1 The Natural Logarithm Function

3. $D_x \ln(x^2-5x+6) = \dfrac{1}{x^2-5x+6}(2x-5) = \dfrac{2x-5}{x^2-5x+6}$

6. $D_x \ln\sqrt{3x-25} = D_x \ln(3x-25)^{1/2} = D_x(1/2)\ln(3x-25) = \dfrac{1}{2}\dfrac{1}{3x-25}(3) = \dfrac{3}{2(3x-25)}$

9. $y = 3 \ln x + (\ln x)^3$

$\dfrac{dy}{dx} = 3\dfrac{1}{x} + 3(\ln x)^2 \dfrac{1}{x} = \dfrac{3(1 + \ln^2 x)}{x}$

12. $f'(x) = \dfrac{1}{x+\sqrt{x^2+1}}\left[1 + \dfrac{1}{2\sqrt{x^2+1}} 2x\right] = \dfrac{\sqrt{x^2+1} + x}{x\sqrt{x^2+1} + (x^2+1)} = \dfrac{1}{\sqrt{x^2+1}}$

[The last step was obtained by multiplying numerator and denominator by $(\sqrt{x^2+1} - x)$; i.e., by rationalizing the numerator.]

15. $\displaystyle\int \dfrac{4}{2x+1} dx = \int 2 \dfrac{1}{2x+1} 2dx$ $\quad\boxed{\text{Let } u = 2x+1 \\ \text{Then } du = 2dx}$

$= \displaystyle\int 2 \dfrac{1}{u} du = 2 \ln|u| + C = 2 \ln|2x+1| + C$

18. $\displaystyle\int \dfrac{x}{x^2+4} dx = \int \dfrac{1}{2}\dfrac{1}{x^2+4} 2xdx$ $\quad\boxed{\text{Let } u = x^2+4 \\ \text{Then } du = 2xdx}$

$= \displaystyle\int \dfrac{1}{2}\dfrac{1}{u} du = \dfrac{1}{2}\ln|u| + C = \dfrac{1}{2}\ln|x^2+4| + C = \dfrac{1}{2}\ln(x^2+4) + C$

Problem Set 7.1

21. $\int_0^3 \frac{x^3}{x^4+1} dx = \frac{1}{4}\int_0^3 \frac{1}{x^4+1} 4x^3 dx$

$\boxed{\text{Let } u = x^4+1 \\ \text{Then } du = 4x^3 dx \\ x=3 \Rightarrow u=82 \\ x=0 \Rightarrow u=1}$

$= \frac{1}{4}\int_1^{82} \frac{1}{u} du = \frac{1}{4}\left[\ln|u|\right]_1^{82} = \frac{\ln 82 - \ln 1}{4} = \frac{\ln 82}{4}$

≈ 1.1017

24. $(1/2)\ln(x-9) + (1/2)\ln x = \ln(x-9)^{1/2} + \ln x^{1/2} = \ln[(x-9)^{1/2} x^{1/2}]$
$= \ln\sqrt{x(x-9)} \quad [x > 9]$

27. $\ln y = \ln(x+11) - (1/2)\ln(x^3-4)$

Then, taking the derivative of each side with respect to x, obtain:

$\frac{1}{y}\frac{dy}{dx} = \frac{1}{x+11} - \frac{1}{2}\frac{1}{x^3-4} 3x^2 = \frac{1}{x+11} - \frac{3x^2}{2(x^3-4)} = \frac{-x^3-33x^2-8}{2(x+11)(x^3-4)}$

Therefore, $\frac{dy}{dx} = y \frac{(-x^3-33x^2-8)}{2(x+11)(x^3-4)} = \frac{(x+11)(-x^3-33x^3-8)}{2(x+11)(x^3-4)^{3/2}} = \frac{x^3+33x^2+8}{-2(x^3-4)^{3/2}}$

30. $\ln y = (2/3)\ln(x^2+3) + 2\ln(3x+2) - (1/2)\ln(x+1)$

Then $\frac{1}{y}\frac{dy}{dx} = \frac{2}{3(x^2+3)}(2x) + \frac{2}{3x+2}(3) - \frac{1}{2(x+1)}$

$= \frac{8x(3x+2)(x+1) + 36(x^2+3)(x+1) - 3(x^2+3)(3x+2)}{6(x^2+3)(3x+2)(x+1)}$

$= \frac{51x^3+70x^2+97x+90}{6(x^2+3)(3x+2)(x+1)}$, so $\frac{dy}{dx} = y \frac{51x^3+70x^2+97x+90}{6(x^2+3)(3x+2)(x+1)}$

$= \frac{(x^2+3)^{2/3}(3x+2)^2(51x^3+70x^2+97x+90)}{\sqrt{x+1}\, 6(x^2+3)(3x+2)(x+1)} = \frac{(51x^3+70x^2+97x+90)(3x+2)}{6(x^2+3)^{1/3}(x+1)^{3/2}}$

33. $y = \ln(1/x) = \ln 1 - \ln x = -\ln x$.

$y = -\ln x$ is a reflection of $y = \ln x$.

with respect to the x-axis.

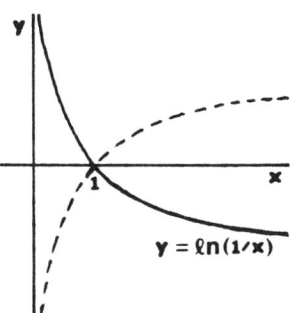

36. $\lim\limits_{x \to 0} \ln \dfrac{\sin x}{x} = \ln\left(\lim\limits_{x \to 0} \dfrac{\sin x}{x}\right)$ [since $\ln$ is continuous at $\lim\limits_{x \to 0} \dfrac{\sin x}{x} = 1$]

$= \ln 1 = 0$.

39. $\ln 4 > 1 \iff m \ln 4 > m \iff \ln 4^m > m$. [$m > 0$]

Thus, given any $m > 0$, $\ln x > m$ for all $x > 4^m$, since $\ln$ is an increasing function.

Therefore, $\lim\limits_{x \to \infty} \ln x = \infty$.

42. (a) $t > 1 \Rightarrow \dfrac{1}{t} < \dfrac{1}{\sqrt{t}} \Rightarrow \displaystyle\int_1^x \dfrac{1}{t}\,dt < \int_1^x t^{-1/2}\,dt \Rightarrow \ln x < \left[2\sqrt{t}\right]_1^x$

$\Rightarrow \ln x < 2(\sqrt{x}-1)$. [$x > 1$]

(b) For $x > 1$, $0 < \dfrac{\ln x}{x} < \dfrac{2(\sqrt{x}-1)}{x} = \dfrac{2}{\sqrt{x}} - \dfrac{2}{x}$.

$\lim\limits_{x \to \infty}(0) = 0$ and $\lim\limits_{x \to \infty}\left(\dfrac{2}{\sqrt{x}} - \dfrac{2}{x}\right) = 0$, so $\lim\limits_{x \to \infty} \dfrac{\ln x}{x} = 0$. [Squeeze Thm.]

Problem Set 7.2   Inverse Functions and Their Derivatives

3. $f(x) = 2$ for three values of x, so f does not have an inverse.

6. f has an inverse. $f^{-1}(2) \approx 1$.

9. [See Example 2, Page 109] $f'(x) = \sec^2 x$ which exists and is greater than zero on $(-\pi/2, \pi/2)$, so f is increasing on that interval.

12. $f'(x) = 2x+1 > 0$ for $x > -1/2$, so f is increasing on $[-1/2, \infty)$.

15. $f(x) = 3x-1$.

Step 1: Let $y = 3x-1$. Then $x = \frac{y+1}{3}$.

Step 2: $f^{-1}(y) = \frac{y+1}{3}$.

Step 3: $f^{-1}(x) = \frac{x+1}{3}$.

$f^{-1}(f(x)) = f^{-1}(3x-1) = \frac{(3x-1)+1}{3} = x$.

$f(f^{-1}(x)) = f(\frac{x+1}{3}) = 3(\frac{x+1}{3}) - 1 = x$.

18. $f(x) = -\sqrt{2-x}$.

Step 1: Let $y = -\sqrt{2-x}$. Then $y^2 = 2-x$, and $x = 2-y^2$.
Step 2: $f^{-1}(y) = 2-y^2$.
Step 3: $f^{-1}(x) = 2-x^2$.

$f^{-1}(f(x)) = f^{-1}(-\sqrt{2-x}) = 2 - [-\sqrt{2-x}]^2 = 2 - (2-x) = x$.

$f(f^{-1}(x)) = f(2-x^2) = -\sqrt{2 - [2-x^2]} = -\sqrt{x^2} = -|x| = -(-x) = x$ [since the domain of f^{-1}, which equals the range of f, contains only non-positive numbers].

21. $f(x) = x^2$, $x \leq 0$.

Step 1: Let $y = x^2$. Then $x = -\sqrt{y}$ [since $x \leq 0$].
Step 2: $f^{-1}(y) = -\sqrt{y}$.
Step 3: $f^{-1}(x) = -\sqrt{x}$.

$f^{-1}(f(x)) = f^{-1}(x^2) = -\sqrt{x^2} = -|x| = -(-x) = x$ [since the domain of f contains only nonpositive numbers].

$f(f^{-1}(x)) = f(-\sqrt{x}) = (-\sqrt{x})^2 = x$.

24. $f(x) = x^{3/2}$, $x \geq 0$.

  Step 1: Let $y = x^{3/2}$. Then $x = y^{2/3}$.
  Step 2: $f^{-1}(y) = y^{2/3}$.
  Step 3: $f^{-1}(x) = x^{2/3}$.

  $f^{-1}(f(x)) = f^{-1}(x^{3/2}) = (x^{3/2})^{2/3} = x$.

  $f(f^{-1}(x)) = f(x^{2/3}) = (x^{2/3})^{3/2} = x$ [since the domain of f^{-1}, which is the range of f, contains only nonnegative numbers].

27. $f(x) = \dfrac{x^3+1}{x^3+2}$.

  Step 1: Let $y = \dfrac{x^3+1}{x^3+2}$. Then $y(x^3+2) = x^3+1$.
  $$yx^3 + 2y = x^3 + 1.$$
  $$yx^3 - x^3 = -2y + 1.$$
  $$x^3(y-1) = -2y + 1.$$
  $$x^3 = \dfrac{-2y+1}{y-1}, \text{ so } x = \sqrt[3]{\dfrac{-2y+1}{y-1}}.$$

  Step 2: $f^{-1}(y) = \sqrt[3]{\dfrac{-2y+1}{y-1}}$

  Step 3: $f^{-1}(x) = \sqrt[3]{\dfrac{-2x+1}{x-1}}$

  $f^{-1}(f(x)) = f^{-1}\left(\dfrac{x^3+1}{x^3+2}\right) = \sqrt[3]{\dfrac{-2\dfrac{x^3+1}{x^3+2} + 1}{\dfrac{x^3+1}{x^3+2} - 1}} = \sqrt[3]{\dfrac{-2(x^3+1)+(x^3+2)}{(x^3+1)-(x^3+2)}} = \sqrt[3]{x^3} = x$.

  Similar work with a complex fraction yields $f(f^{-1}(x)) = \dfrac{-x}{-1} = x$.

30. $f(x) = x^2 - 3x + 1$.

  $f'(x) = 2x - 3$. Auxiliary axis for f':

  ```
           Dec         Inc
           (-)   (0)   (+)
  ─────────────┼───────────── x
                3/2
  ```

  Restrict the domain to $[3/2, \infty)$. (Could use $(-\infty, 3/2]$.)

  Step 1: Let $y = x^2 - 3x + 1$. Then $x^2 - 3x + (1-y) = 0$ is a quadratic equation in x. To solve for x use the quadratic formula.

Problem Set 7.2

30. (continued)

$$x = \frac{-(-3) + \sqrt{(-3)^2 - 4(1)(1-y)}}{2}$$ [Use only the "+" since x is greater than or equal to 3/2]

$$= \frac{3 + \sqrt{5+4y}}{2}$$

Step 2: $f^{-1}(y) = \frac{3 + \sqrt{5+4y}}{2}$ Step 3: $f^{-1}(x) = \frac{3 + \sqrt{5+4x}}{2}$

33. $(f^{-1})'(3) \approx -1/4$

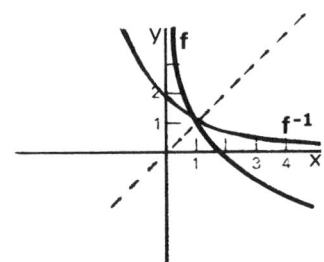

36. Let $y = f(x)$. Then $y = 2$ when $x = 1$.
$f'(x) = 5x^4 + 5$, so $f'(1) = 10$, and then $(f^{-1})'(2) = \frac{1}{f'(1)} = \frac{1}{10}$

39. Let $y = g(x)$ and $z = f(y)$. Then $z = f \circ g(x)$.

For each z, there is a unique y since f^{-1} exists.

For each y, there is a unique x since g^{-1} exists.

Therefore, for each z, there is a unique x, so $(f \circ g)^{-1} = h^{-1}$ exists.

We need to show that $g^{-1} \circ f^{-1}$ is $(f \circ g)^{-1}$.

$[(g^{-1} \circ f^{-1}) \circ (f \circ g)](x) = (g^{-1} \circ f^{-1} \circ f \circ g)(x) = (g^{-1} \circ g)(x) = x$

$[(f \circ g) \circ (g^{-1} \circ f^{-1})](x) = (f \circ g \circ g^{-1} \circ f^{-1})(x) = (f \circ f^{-1})(x) = x$

Therefore, $h^{-1} = g^{-1} \circ f^{-1}$.

Problem Set 7.3 The Natural Exponential Function

3. $e^{2 \ln x} = e^{\ln x^2} = x^2$ $[x > 0]$

6. $\ln e^{-x+2} = -x+2$

9. $e^{\ln 2 + \ln x} = e^{\ln(2x)} = 2x$

12. $D_x y = e^{3x^2-x}(6x-1) = (6x-1)e^{3x^2-x}$

15. $y = e^{\ln x} = x$, so $D_x y = 1$

18. $e^{x^2 \ln x}[(x^2)(1/x) + (\ln x)(2x)] = e^{x^2 \ln x}(x + 2x\ln x)$

21. $\dfrac{d}{dx}(e^{xy} + y) = \dfrac{d}{dx}(2)$

    $e^{xy}(x\dfrac{dy}{dx} + y) + \dfrac{dy}{dx} = 0$; $\dfrac{dy}{dx}(xe^{xy} + 1) = -ye^{xy}$; $\dfrac{dy}{dx} = \dfrac{-ye^{xy}}{xe^{xy} + 1}$

    Therefore, $\dfrac{dy}{dx} = \dfrac{-y(2-y)}{x(2-y) + 1} = \dfrac{-y(2-y)}{2x-xy+1}$ [since $e^{xy} = 2-y$]

24. $f(x) = e^{-x}$ is a decreasing function, so $a < b \Rightarrow e^{-a} > e^{-b}$.

27. $f(x) = e^{-x^2}$

    Domain: R

    Symmetry: With respect to the y-axis [f is an even function.]

    Intercepts: No x-intercepts; y-intercept is $f(0) = 1$.

27. (continued)

Monotonicity: $f'(x) = e^{-x^2}(-2x) = -2xe^{-x^2}$.

```
   Inc      Dec
   (+)  (0) (-)
   ─────┼─────── x
        0
```

Concavity: $f''(x) = e^{-x^2}(-2) + (-2x)e^{-x^2}(-2x) = e^{-x^2}(4x^2-2)$

$= e^{-x^2}(2x+\sqrt{2})(2x-\sqrt{2})$.

```
   CU        CD        CU
   (+)  (0)  (-)  (0)  (+)
   ─────┼─────────┼───────── x
      -0.71      0.71
```

Asymptotes: $y=0$ (horizontal) since $\lim_{x \to \pm\infty} e^{-x^2} = \lim_{x \to \pm\infty} \frac{1}{e^{x^2}} = 0$.

Summary	
$(-\infty,-0.71)$	Inc and CU
At -0.71	Inflection
$(-0.71,0)$	Inc and CD
At 0	Levels off
$(0,0.71)$	Dec and CD
At 0.71	Inflection
$(0.71,\infty)$	Dec and CU

x	f(x)	
0	1	(Maximum)
0.71	0.61	(Inflection Point)
1	0.37	
2	0.02	

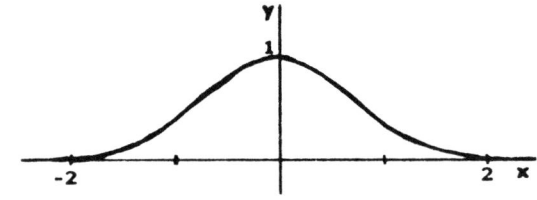

30. $\int xe^{x^2-3}dx = \int \frac{1}{2} e^{x^2-3} 2x\,dx$ $\boxed{\text{Let } u = x^2-3 \\ \text{Then } du = 2x\,dx}$

$= \int \frac{1}{2} e^u du = \frac{1}{2} e^u + C = \frac{1}{2} e^{x^2-3} + C.$

33. $\int \frac{e^{-1/x}}{x^2} dx = \int e^{-x^{-1}}(x^{-2})dx$ $\boxed{\text{Let } u = -x^{-1} \\ \text{Then } du = x^{-2}dx}$

$= \int e^u du = e^u + C = e^{-1/x} + C.$

36. $\int_1^2 \frac{e^{3/x}}{x^2}dx = \frac{-1}{3}\int_1^2 e^{3x^{-1}}(-3x^{-2})dx$

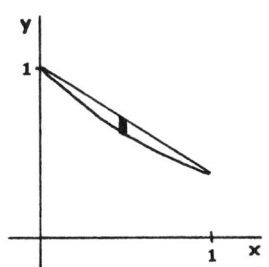
Let $u = 3x^{-1}$
Then $du = -3x^{-2}dx$
$x=2 \Rightarrow u=1.5$
$x=1 \Rightarrow u=3$

$= \frac{-1}{3}\int_3^{1.5} e^u du = \frac{-1}{3}\left[e^u\right]_3^{1.5} = \frac{-1}{3}(e^{1.5}-e^3) \approx 5.2013$

39. The line has slope $\frac{1/e - 1}{1-0} = \frac{1-e}{e}$, and y-intercept 1. An equation of the line is $y = (\frac{1-e}{e})x + 1$

$\Delta A \approx \left([(\frac{1-e}{e})x + 1] - e^{-x}\right)\Delta x$

$A = \int_0^1 [(\frac{1-e}{e})x + 1 - e^{-x}]dx = \left[(\frac{1-e}{e})\frac{x^2}{2} + x + e^{-x}\right]_0^1 = \frac{3-e}{2e} \approx 0.0518$

42. $e^{0.3} \approx 1 + (0.3) + \frac{(0.3)^2}{2!} + \frac{(0.3)^3}{3!} + \frac{(0.3)^4}{4!} = 1.3498375$

$e^{0.3} \approx 1.3498588$ (by calculator). They differ by 0.0000213.

Problem Set 7.4 General Exponential and Logarithmic Functions

3. $x = 9^{3/2} = 27$

6. $\frac{1}{x} = 4^3 = 64$, so $x = 1/64$

9. $\log_5 13 = \frac{\ln 13}{\ln 5} \approx 1.5937$

12. $\log_{10}(91.2)^3 = 3\log_{10}91.2 = 3\dfrac{\ln 91.2}{\ln 10} \approx 5.8800$

15. $\ln 4^{3x-1} = \ln 5;\ (3x-1)\ln 4 = \ln 5;\ 3x\ln 4 - \ln 4 = \ln 5;\ x = \dfrac{\ln 5 + \ln 4}{3\ln 4}$

$= \dfrac{\ln 20}{\ln 64} \approx 0.7203$

18. $D_x(3^{2x^4-4x}) = 3^{2x^4-4x}(\ln 3)(8x^3-4) = 3^{2x^4-4x}(8x^3-4)\ln 3$

21. $D_x[2^x \ln(x+5)] = (2^x)\left(\dfrac{1}{x+5}\right) + [\ln(x+5)](2^x \ln 2) = \dfrac{2^x[1 + (x+5)\ln 2\, \ln(x+5)]}{x+5}$

24. $\displaystyle\int 10^{5x-1}dx = \int \dfrac{1}{5} 10^{5x-1}\, 5\,dx$ | Let $u = 5x-1$
 Then $du = 5\,dx$ |

$= \displaystyle\int \dfrac{1}{5} 10^u du = \dfrac{1}{5}\dfrac{10^u}{\ln 10} + C = \dfrac{10^{5x-1}}{5\ln 10} + C$

27. $y = 10^{(x^2)} + x^{20}$

$\dfrac{dy}{dx} = 10^{(x^2)}(\ln 10)(2x) + 20x^{19} = 2x(\ln 10)10^{(x^2)} + 20x^{19}$

30. $y = 2^{(e^x)} + 2^{ex}$

$\dfrac{dy}{dx} = 2^{(e^x)}(\ln 2)(e^x) + 2^{ex}(\ln 2)(e) = (2^{e^x}e^x + 2^{ex}e)\ln 2$

33. $f(x) = x^{\sin x} = e^{(\sin x)\ln x}$

$f'(x) = e^{(\sin x)\ln x}[(\sin x)(1/x) + (\ln x)(\cos x)] = x^{\sin x}\left(\dfrac{\sin x}{x} + \ln x \cos x\right)$

Then $f'(1) = 1^{\sin 1}\left(\dfrac{\sin 1}{1} + \ln 1 \cos 1\right) = \sin 1 \approx 0.8415.$

36. $\log_{1/3} x = \dfrac{\ln x}{\ln(1/3)} = \dfrac{\ln x}{\ln 1 - \ln 3}$

$= \dfrac{\ln x}{-\ln 3} = -\dfrac{\ln x}{\ln 3} = -\log_3 x$

Hence, the graphs are symmetric with respect to the x-axis.

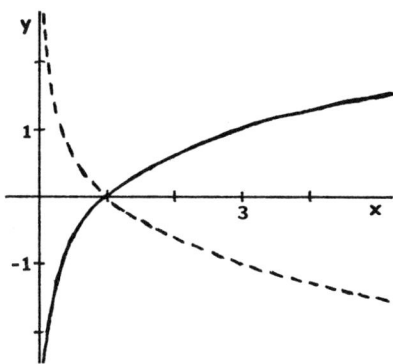

39. $\overline{C}$ is the 12th note after C, so the frequency of $\overline{C}$ is $262r^{12}$. We are also given that the frequency of $\overline{C}$ is $2(262) = 524$.

Therefore, $262r^{12} = 524$; $r^{12} = 2$; $12 \ln r = \ln 2$; $\ln r = (\ln 2)/12$;

$r = e^{(\ln 2)/12} \approx 1.0595$

A is the 9th note after C, so the frequency of A is $262r^9 \approx 440.6297$ [In 1938 a frequency of 440 for A was internationally adopted as the standard.]

42. WRONG 1: $y' = g(x) f(x)^{g(x)-1} f'(x)$ [Misapplying the Power Rule]

WRONG 2: $y' = f(x)^{g(x)} \ln[f(x)] g'(x)$ [Misapplying the Exponential Function Rule]

RIGHT: $y = f(x)^{g(x)} = e^{g(x)[\ln f(x)]}$

$y' = e^{g(x)[\ln f(x)]} \left([g(x)][\dfrac{1}{f(x)} f'(x)] + [\ln f(x)][g'(x)] \right)$

$= f(x)^{g(x)} \left(\dfrac{g(x) f'(x)}{f(x)} + [\ln f(x)] g'(x) \right)$

$= f(x)^{g(x)-1} g(x) f'(x) + f(x)^{g(x)} [\ln f(x)] g'(x)$

Thus, WRONG 1 + WRONG 2 = RIGHT

Problem Set 7.5 Exponential Growth and Decay

3. The solution is $y = y_0 e^{0.006t}$ [y_0 is the value of y when $t=0$ -- see the first derivation in this section.] Although we don't know the value of y when $t=0$, we do know that $y=2$ if $t=10$. Substituting those values we can solve for y_0.

$$2 = y_0 e^{0.006(10)}, \text{ so } y_0 = 2e^{-0.06} \approx 1.8835$$

Therefore, $y = 2e^{-0.06} e^{0.006t}$, or $y \approx 1.8835 e^{0.006t}$

6. $y = 10000(2.4)^{t/10}$ [from Problem 5]. Find t when $y = 20000$.

$$20000 = 10000(2.4)^{t/10}; \quad 2 = (2.4)^{t/10}; \quad \ln 2 = (t/10)\ln(2.4)$$

$$t = \frac{10 \ln 2}{\ln(2.4)} \approx 7.9174 \text{ [days]} \quad [7 \text{ days, } 22 \text{ hrs, } 1 \text{ min, } 6.88 \text{ secs}]$$

9. End of 1 year: $(1.032)(4.5) \approx 4.644$ [million]

End of 2 years: $(1.032)(\text{pop at end of 1 year}) = (1.032)^2 (4.5) \approx 4.793$ [million]

End of 10 years: $(1.032)^{10}(4.5) \approx 6.166$ [million]

End of 100 years: $(1.032)^{100}(4.5) \approx 105.0$ [million]

12. $y = y_0 e^{kt}$ where y_0 is the amount of substance when $t = 0$.
Let $y_0 = 1$ unit of the substance, so $y = e^{kt}$.

85% of the 1 unit is left when $t = 3$, so $0.85 = e^{k(3)}$.

Solving for k, $k = \frac{\ln(0.85)}{3}$, so $y = e^{(\ln 0.85)t/3} = 0.85^{t/3}$

At half-life, $y=0.5$: $0.5 = 0.85^{t/3}$, so $\ln 0.5 = (t/3)\ln 0.85$

$$t = \frac{3 \ln 0.5}{\ln 0.85} \approx 12.7951 \text{ [days]}$$

15. $\frac{dT}{dt} = k(T-75); \quad \frac{1}{T-75} dT = k\,dt; \quad \ln(T-75) = kt + C$

$T=300$ when $t=0$: $\ln(300-75) = k(0) + C \Rightarrow C = \ln 225$

$\ln(T-75) = kt + \ln 225$

$T-75 = e^{kt+\ln 225} = e^{kt} e^{\ln 225} = e^{kt} 225$; thus, $T = 225 e^{kt} + 75$

15. (continued)

T=200 when t=0.5: $200 = 225e^{k(0.5)} + 75 \Rightarrow k = 2\ln(\frac{125}{225}) = 2\ln(5/9)$.

Therefore, $T = 225e^{[2\ln(5/9)]t} + 75 = 225(25/81)^t + 75$.

After 3 hours: $T = 225(25/81)^3 + 75 \approx 81.62\ °F$.

18. $A(2) = 375(1 + \frac{0.144}{n})^{2n}$ for n compounding periods per year.

$A(2) = 375e^{(0.144)(2)}$ for compounding continuously.

(a) n=1: $A(2) = 375(1 + 0.144)^2 \approx 490.78$ dollars.

(b) n=12: $A(2) = 375(1 + \frac{0.144}{12})^{24} \approx 499.30$ dollars.

(c) n=365: $A(2) = 375(1 + \frac{0.144}{365})^{365} \approx 500.13$ dollars.

(d) continuous: $A(2) = 375e^{0.288} \approx 500.16$ dollars.

21. 1984 - 1626 = 358 years.

$A(358) = 24e^{0.06(358)} \approx 51{,}513{,}043{,}900$. [51.5 billion dollars]

24. $\frac{dy}{dt} = ky(L-y)$; $\frac{1}{y(L-y)}dy = kdt$; $[\frac{1}{Ly} + \frac{1}{L(L-y)}]dy = kdt$.

Then $\frac{\ln y}{L} - \frac{\ln(L-y)}{L} = kt + C$.

$\ln[y/(L-y)] = Lkt + CL$; $y/(L-y) = e^{Lkt}\overline{C}$ [where $\overline{C} = e^{CL}$].

$y = y_0$ when $t=0 \Rightarrow \overline{C} = y_0/(L-y_0)$.

Then $\frac{y}{L-y} = \frac{e^{Lkt}y_0}{L-y_0}$; $\frac{L-y}{y} = \frac{(L-y_0)e^{-Lkt}}{y_0}$; $\frac{L}{y} - 1 = \frac{(L-y_0)e^{-Lkt}}{y_0}$.

$\frac{L}{y} = \frac{(L-y_0)e^{-Lkt} + y_0}{y_0}$; $y = \frac{y_0}{y_0 + (L-y_0)e^{-Lkt}} \cdot L$; $y = \frac{Ly_0}{y_0 + (L-y_0)e^{-Lkt}}$.

Problem Set 7.5

27. (a) $\lim_{x\to 0} (1-x)^{1/x} = \lim_{z\to 0} \left((1+z)^{1/z}\right)^{-1} = e^{-1} = 1/e.$ $\quad [z = -x]$

(b) $\lim_{x\to 0} (1+3x)^{1/x} = \lim_{z\to 0} (1+z)^{3/z} = \lim_{z\to 0} \left((1+z)^{1/z}\right)^3 = e^3.$ $\quad [z=3x]$

(c) $\lim_{n\to\infty} \left(\frac{n+2}{n}\right)^n = \lim_{z\to 0^+} (1+z)^{2/z} = \lim_{z\to 0^+} \left((1+z)^{1/z}\right)^2 = e^2.$ $\quad [z = 2/n]$

(d) $\lim_{n\to\infty} \left(\frac{n-1}{n}\right)^{2n} = \lim_{z\to 0^-} (1+z)^{-2/x} = \lim_{z\to 0^-} \left((1+z)^{1/z}\right)^{-2} = e^{-2}.$ $\quad [z = -1/n]$

Problem Set 7.6 The Inverse Trigonometric Functions

3. $-\pi/4$, since $\sin(-\pi/4) = -\sqrt{2}/2$ and $-\pi/4$ is on the interval $[-\pi/2, \pi/2]$.

6. $\sec^{-1}(-1/2)$ doesn't exist since $\sec x \neq -1/2$ for any x.

9. $\sec^{-1}(-2) = \cos^{-1}(-1/2) = 2\pi/3$ since $\cos(2\pi/3) = -1/2$ and $2\pi/3 \in [0,\pi]$.

12. $\sin(\cos^{-1} 0.6) = \sqrt{1-\cos^2(\cos^{-1} 0.6)} = \sqrt{1-(0.6)^2} = 0.8.$

15. $-1.3780.$

18. $1.1746.$

21. $0.6075.$

24. $-6.4315.$

27. $\theta = \sin^{-1}(5/x) = \cos^{-1}(\sqrt{x^2-25}/x)$
    $= \tan^{-1}(5/\sqrt{x^2-25}) = \sec^{-1}(x/\sqrt{x^2-25}).$

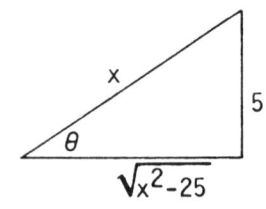

30. Let $\beta = \alpha + \theta$, and note that $x > 5$

$\sin\theta = \sin(\beta-\alpha) = \sin\beta\cos\alpha - \cos\beta\sin\alpha$

$= \dfrac{5}{x} \dfrac{\sqrt{x^2-25}}{\sqrt{x^2-21}} - \dfrac{\sqrt{x^2-25}}{x} \dfrac{2}{\sqrt{x^2-21}} = \dfrac{3\sqrt{x^2-25}}{x\sqrt{x^2-21}}$

$\cos\theta = \cos(\beta-\alpha) = \cos\beta\cos\alpha + \sin\beta\sin\alpha$

$= \dfrac{\sqrt{x^2-25}}{x} \dfrac{\sqrt{x^2-25}}{\sqrt{x^2-21}} + \dfrac{5}{x} \dfrac{2}{\sqrt{x^2-21}} = \dfrac{x^2-15}{x\sqrt{x^2-21}}$

$\tan\theta = \dfrac{\sin\theta}{\cos\theta} = \dfrac{3\sqrt{x^2-25}}{x^2-15}$

Then, $\theta = \sin^{-1} \dfrac{3\sqrt{x^2-25}}{x\sqrt{x^2-21}} = \cos^{-1} \dfrac{x^2-15}{x\sqrt{x^2-21}} = \tan^{-1} \dfrac{3\sqrt{x^2-25}}{x^2-15} = \sec^{-1} \dfrac{x\sqrt{x^2-21}}{x^2-15}$

33. $\sin[\cos^{-1}(3/5) + \cos^{-1}(5/13)]$

$= \sin[\cos^{-1}(3/5)] \cos[\cos^{-1}(5/13)] + \cos[\cos^{-1}(3/5)] \sin[\cos^{-1}(5/13)]$

$= \sqrt{1-(3/5)^2}\,(5/13) + (3/5)\sqrt{1-(5/13)^2} = \dfrac{4}{5}\dfrac{5}{13} + \dfrac{3}{5}\dfrac{12}{13} = \dfrac{56}{65} \approx 0.8615$

36. $\sin(\tan^{-1}x) = \dfrac{\tan(\tan^{-1}x)}{\sec(\tan^{-1}x)} = \dfrac{x}{\sqrt{1+x^2}}$

39. $\cos(\tan^{-1}x) = \dfrac{1}{\sec(\tan^{-1}x)} = \dfrac{1}{\sqrt{1+x^2}}$

42. Let $y = \sec^{-1}x$, so $\sec y = x$

Note the graph of $x = \sec y$.

(a) $\lim\limits_{x \to \infty} \sec^{-1}x = \lim\limits_{y \to (\pi/2)^-} y = \pi/2$

(b) $\lim\limits_{x \to -\infty} \sec^{-1}x = \lim\limits_{y \to (\pi/2)^+} y = \pi/2$

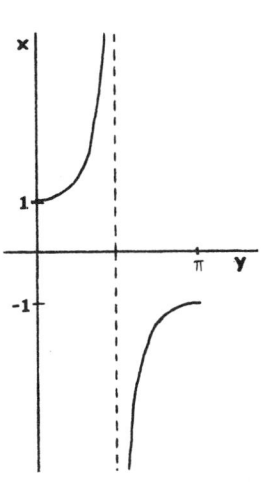

Problem Set 7.6

45. In each case, use the three-step approach (introduced in Section 7.2) for finding the inverse of a function.

(a) Restrict $2x$ so that $0 \leq 2x \leq \pi$; i.e., so that x is in $[0, \pi/2]$.

Let $y = 3\cos 2x$.

$y/3 = \cos 2x$; $2x = \cos^{-1}(y/3)$; $x = (1/2)\cos^{-1}(y/3)$ [Step 1]

$f^{-1}(y) = (1/2)\cos^{-1}(y/3)$ [Step 2]

$f^{-1}(x) = (1/2)\cos^{-1}(x/3)$ [Step 3]

(b) Restrict $3x$ so that $-\pi/2 \leq 3x \leq \pi/2$; i.e., x is in $[-\pi/6, \pi/6]$.

Let $y = 2\sin 3x$

$y/2 = \sin 3x$; $3x = \sin^{-1}(y/2)$; $x = (1/3)\sin^{-1}(y/2)$ [Step 1]

$f^{-1}(y) = (1/3)\sin^{-1}(y/2)$ [Step 2]

$f^{-1}(x) = (1/3)\sin^{-1}(x/2)$ [Step 3]

(c) Restrict x to $(-\pi/2, \pi/2)$.

Let $y = (1/2)\tan x$

$2y = \tan x$; $x = \tan^{-1} 2y$ [Step 1]

$f^{-1}(y) = \tan^{-1} 2y$ [Step 2]

$f^{-1}(x) = \tan^{-1} 2x$ [Step 3]

(d) Restrict x so that $-\pi/2 \leq (1/x) \leq \pi/2$; i.e., $x \leq -2/\pi$ or $x \geq 2/\pi$; i.e. x is in $(-\infty, -2/\pi] \cup [2/\pi, \infty)$.

Let $y = \sin(1/x)$

$\sin^{-1} y = 1/x$; $x = 1/\sin^{-1} y$ [Step 1]

$f^{-1}(y) = 1/\sin^{-1} y$ [Step 2]

$f^{-1}(x) = 1/\sin^{-1} x$ [Step 3]

Problem Set 7.7 Derivatives of Trigonometric Functions

3. $\dfrac{dy}{dx} = (\cot x)(-\csc x \cot x) + (\csc x)(-\csc^2 x) = -\csc x (\cot^2 x + \csc^2 x)$

6. $\dfrac{dy}{dx} = (e^x)(-\sin x) + (\cos x)(e^x) = e^x(\cos x - \sin x)$

9. $\dfrac{dy}{dx} = \dfrac{1}{\sqrt{1-(x^2)^2}} (2x) = \dfrac{2x}{\sqrt{1-x^4}}$

12. $\dfrac{dy}{dx} = (3x-1) \dfrac{-1}{\sqrt{1-(x^2)^2}} (2x) + (\cos^{-1}(x^2))(3) = \dfrac{-2x(3x-1)}{\sqrt{1-x^4}} + 3\cos^{-1}(x^2)$

15. $\dfrac{dy}{dx} = (x) \dfrac{1}{1+x^2} + (\tan^{-1} x)(1) = \dfrac{x}{1+x^2} + \tan^{-1} x$

18. $\dfrac{dy}{dx} = \sec^2(\sin^{-1} x) \dfrac{1}{\sqrt{1-x^2}} = \dfrac{1}{\cos^2(\sin^{-1} x)\sqrt{1-x^2}} = \dfrac{1}{(1-x^2)\sqrt{1-x^2}} = \dfrac{1}{(1-x^2)^{3/2}}$

[In the next to last step, identity (2) from Section 7.6 was used.]

21. $\dfrac{dy}{dx} = \dfrac{1}{1 + \left(\dfrac{1-x}{1+x}\right)^2} \dfrac{(1+x)(-1) - (1-x)(1)}{(1+x)^2} = \dfrac{-2}{(1+x)^2 + (1-x)^2} = \dfrac{-2}{2+2x^2} = \dfrac{-1}{1+x^2}$

24. $\displaystyle\int \sin 2x \cos 2x \, dx = \int \dfrac{1}{2} \sin 4x \, dx = \dfrac{-1}{8} \cos 4x + C$ [Use u=4x substitution if necessary to see result.]

[Note: Could also use u = sin2x substitution as the first step but the use of the double-angle identity is a simpler approach.]

27. $\displaystyle\int \dfrac{\sec^2 x}{\tan x} dx = \int \dfrac{1}{u} du = \ln|u| + C = \ln|\tan x| + C$

| Let u = tan x |
| Then du = $\sec^2 x$ dx |

30. $\int_0^{\pi/2} \sin^2 x \cos x \, dx = \int_0^1 u^2 \, du = \left[\frac{u^3}{3}\right]_0^1 = \frac{1}{3}$

| Let $u = \sin x$ |
| Then $du = \cos x \, dx$ |
| $x = \pi/2 \Rightarrow u = 1$ |
| $x = 0 \Rightarrow u = 0$ |

33. $\int_{-1}^{1} \frac{1}{1+x^2} \, dx = \left[\tan^{-1} x\right]_{-1}^{1} = \tan^{-1}(1) - \tan^{-1}(-1) = \pi/4 - (-\pi/4) = \pi/2$

36. $\int \frac{e^x}{1+e^{2x}} \, dx = \int \frac{1}{1+(e^x)^2} e^x \, dx$

| Let $u = e^x$ |
| Then $du = e^x \, dx$ |

$= \int \frac{1}{1+u^2} \, du = \tan^{-1} u + C = \tan^{-1}(e^x) + C$

39. $\frac{d}{dx}\left[\frac{1}{a} \tan^{-1}(x/a) + C\right] = \frac{1}{a} \frac{1}{1+(x/a)^2} \frac{1}{a} = \frac{1}{a^2} \frac{1}{1+(x^2/a^2)} = \frac{1}{a^2+x^2}$

42. $\int_{-a}^{a} \sqrt{a^2-x^2} \, dx = \left[\frac{x}{2}\sqrt{a^2-x^2} + \frac{a^2}{2} \sin^{-1}(x/a)\right]_{-a}^{a} = \frac{a^2}{2} \sin^{-1}(1) - \frac{a^2}{2} \sin^{-1}(-1)$

$= \frac{a^2}{2} \frac{\pi}{2} - \frac{a^2}{2} \frac{-\pi}{2} = \frac{\pi a^2}{2}$ [This is the area in the upper semicircle of the graph of $x^2 + y^2 = a^2$.]

45. $\beta(b) = \tan^{-1}(12/b) - \tan^{-1}(2/b)$, $b > 0$

$\beta'(b) = \frac{1}{1+(144/b^2)} \frac{-12}{b^2} - \frac{1}{1+(4/b^2)} \frac{-2}{b^2}$

$= \frac{-12}{b^2+144} + \frac{2}{b^2+4} = \frac{-12b^2-48+2b^2+288}{(b^2+144)(b^2+4)}$

$= \frac{-10(b^2-24)}{(b^2+144)(b^2+4)}$ Auxiliary axis for β':

Inc (+)   (0)   Dec (-)
         $\sqrt{24}$

β is maximum where $b = \sqrt{24} \approx 4.8990$ [ft] [about 4'10.8"]

48. Let β be the angle of elevation.

$\frac{dx}{dt} = -600$ [mph] is given.

Determine $\frac{d\beta}{dt}$ in radians per minute.

$\beta = \tan^{-1}(2/x)$

Then $\frac{d\beta}{dt} = \frac{1}{1 + (2/x)^2} \frac{-2}{x^2} \frac{dx}{dt} = \frac{-2}{x^2+4}(-600) = \frac{1200}{x^2+4}$

When the distance from the observor to the plane is 3, $x = \sqrt{5}$, so

$\frac{d\beta}{dt} = \frac{1200}{9} \approx 133.3333$ [rad/hour]. Divide by 60 to get rad/min.

Then, the angle of elevation is increasing at about 2.2222 rad/min.

Problem Set 7.8 The Hyperbolic Functions and Their Inverses

3. $\sinh x \cosh y + \cosh x \sinh y = \frac{e^x - e^{-x}}{2} \frac{e^y + e^{-y}}{2} + \frac{e^x + e^{-x}}{2} \frac{e^y - e^{-y}}{2}$

$= \frac{e^{x+y} + e^{x-y} - e^{-x+y} - e^{-x-y} + e^{x+y} - e^{x-y} + e^{-x+y} - e^{-x-y}}{4}$

$= \frac{2e^{x+y} - 2e^{-(x+y)}}{4} = \frac{e^{x+y} - e^{-(x+y)}}{2} = \sinh(x+y)$

6. $\sinh 2x = \sinh(x+x) = \sinh x \cosh x + \cosh x \sinh x = 2 \sinh x \cosh x$

9. $D_x y = [\sinh(x^2-1)](2x) = 2x \sinh(x^2-1)$

12. $D_x y = (e^x)(\sinh x) + (\cosh x)(e^x) = e^x(\sinh x + \cosh x) = e^x e^x = e^{2x}$
[For the next to last step, see Problem 1.]

15. $D_x y = \frac{1}{\sqrt{(x^3)^2-1}}(3x^2) = \frac{3x^2}{\sqrt{x^6-1}}$, $x > 1$

18. $D_x y = \dfrac{1}{\sinh^{-1} x} \dfrac{1}{\sqrt{x^2+1}} = \dfrac{1}{\sqrt{x^2+1}\ \sinh^{-1} x}$

21. $\displaystyle\int x\ \cosh(x^2+3)\ dx = \int \dfrac{1}{2} \cosh(x^2+3)\ 2x\,dx$

$\boxed{\text{Let}\quad u = x^2+3 \\ \text{Then}\ du = 2x\,dx}$

$\quad = \displaystyle\int \dfrac{1}{2} \cosh u\ du = \dfrac{1}{2} \sinh u + C = \dfrac{1}{2} \sinh(x^2+3) + C$

24. $\displaystyle\int \tanh x\ \ell n(\cosh x)\ dx$

$\boxed{\text{Let}\quad u = \ell n(\cosh x) \\ \text{Then}\ du = \dfrac{1}{\cosh x}\ \sinh x\ dx \\ \phantom{\text{Then}\ du} = \tanh x\ dx}$

$\quad = \displaystyle\int u\ du = \dfrac{u^2}{2} + C = \dfrac{1}{2} \ell n^2(\cosh x)$

27. Using the Method of Disks:

$\text{Volume} = \displaystyle\int_0^1 \pi (\cosh x)^2\,dx$

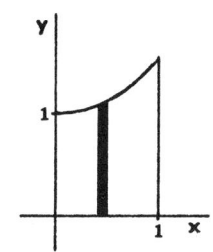

$= \dfrac{\pi}{2} \displaystyle\int_0^1 (1 + \cosh 2x)\,dx \quad \text{[from hint]}$

$= \dfrac{\pi}{2}\left[x + \dfrac{1}{2} \sinh 2x \right]_0^1 = \dfrac{\pi}{2}\left(1 + \dfrac{1}{2} \sinh 2\right) \approx 4.4193$

30. $y = b - a\ \cosh(x/a) \quad (*)$

(a) $b - a\ \cosh(x/a)$ defines an even function. Therefore, if the width of the arch along the x-axis is 2a, the curve intersects the x-axis at -a and at a. Then letting y=0 and x=a in (*), $0 = b - a\ \cosh(1)$. Therefore, $b = a\ \cosh(1) \approx 1.54308a$

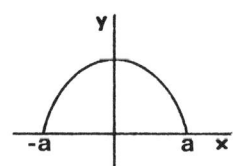

(b) The height of the arch is the value of y when x=0. That is $b - a\ \cosh(0/a) = a\ \cosh 1 - a(1) = a(\cosh 1 - 1) \approx 0.54308a$.

(c) Width = 48 => a = 24 => height = 24($\cosh 1 - 1$) $\approx$ 13 [feet]

Problem Set 7.9   Chapter Review Problems

True-False Quiz

3. True.   The integral equals $\ln(e^3) = 3\ln(e) = 3(1) = 3$.

6. False. Cex: $\ln(e)/\ln(e) = 1/1 = 1$, but $\ln(e) - \ln(e) = 1-1 = 0$

   [Note: $\ln(x/y) = \ln x - \ln y$ is true if $x > 0$ and $y > 0$]

9. True. $(f \circ g)(x) = f(\ln[x-4]) = 4 + e^{\ln(x-4)} = 4 + (x-4) = x$

   $(g \circ f)(x) = g(4+e^x) = \ln([4+e^x]-4) = \ln(e^x) = x$

12. True.   Definition of raising a positive number to a real power.

15. True.   Direct application of the Power Rule (base is a variable and the exponent is a constant).

18. False.   They intersect at $\pi/4$, $-3\pi/4$, $5\pi/4$, $-7\pi/4$, $\cdots$.

   The product of their slopes at these points is $(\frac{-\sqrt{2}}{2})(\frac{\sqrt{2}}{2}) = \frac{-1}{2}$.

21. False. $\cosh(\ln 3) = \frac{e^{\ln 3} + e^{-\ln 3}}{2} = \frac{3 + 3^{-1}}{2} = \frac{5}{3}$

24. True. $f(x) = \frac{\sinh x}{\cosh x}$, ratio of an odd function and an even function.

27. False. Cex: $\ln(x-3)$ is not defined for $x=0$.

   [Note: It is true, however, for all x such that $|x| > 3$.]

30. True. $D_x(a^x) = a^x \ln a$. $a^x \ln a = a^x \Rightarrow \ln a = 1 \Rightarrow a = e$

Miscellaneous Problems

3. $D_x\left(e^{x^2-4x}\right) = e^{x^2-4x}(2x-4) = (2x-4)e^{x^2-4x}$

6. $D_x(e^{\ln \cot x}) = D_x(\cot x) = -\csc^2 x$

9. $D_x[\sinh^{-1}(\tan x)] = \dfrac{1}{\sqrt{\tan^2 x + 1}}(\sec^2 x) = \dfrac{\sec^2 x}{|\sec x|} = |\sec x|$

12. $D_x[\ln \sin^2(x/2)] = D_x[2 \ln \sin(x/2)] = \dfrac{2}{\sin(x/2)} \cos(x/2) \cdot \dfrac{1}{2} = \cot(x/2)$

15. $D_x(\cos e^{\sqrt{x}}) = (-\sin e^{\sqrt{x}})(e^{\sqrt{x}})\left(\dfrac{1}{2\sqrt{x}}\right) = \dfrac{-e^{\sqrt{x}} \sin e^{\sqrt{x}}}{2\sqrt{x}}$

18. $D_x[4^{3x} + (3x)^4] = D_x(64^x + 81x^4) = 64^x \ln 64 + 324x^3$ [Used $4^{3x} = (4^3)^x$.]

21. $D_x(4 \tan 5x \sec 5x) = (4 \tan 5x)[(\sec 5x \tan 5x)(5)] + (\sec 5x)[(4\sec^2 5x)(5)]$
$= 20 \sec 5x \tan^2 5x + 20 \sec^3 5x$

24. $D_x[(1+x^2)^e] = e(1+x^2)^{e-1}(2x) = 2ex(1+x^2)^{e-1}$

27. $\int e^x \sin e^x \, dx = \int \sin u \, du = -\cos u + C = -\cos e^x + C$ Let $u = e^x$
    $D_x(-\cos e^x + C) = (\sin e^x)(e^x) = e^x \sin e^x$ Then $du = e^x dx$

30. $\int 4x \cos x^2 \, dx = \int 2 \cos x^2 \cdot 2x \, dx$ Let $u = x^2$
    $= \int 2 \cos u \, du = 2 \sin u + C = 2 \sin x^2 + C$ Then $du = 2x dx$
    $D_x(2 \sin x^2 + C) = 2(\cos x^2)(2x) = 4x \cos x^2$

33. $\int \frac{-1}{x + x(\ln x)^2} dx = \int \frac{-1}{1 + (\ln x)^2} \cdot \frac{1}{x} dx$ | Let $u = \ln x$
 Then $du = (1/x)dx$

$= \int \frac{-1}{1+u^2} du = -\tan^{-1} u + C = -\tan^{-1}(\ln x) + C.$

$D_x[-\tan^{-1}(\ln x) + C] = \frac{-1}{1 + (\ln x)^2}\left(\frac{1}{x}\right) = \frac{-1}{x + x(\ln x)^2}.$

36. $f(x) = \frac{x^2}{e^x}.$

$f'(x) = \frac{(e^x)(2x) - (x^2)(e^x)}{(e^x)^2} = \frac{x(2-x)}{e^x}.$

```
       Dec     Inc     Dec
       (-) (0) (+) (0) (-)
    ─────┼───────┼──────── x
         0       2
```

$f''(x) = \frac{(e^x)(2-2x) - (2x-x^2)(e^x)}{(e^x)^2} = \frac{x^2 - 4x + 2}{e^x}.$ Use the quadratic formula to find that $f''(x) = 0$ if $x = 2 \pm \sqrt{2}$ (about 0.59 or 3.41).

```
       CU          CD           CU
       (+)  (0)    (-)   (0)    (+)
    ────┼──────────┼────────────── x
        0.59       3.41
```

Summary	
$(-\infty, 0)$	Dec and CU
At 0	Local Minimum
$(0, 0.59)$	Inc and CU
At 0.59	Inflection
$(0.59, 2)$	Inc and CD
At 2	Local Maximum
$(2, 3.41)$	Dec and CD
At 3.41	Inflection
$(3.41, \infty)$	Dec and CU

x	f(x)
0	0
0.59	0.19
2	0.54
3.41	0.38
-0.5	0.41
-1	2.72

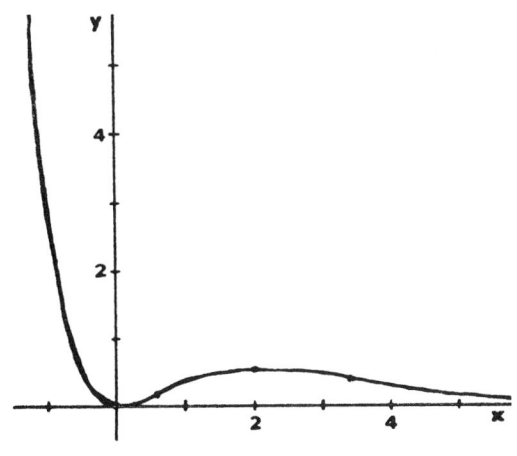

Problem Set 7.9

39. $A(1) = 100(1 + \frac{0.12}{n})^n$ for n compounding periods per year for 1 year.

$A(1) = 100e^{0.12}$ for continuous compounding for 1 year.

(a) n=1: $A(1) = 100(1 + \frac{0.12}{1})^1 = 112$ dollars.

(b) n=12: $A(1) = 100(1 + \frac{0.12}{12})^{12} = 112.68$ dollars.

(c) n=365: $A(1) = 100(1 + \frac{0.12}{365})^{365} = 112.75$ dollars.

(d) continuous: $A(1) = 100e^{0.12} = 112.75$ dollars.

42. Let t denote the time and A the population after 1970.

Since A=10000 when t=0, $A = 10000e^{kt}$ for some constant k.

A=14000 when t=10, so $14000 = 10000e^{10k}$. Therefore, $k = \frac{\ln(1.4)}{10}$.

Therefore, $A = 10000e^{(\ln 1.4)t/10} = 10000(1.4)^{t/10}$.

When t=30, $A = 10000(1.4)^{(30)/10}$

$= 10000(1.4)^3$

$= 27440$.

Therefore, in the year 2000 the population will be about 27,440.

CHAPTER 8   TECHNIQUES OF INTEGRATION

Problem Set 8.1   Integration by Substitution

3. $\int x(x^2+1)^4 dx = \int \frac{1}{2}(x^2+1)^4 \, 2x\,dx$

   $\boxed{\text{Let } u = x^2+1 \\ \text{Then } du = 2x\,dx}$

   $= \int \frac{1}{2} u^4 du = \frac{u^5}{10} + C = \frac{(x^2+1)^5}{10} + C$

6. $\int \frac{e^x}{1+2e^x} dx = \int \frac{1}{2} \frac{1}{1+2e^x} 2e^x dx$

   $\boxed{\text{Let } u = 1+2e^x \\ \text{Then } du = 2e^x dx}$

   $= \int \frac{1}{2} \frac{1}{u} du = \frac{\ln|u|}{2} + C = \frac{\ln(1+2e^x)}{2} + C$

9. $\int 3t\sqrt{2+t^2}\, dt = \int \frac{3}{2}(2+t^2)^{1/2}\, 2t\,dt$

   $\boxed{\text{Let } u = 2+t^2 \\ \text{Then } du = 2t\,dt}$

   $= \int \frac{3}{2} u^{1/2} du = u^{3/2} + C = (2+t^2)^{3/2} + C$

12. $\int \frac{e^{\sin z}}{\sec z} dz = \int e^{\sin z} \cos z\, dz = \int e^u du = e^u + C = e^{\sin z} + C$

   $\boxed{\text{Let } u = \sin z \\ \text{Then } du = \cos z\, dz}$

15. $\int_0^{\pi/2} \frac{\cos x}{1+\sin^2 x} dx = \int_0^1 \frac{1}{1+u^2} du$

   $\boxed{\text{Let } u = \sin x \\ \text{Then } du = \cos x\, dx \\ x=\pi/2 \Rightarrow u=1 \\ x=0 \quad \Rightarrow u=0}$

   $= \left[\tan^{-1} u\right]_0^1 = \tan^{-1}(1) - \tan^{-1}(0) = \frac{\pi}{4} - 0 = \frac{\pi}{4} \approx 0.7854$

18. $\int \frac{x^3+2x^2}{x-2} dx = \int (x^2+4x+8 + \frac{16}{x-2}) dx = \frac{x^3}{3} + 2x^2 + 8x + 16\ln|x-2| + C$

   [The first step was obtained by doing a long division.]

21. $\int \frac{\cos(\ln 4x^2)}{x} dx = \int \frac{1}{2} \cos(\ln 4x^2) \frac{2}{x} dx$

$\boxed{\text{Let } u = \ln(4x^2) \\ \text{Then } du = \frac{1}{4x^2} 8x\, dx \\ = (2/x)\, dx}$

$= \int \frac{1}{2} \cos u\, du = \frac{\sin u}{2} + C = \frac{\sin(\ln 4x^2)}{2} + C$

24. $\int \frac{x}{x^4+1} dx = \int \frac{1}{2} \frac{1}{1+(x^2)^2} 2x\, dx$

$\boxed{\text{Let } u = x^2 \\ \text{Then } du = 2x\, dx}$

$= \int \frac{1}{2} \frac{1}{1+u^2} du = \frac{\tan^{-1} u}{2} + C = \frac{\tan^{-1}(x^2)}{2} + C$

27. $\int_0^1 x\, 10^{x^2} dx = \frac{1}{2} \int_0^1 10^{x^2} 2x\, dx$

$\boxed{\text{Let } u = x^2 \\ \text{Then } du = 2x\, dx \\ x=1 \Rightarrow u=1 \\ x=0 \Rightarrow u=0}$

$= \frac{1}{2} \int_0^1 10^u\, du = \frac{1}{2} \left[\frac{10^u}{\ln 10}\right]_0^1 = \frac{10-1}{2\ln 10} \approx 1.9543$

30. $\int \frac{\sin(4t-1)}{1-\sin^2(4t-1)} dt = \int \frac{\sin(4t-1)}{\cos^2(4t-1)} dt$

$\boxed{\text{Let } u = \cos(4t-1) \\ \text{Then } du = -4\sin(4t-1) dt}$

$= \int \frac{-1}{4} [\cos(4t-1)]^{-2} [-4\sin(4t-1)] dt = \int \frac{-1}{4} u^{-2} du = \frac{u^{-1}}{4} + C$

$= \frac{[\cos(4t-1)]^{-1}}{4} + C = \frac{\sec(4t-1)}{4} + C$

33. $\int e^x \sec e^x dx = \int \sec u\, du = \ln|\sec u + \tan u| + C$

$\boxed{\text{Let } u = e^x \\ \text{Then } du = e^x dx}$

$= \ln|\sec e^x + \tan e^x| + C$ [See Example 9 for 2nd step.]

36. $\int \frac{(6t-1)\sin\sqrt{3t^2-t-1}}{\sqrt{3t^2-t-1}} dt$

$\boxed{\text{Let } u = \sqrt{3t^2-t-1} \\ \text{Then } du = \frac{(6t-1)}{2\sqrt{3t^2-t-1}} dt}$

$= \int 2 \sin u\, du = -2\cos u + C = -2\cos\sqrt{3t^2-t-1} + C$

39. $\int \dfrac{t^2 \cos^2(t^3-2)}{\sin^2(t^3-2)} \, dt = \int \dfrac{1}{3} \cot^2(t^3-2) \, 3t^2 \, dt$ $\boxed{\text{Let } u = t^3-2 \\ \text{Then } du = 3t^2 dt}$

$= \int \dfrac{1}{3} \cot^2 u \, du = \int \dfrac{1}{3}(\csc^2 u - 1) \, du = \dfrac{-\cot u - u}{3} + C = \dfrac{\cot(t^3-2) + t^3 - 2}{-3} + C$

$= \dfrac{t^3 + \cot(t^3-2)}{-3} + C_1 \quad$ [where $C_1 = C + \dfrac{2}{3}$]

42. $\int (t+1) e^{-t^2-2t-5} \, dt = \int \dfrac{-1}{2} e^{-t^2-2t-5}[-2(t+1)] \, dt$ $\boxed{\text{Let } u = -t^2-2t-5 \\ \text{Then } du = (-2t-2)dt \\ \phantom{\text{Then } du} = -2(t+1)dt}$

$= \int \dfrac{-1}{2} e^u \, du = \dfrac{-e^u}{2} + C = \dfrac{-e^{-t^2-2t-5}}{2} + C$

45. $\int \dfrac{\sec x \tan x}{1+\sec^2 x} \, dx = \int \dfrac{1}{1+u^2} \, du = \tan^{-1} u + C = \tan^{-1}(\sec x) + C$ $\boxed{\text{Let } u = \sec x \\ \text{Then } du = \sec x \tan x \, dx}$

48. $\int \dfrac{1}{2t\sqrt{4t^2-1}} \, dt = \int \dfrac{1}{2} \dfrac{1}{(2t)\sqrt{(2t)^2-1}} \, 2dt$ $\boxed{\text{Let } u = 2t \\ \text{Then } du = 2dt}$

$= \int \dfrac{1}{2} \dfrac{1}{u\sqrt{u^2-1}} \, du = \dfrac{1}{2} \sec^{-1}\left(\dfrac{|u|}{1}\right) + C = \dfrac{\sec^{-1}|2t|}{2} + C$

51. $\int \dfrac{1}{x^2+2x+5} \, dx = \int \dfrac{1}{(x+1)^2+4} \, dx = \dfrac{1}{2} \tan^{-1}\left(\dfrac{x+1}{2}\right) + C$ [Could let $u = x+1$.]

54. $\int \dfrac{1}{\sqrt{16+6x-x^2}} \, dx = \int \dfrac{1}{\sqrt{25-(x^2-6x+9)}} \, dx = \int \dfrac{1}{\sqrt{25-(x-3)^2}} \, dx = \sin^{-1}\left(\dfrac{x-3}{5}\right) + C$

57. $\int \dfrac{1}{t\sqrt{2t^2-9}} \, dt = \int \dfrac{1}{(\sqrt{2}t)\sqrt{(\sqrt{2}t)^2-9}} \, \sqrt{2} \, dt$ $\boxed{\text{Let } u = \sqrt{2}t \\ \text{Then } du = \sqrt{2} \, dt}$

$= \int \dfrac{1}{u\sqrt{u^2-9}} \, du = \dfrac{1}{3} \sec^{-1}\left(\dfrac{|u|}{3}\right) + C = \dfrac{1}{3} \sec^{-1}\left(\dfrac{|\sqrt{2}t|}{3}\right) + C$

Problem Set 8.1

60. $\int 2t\sqrt{3-4t}\,dt = 2\dfrac{2(-12t-6)}{15(16)}(3-4t)^{3/2}+C = \dfrac{(2t+1)(3-4t)^{3/2}}{-10}+C.$ [Formula 96: a=-4, b=3]

63. $\int x^2\sqrt{9-2x^2}\,dx = \dfrac{1}{2\sqrt{2}}\int(\sqrt{2}x)^2\sqrt{9-(\sqrt{2}x)^2}\,\sqrt{2}\,dx$ [Formula 57: a=3, u=$\sqrt{2}$x]

$= \dfrac{1}{2\sqrt{2}}\left[\dfrac{\sqrt{2}x}{8}(4x^2-9)\sqrt{9-2x^2} + \dfrac{81}{8}\sin^{-1}(\dfrac{\sqrt{2}x}{3}) + C\right]$

$= \dfrac{x(4x^2-9)\sqrt{9-2x^2}}{16} + \dfrac{81}{16\sqrt{2}}\sin^{-1}(\dfrac{\sqrt{2}x}{3}) + K.$

66. $\int t^2\sqrt{3+5t^2}\,dt = \dfrac{1}{5\sqrt{5}}\int(\sqrt{5}t)^2\sqrt{3+(\sqrt{5}t)^2}\,\sqrt{5}\,dt$ [Formula 48: a=$\sqrt{3}$, u=$\sqrt{5}$t]

$= \dfrac{1}{5\sqrt{5}}\left[\dfrac{\sqrt{5}t}{8}(10t^2+3)\sqrt{3+5t^2} - \dfrac{9}{8}\ln(\sqrt{5}t + \sqrt{3+5t^2}) + C\right]$

$= \dfrac{t(10t^2+3)\sqrt{3+5t^2}}{40} - \dfrac{9\ln(\sqrt{5}t + \sqrt{3+5t^2})}{40\sqrt{5}} + K.$

69. $\int\dfrac{\sin t\,\cos t}{\sqrt{3\sin t + 5}}\,dt = \int\dfrac{u}{\sqrt{3u+5}}\,du$ $\boxed{u = \sin t;\ du = \cos t\,dt}$

[Formula 98: a=3, b=5]

$= \dfrac{2}{27}(3u-10)\sqrt{3u+5} + C = \dfrac{2}{27}(3\sin t - 10)\sqrt{3\sin t + 5} + C.$

72. $\dfrac{\sin x}{\cos x} + \dfrac{\cos x}{1+\sin x} = \dfrac{\sin x+\sin^2 x+\cos^2 x}{\cos x(1+\sin x)} = \dfrac{1+\sin x}{\cos x(1+\sin x)} = \dfrac{1}{\cos x} = \sec x.$

$\int\sec x\,dx = \int\left(\dfrac{\sin x}{\cos x} + \dfrac{\cos x}{1+\sin x}\right)dx = \int\tan x\,dx + \int\dfrac{1}{u}\,du$ $\boxed{\text{Let}\quad u = 1+\sin x \\ \text{Then } du = \cos x\,dx}$

$= -\ln|\cos x| + \ln|u| + C = -\ln|\cos x| + \ln|1+\sin x| + C$

$= \ln\left|\dfrac{1+\sin x}{\cos x}\right| + C = \ln\left|\dfrac{1}{\cos x} + \dfrac{\sin x}{\cos x}\right| + C = \ln|\sec x + \tan x| + C.$

Problem Set 8.2 Some Trigonometric Integrals

3. $\int \cos^3 x\, dx = \int (1-\sin^2 x)\cos x\, dx = \int (1-u^2)\, du$

$\quad = u - \dfrac{u^3}{3} + C = \sin x - \dfrac{\sin^3 x}{3} + C$

$\boxed{\text{Let } u = \sin x \\ \text{Then } du = \cos x\, dx}$

6. $\int \cos^6 t\, dt = \int \left[\dfrac{1}{2}(1+\cos 2t)\right]^3 dt = \int \dfrac{1}{8}(1 + 3\cos 2t + 3\cos^2 2t + \cos^3 2t)\, dt$

$\quad = \dfrac{t}{8} + \dfrac{3\sin 2t}{16} + \int \dfrac{3}{8}\left(\dfrac{1}{2}(1+\cos 4t)\right) dt + \int \dfrac{1}{8}(1-\sin^2 2t)\cos 2t\, dt$

$\quad = \dfrac{t}{8} + \dfrac{3\sin 2t}{16} + \dfrac{3t}{16} + \dfrac{3\sin 4t}{64} + \dfrac{1}{16}\left(\sin 2t - \dfrac{\sin^3 2t}{3}\right) + C$

Hence, $\displaystyle\int_0^{\pi/2} \cos^6 t\, dt = \dfrac{\pi}{16} + 0 + \dfrac{3\pi}{32} + 0 + 0 - 0 - (0+0+0+0+0-0) = \dfrac{5\pi}{32} \approx 0.4909$

9. $\int \sin^7 3x \cos^2 3x\, dx = \int (1-\cos^2 3x)^3 \cos^2 3x \sin 3x\, dx$ $\boxed{\text{Let } u = \cos 3x \\ \text{Then } du = -3\sin 3x\, dx}$

$\quad = \int \dfrac{-1}{3}(1-u^2)^3 u^2\, du = \int \left(\dfrac{-u^2}{3} + u^4 - u^6 + \dfrac{u^8}{3}\right) du = \dfrac{-u^3}{9} + \dfrac{u^5}{5} - \dfrac{u^7}{7} + \dfrac{u^9}{27} + C$

$\quad = \dfrac{-\cos^3 3x}{9} + \dfrac{\cos^5 3x}{5} - \dfrac{\cos^7 3x}{7} + \dfrac{\cos^9 3x}{27} + C$

12. $\int \sin^{1/2}\theta \cos^3\theta\, d\theta = \int \sin^{1/2}\theta(1-\sin^2\theta)\cos\theta\, d\theta$ $\boxed{\text{Let } u = \sin\theta \\ \text{Then } du = \cos\theta\, d\theta}$

$\quad = \int u^{1/2}(1-u^2)\, du = \int (u^{1/2} - u^{5/2})\, du$

$\quad = \dfrac{2u^{3/2}}{3} - \dfrac{2u^{7/2}}{7} + C = \dfrac{2\sin^{3/2}\theta}{3} - \dfrac{2\sin^{7/2}\theta}{7} + C$

15. $\int \tan^3 3y \sec^3 3y\, dy$ $\boxed{\text{Let } u = \sec 3y \\ \text{Then } du = 3\sec 3y\tan 3y\, dy}$

$\quad = \int (\sec^2 3y - 1)\sec^2 3y\, \sec 3y \tan 3y\, dy$

$\quad = \int \dfrac{1}{3}(u^2 - 1)u^2\, du = \int \left(\dfrac{u^4}{3} - \dfrac{u^2}{3}\right) du = \dfrac{u^5}{15} - \dfrac{u^3}{9} + C = \dfrac{\sec^5 3y}{15} - \dfrac{\sec^3 3y}{9} + C$

18. $\int \cot x \sec^2 x \, dx = \int \frac{1}{\tan x} \sec^2 x \, dx$

$\boxed{\text{Let } u = \tan x \\ \text{Then } du = \sec^2 x \, dx}$

$= \int \frac{1}{u} \, du = \ln|u| + C = \ln|\tan x| + C$

21. $\int \sin 4y \cos 5y \, dy = \int \frac{1}{2}(\sin 9y + \sin[-1]y) \, dy = \int \frac{1}{2}(\sin 9y - \sin y) \, dy$

$= \frac{1}{2}\left(\frac{-\cos 9y}{9} + \cos y\right) + C = \frac{\cos 9y}{-18} + \frac{\cos y}{2} + C$

24. $\int \csc^4 3y \, dy = \int (\cot^2 3y + 1) \csc^2 3y \, dy$

$\boxed{\text{Let } u = \cot 3y \\ \text{Then } du = -3\csc^2 3y \, dy}$

$= \int \frac{-1}{3}(u^2+1) \, du = \frac{-1}{3}\left(\frac{u^3}{3} + u\right) + C = \frac{\cot^3 3y + 3\cot 3y}{-9} + C$

27. $\int \cot^3 x \, dx = \int \cot x (\csc^2 x - 1) \, dx$

$= \int \cot x \csc^2 x \, dx - \int \cot x \, dx$

$\boxed{\text{Let } u = \cot x \\ \text{Then } du = -\csc^2 x \, dx}$

$= \int -u \, du - \ln|\sin x| = \frac{-u^2}{2} - \ln|\sin x| + C = \frac{-\cot^2 x}{2} - \ln|\sin x| + C$

30. $\int_{-\pi}^{\pi} \cos mx \cos nx \, dx = 2\int_0^{\pi} \frac{1}{2}[\cos(m+n)x + \cos(m-n)x] \, dx$

$= \left[\frac{\sin(m+n)x}{m+n} + \frac{\sin(m-n)x}{m-n}\right]_0^{\pi} = 0$ if m+n and m-n are integers.

Problem Set 8.2

Problem Set 8.3 Rationalizing Substitutions

3. $\int \dfrac{t}{\sqrt{2t+7}}\,dt = \int \dfrac{u^2-7}{2u}\,u\,du$

$\boxed{\text{Let } u = \sqrt{2t+7};\ u^2 = 2t+7;\ t = \dfrac{u^2-7}{2} \\ \text{Then } 2u\,du = 2\,dt \\ u\,du = dt}$

$= \int\left(\dfrac{u^2}{2} - \dfrac{7}{2}\right)du = \dfrac{u^3}{6} - \dfrac{7u}{2} + C = \dfrac{(2t+7)^{3/2}}{6} - \dfrac{7(2t+7)^{1/2}}{2} + C$

6. $\int_0^4 \dfrac{\sqrt{t}}{t+1}\,dt = \int_0^2 \dfrac{u}{u^2+1}\,2u\,du = 2\int_0^2 \dfrac{u^2}{u^2+1}\,du$

$\boxed{\text{Let } u = \sqrt{t};\ u^2 = t \\ \text{Then } 2u\,du = dt \\ t=4 \Rightarrow u = 2 \\ t=0 \Rightarrow u = 0}$

$= 2\int_0^2 \left(1 - \dfrac{1}{u^2+1}\right)du = 2\left[u - \tan^{-1}u\right]_0^2 = 2(2-\tan^{-1}2) \approx 1.7857$

9. $\int \dfrac{\sqrt{1-x^2}}{x}\,dx = \int \dfrac{\cos t}{\sin t}\cos t\,dt$

$\boxed{\text{Let } x = \sin t,\ t \neq 0,\ t \text{ in } [-\pi/2,\pi/2] \\ \text{Then } dx = \cos t\,dt}$

$= \int \dfrac{1-\sin^2 t}{\sin t}\,dt = \int(\csc t - \sin t)\,dt$

$= -\ln|\csc t + \cot t| + \cos t + C$

$= -\ln\left|\dfrac{1}{x} + \dfrac{\sqrt{1-x^2}}{x}\right| + \sqrt{1-x^2} + C = \sqrt{1-x^2} - \ln\left(\dfrac{1+\sqrt{1-x^2}}{|x|}\right) + C$

12. $\int \dfrac{1}{(x^2+9)^{3/2}}\,dx = \int \dfrac{1}{(9\tan^2 u + 9)^{3/2}}\,3\sec^2 u\,du$

$\boxed{\text{Let } x = 3\tan u,\ u \text{ in } (\tfrac{-\pi}{2},\tfrac{\pi}{2}) \\ \text{Then } dx = 3\sec^2 u\,du}$

$= \int \dfrac{3\sec^2 u}{27\sec^3 u}\,du = \int \dfrac{1}{9}\cos u\,du$

$= \dfrac{\sin u}{9} + C = \dfrac{x}{9\sqrt{x^2+9}} + C$

15. $\int \dfrac{t}{\sqrt{4-t^2}}\,dt = \int \dfrac{-1}{2}\,u^{-1/2}\,du = -u^{1/2} + C = -\sqrt{4-t^2} + C$

$\boxed{\text{Let } u = 4-t^2 \\ \text{Then } du = -2t\,dt}$

18. $\int \dfrac{2x+1}{\sqrt{x^2+9}}\, dx = \int \dfrac{6\tan u + 1}{\sqrt{9\tan^2 u + 9}}\, 3\sec^2 u\, du$

$\boxed{\text{Let } x = 3\tan u,\ u \text{ in } (-\pi/2, \pi/2) \\ \text{Then } dx = 3\sec^2 u\, du}$

$= \int \dfrac{3(6\tan u + 1)\sec^2 u}{3\sec u}\, du = \int (6\sec u \tan u + \sec u)\, du$

$= 6\sec u + \ln|\sec u + \tan u| + C$

$= 6\,\dfrac{\sqrt{x^2+9}}{3} + \ln\left|\dfrac{\sqrt{x^2+9}}{3} + \dfrac{x}{3}\right| + C$

$= 2\sqrt{x^2+9} + \ln(\sqrt{x^2+9} + x) - \ln 3 + C = 2\sqrt{x^2+9} + \ln(\sqrt{x^2+9} + x) + K$

(Triangle: hypotenuse $\sqrt{x^2+9}$, opposite x, adjacent 3, angle u)

21. $\int \dfrac{1}{\sqrt{x^2+2x+5}}\, dx = \int \dfrac{1}{\sqrt{(x+1)^2+4}}\, dx$

$\boxed{\text{Let } x+1 = 2\tan u,\ u \text{ in } (-\pi/2, \pi/2) \\ \text{Then } dx = 2\sec^2 u\, du}$

$= \int \dfrac{1}{\sqrt{4\tan^2 u + 4}}\, 2\sec^2 u\, du = \int \dfrac{2\sec^2 u}{2\sec u}\, du$

$= \int \sec u\, du = \ln|\sec u + \tan u| + C$

$= \ln\left|\dfrac{\sqrt{x^2+2x+5}}{2} + \dfrac{x+1}{2}\right| + C$

$= \ln(\sqrt{x^2+2x+5} + x+1) - \ln 2 + C = \ln(\sqrt{x^2+2x+5} + x+1) + K$

(Triangle: hypotenuse $\sqrt{x^2+2x+5}$, opposite $x+1$, adjacent 2, angle u)

24. $\int \dfrac{2x-1}{\sqrt{x^2+4x+5}}\, dx = \int \dfrac{2x-1}{\sqrt{(x+2)^2+1}}\, dx$

$\boxed{\text{Let } x+2 = \tan u,\ u \text{ in } (-\pi/2, \pi/2) \\ \text{Then } dx = \sec^2 u\, du}$

$= \int \dfrac{2(\tan u - 2) - 1}{\sqrt{\tan^2 u + 1}}\, \sec^2 u\, du$

$= \int \dfrac{(2\tan u - 5)\sec^2 u}{\sec u}\, du$

(Triangle: hypotenuse $\sqrt{x^2+4x+5}$, opposite $x+2$, adjacent 1, angle u)

$= \int (2\tan u \sec u - 5\sec u)\, du = 2\sec u - 5\ln|\sec u + \tan u| + C$

$= 2\sqrt{x^2+4x+5} - 5\ln(\sqrt{x^2+4x+5} + x+2) + C$

27. $\int \dfrac{1}{\sqrt{4x-x^2}}\, dx = \int \dfrac{1}{\sqrt{4-(x-2)^2}}\, dx = \int \dfrac{1}{\sqrt{4-u^2}}\, du$

$\boxed{\text{Let } u = x-2 \\ \text{Then } du = dx}$

$= \sin^{-1}\left(\dfrac{u}{2}\right) + C = \sin^{-1}\left(\dfrac{x-2}{2}\right) + C$

30. $\int \dfrac{2x-1}{x^2-6x+18}\,dx = \int \dfrac{(2x-6)+5}{x^2-6x+18}\,dx$

$= \int \dfrac{2x-6}{x^2-6x+18}\,dx + \int \dfrac{5}{(x-3)^2+9}\,dx$

1st integral: Let $u = x^2-6x+18$
Then $du = (2x-6)dx$

$= \int \dfrac{1}{u}\,du + \int \dfrac{5}{v^2+9}\,dv$

2nd integral: Let $v = x-3$
Then $dv = dx$

$= \ln|u| + \dfrac{5}{3}\tan^{-1}\left(\dfrac{v}{3}\right) + C = \ln(x^2-6x+18) + 5\tan^{-1}\left(\dfrac{x-3}{3}\right) + C$

33. (a) $\int \dfrac{x}{x^2+9}\,dx = \int \dfrac{1}{2}\dfrac{1}{u}\,du = \dfrac{1}{2}\ln|u| + C = \ln(\sqrt{x^2+9}) + C$

Let $u = x^2+9$
Then $du = 2xdx$

(b) $\int \dfrac{x}{x^2+9}\,dx = \int \dfrac{3\tan u}{9\tan^2 u + 9}\,3\sec^2 u\,du$

Let $x = 3\tan u$
Then $dx = 3\sec^2 u\,du$

$= \int \dfrac{9\tan u \sec^2 u}{9\sec^2 u}\,du = \int \tan u\,du$

$= -\ln|\cos u| + K = -\ln \dfrac{3}{\sqrt{x^2+9}} + K$

$= -\ln 3 + \ln(\sqrt{x^2+9}) + K = \ln(\sqrt{x^2+9}) + C$

Problem Set 8.4 Integration by Parts

3. $\int x \sin 3x\,dx$

Let $u = x$ and $dv = \sin 3x\,dx$
Then $du = dx$ and $v = \dfrac{-1}{3}\cos 3x$

$= \dfrac{-x\cos 3x}{3} - \int \dfrac{-1}{3}\cos 3x\,dx$

$= \dfrac{-x\cos 3x}{3} + \dfrac{\sin 3x}{9} + C$

6. $\int x\sqrt{x+1}\,dx$

Let $u = x$ and $dv = (x+1)^{1/2}dx$
Then $du = dx$ and $v = \dfrac{2(x+1)^{3/2}}{3}$

$= \dfrac{2x(x+1)^{3/2}}{3} - \int \dfrac{2(x+1)^{3/2}}{3}\,dx$

$= \dfrac{2x(x+1)^{3/2}}{3} - \dfrac{4(x+1)^{5/2}}{15} + C$

9. $\int \sqrt{x} \, \ln x \, dx$

| Let $u = \ln x$ and $dv = x^{1/2} dx$ |
| Then $du = \frac{1}{x} dx$ and $v = \frac{2x^{3/2}}{3}$ |

$= \frac{2x^{3/2} \ln x}{3} - \int \frac{2}{3} x^{1/2} dx$

$= \frac{2x^{3/2} \ln x}{3} - \frac{4x^{3/2}}{9} + C$

12. $\int t \cos 4t \, dt$

| Let $u = t$ and $dv = \cos 4t \, dt$ |
| Then $du = dt$ and $v = \frac{1}{4} \sin 4t$ |

$= \frac{t \sin 4t}{4} - \int \frac{1}{4} \sin 4t \, dt$

$= \frac{t \sin 4t}{4} + \frac{\cos 4t}{16} + C$

15. $\int_{\pi/6}^{\pi/2} x \csc^2 x \, dx$

| Let $u = x$ and $dv = \csc^2 x \, dx$ |
| Then $du = dx$ and $v = -\cot x$ |

$= \left[-x \cot x \right]_{\pi/6}^{\pi/2} - \int_{\pi/6}^{\pi/2} -\cot x \, dx = \left[-x \cot x + \ln|\sin x| \right]_{\pi/6}^{\pi/2}$

$= (0 + 0) - \left(\frac{-\pi\sqrt{3}}{6} + \ln[1/2] \right) = \frac{\pi\sqrt{3}}{6} + \ln 2 \approx 1.6000$

18. $\int x \sin^3 x \, dx = \int x(1 - \cos^2 x) \sin x \, dx = \int x \sin x \, dx - \int x \cos^2 x \sin x \, dx$

$= \left(-x \cos x - \int -\cos x \, dx \right) - \left(\frac{-x}{3} \cos^3 x + \frac{1}{3} \sin x - \frac{1}{9} \sin^3 x \right)$

[Integration by parts with $u = x$ and $dv = \sin x \, dx$ was used on the first integral; the second integral was evaluated in Problem 14.]

$= -x \cos x + \sin x + \frac{x}{3} \cos^3 x - \frac{1}{3} \sin x + \frac{1}{9} \sin^3 x + C$

$= -x \cos x + \frac{2}{3} \sin x + \frac{x}{3} \cos^3 x + \frac{1}{9} \sin^3 x + C$

21. $\int e^t \cos t \, dt = e^t \sin t - \int e^t \sin t \, dt$

| Let $u = e^t$ and $dv = \cos t \, dt$ |
| Then $du = e^t dt$ and $v = \sin t$ |

$\quad = e^t \sin t - (-e^t \cos t + \int e^t \cos t \, dt)$

| Let $u = e^t$ and $dv = \sin t \, dt$ |
| Then $du = e^t dt$ and $v = -\cos t$ |

$\quad = e^t \sin t + e^t \cos t - \int e^t \cos t \, dt$

Therefore, $2\int e^t \cos t \, dt = e^t \sin t + e^t \cos t + C$

Hence, $\int e^t \cos t \, dt = \dfrac{e^t(\sin t + \cos t)}{2} + K$

24. $\int (\ln x)^3 dx = x(\ln x)^3 - \int 3(\ln x)^2 dx$

| Let $u = (\ln x)^3$ and $dv = dx$ |
| Then $du = \dfrac{3(\ln x)^2}{x}$ and $v = x$ |

$\quad = x\ln^3 x - 3(x\ln^2 x - 2x\ln x + 2x) + C$

$\quad = x\ln^3 x - 3x\ln^2 x + 6x\ln x - 6x + C$

27. Area $= \int_0^9 3xe^{-x/3} dx$

| Let $u = 3x$ and $dv = e^{-x/3} dx$ |
| Then $du = 3dx$ and $v = -3e^{-x/3}$ |

$\quad = \left[-9xe^{-x/3}\right]_0^9 - \int_0^9 -9e^{-x/3} dx$

$\quad = \left[-9xe^{-x/3} - 27e^{-x/3}\right]_0^9$

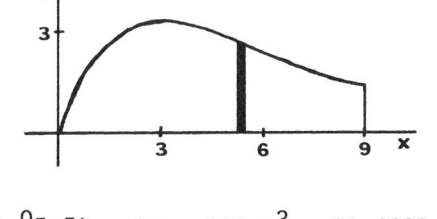

$\quad = \left[-9e^{-x/3}(x+3)\right]_0^9 = (-9e^{-3}[12]) - (-9e^0[3]) = 27 - 108e^{-3} \approx 21.6230$

[sq. units]

30. $\int_0^{\pi/2} \cos^n x \, dx = \left[\dfrac{\cos^{n-1} x \sin x}{n}\right]_0^{\pi/2} + \dfrac{n-1}{n}\int_0^{\pi/2} \cos^{n-2} x \, dx = \dfrac{n-1}{n}\int_0^{\pi/2} \cos^{n-2} x \, dx$

33. $\int x^n e^x dx = x^n e^x - n\int x^{n-1} e^x dx$

| Let $u = x^n$ and $dv = e^x dx$ |
| Then $du = nx^{n-1} dx$ and $v = e^x$ |

$\int x^3 e^x dx = x^3 e^x - 3\int x^2 e^x dx = x^3 e^x - 3(x^2 e^x - 2\int xe^x dx)$

$\quad = x^3 e^x - 3x^2 e^x + 6(xe^x - 1\int e^x dx) = x^3 e^x - 3x^2 e^x + 6xe^x - 6e^x + C$

Problem Set 8.4

Problem Set 8.5 Integration of Rational Functions

3. $\dfrac{5x+3}{x^2-9} = \dfrac{5x+3}{(x+3)(x-3)} = \dfrac{A}{x+3} + \dfrac{B}{x-3} = \dfrac{A(x-3) + B(x+3)}{(x+3)(x-3)}$

   Then $5x+3 = A(x-3) + B(x+3)$; $x=-3 \Rightarrow A=2$; $x=3 \Rightarrow B=3$.

   Therefore, $\displaystyle\int \dfrac{5x+3}{x^2-9}\,dx = \int\left(\dfrac{2}{x+3} + \dfrac{3}{x-3}\right)dx = 2\ln|x+3| + 3\ln|x-3| + C$

6. $\dfrac{3x-13}{x^2+3x-10} = \dfrac{3x-13}{(x+5)(x-2)} = \dfrac{A}{x+5} + \dfrac{B}{x-2} = \dfrac{A(x-2) + B(x+5)}{(x+5)(x-2)}$

   Then $3x-13 = A(x-2) + B(x+5)$; $x=-5 \Rightarrow A=4$; $x=2 \Rightarrow B=-1$

   Therefore, $\displaystyle\int \dfrac{3x-13}{x^2+3x-10}\,dx = \int\left(\dfrac{4}{x+5} - \dfrac{1}{x-2}\right)dx = 4\ln|x+5| - \ln|x-2| + C$

9. $\dfrac{2x^2+x-4}{x^3-x^2-2x} = \dfrac{2x^2+x-4}{x(x-2)(x+1)} = \dfrac{A}{x} + \dfrac{B}{x-2} + \dfrac{C}{x+1} = \dfrac{A(x-2)(x+1) + Bx(x+1) + Cx(x-2)}{x(x-2)(x+1)}$

   Then $2x^2+x-4 = A(x-2)(x+1) + Bx(x+1) + Cx(x-2)$
   $x=0 \Rightarrow A=2$; $x=2 \Rightarrow B=1$; $x=-1 \Rightarrow C=-1$

   $\displaystyle\int \dfrac{2x^2+x-4}{x^3-x^2-2x}\,dx = \int\left(\dfrac{2}{x} + \dfrac{1}{x-2} - \dfrac{1}{x+1}\right)dx = 2\ln|x| + \ln|x-2| - \ln|x+1| + C$

12. $\dfrac{x^4+8x^2+8}{x^3-4x} = x + \dfrac{12x^2+8}{x^3-4x} = x + \dfrac{12x^2+8}{x(x+2)(x-2)} = x + \dfrac{A}{x} + \dfrac{B}{x+2} + \dfrac{C}{x-2}$

    $= x + \dfrac{A(x+2)(x-2) + Bx(x-2) + Cx(x+2)}{x(x+2)(x-2)}$

    Then $12x^2+8 = A(x+2)(x-2) + Bx(x-2) + Cx(x+2)$
    $x=0 \Rightarrow A=-2$; $x=-2 \Rightarrow B=7$; $x=2 \Rightarrow C=7$

    $\displaystyle\int \dfrac{x^4+8x^2+8}{x^3-4x}\,dx = \int\left(x - \dfrac{2}{x} + \dfrac{7}{x+2} + \dfrac{7}{x-2}\right)dx = \dfrac{x^2}{2} - 2\ln|x| + 7\ln|x+2| + 7\ln|x-2| + C$

15. $\dfrac{3x^2-21x+32}{x^3-8x^2+16x} = \dfrac{3x^2-21x+32}{x(x-4)^2} = \dfrac{A}{x} + \dfrac{B}{x-4} + \dfrac{C}{(x-4)^2} = \dfrac{A(x-4)^2 + Bx(x-4) + Cx}{x(x-4)^2}$.

Then $3x^2-21x+32 = A(x-4)^2 + Bx(x-4) + Cx$.
 $x=0 \Rightarrow A=2$; $x=4 \Rightarrow C=-1$; $x=1 \Rightarrow B=1$ [using $A=2, C=-1$].

$\displaystyle\int \dfrac{3x^2-21x+32}{x^3-8x^2+16x}\,dx = \int\left(\dfrac{2}{x} + \dfrac{1}{x-4} - \dfrac{1}{(x-4)^2}\right)dx = 2\ln|x| + \ln|x-4| + \dfrac{1}{x-4} + C$.

18. $\dfrac{2x^2-3x-36}{(2x-1)(x^2+9)} = \dfrac{A}{2x-1} + \dfrac{Bx+C}{x^2+9} = \dfrac{A(x^2+9) + (Bx+C)(2x-1)}{(2x-1)(x^2+9)}$.

Then $2x^2-3x-36 = A(x^2+9) + (Bx+C)(2x-1)$.
 $x=1/2 \Rightarrow A=-4$; $x=0 \Rightarrow C=0$ [using $A=-4$]; $x=1 \Rightarrow B=3$ [using $A=-4, C=0$].

$\displaystyle\int \dfrac{2x^2-3x-36}{(2x-1)(x^2+9)}\,dx = \int\left(\dfrac{-4}{2x-1} + \dfrac{3x}{x^2+9}\right)dx = -2\ln|2x-1| + \dfrac{3}{2}\ln|x^2+9| + C$.

21. $\dfrac{x^3-4x}{(x^2+1)^2} = \dfrac{Ax+B}{x^2+1} + \dfrac{Cx+D}{(x^2+1)^2} = \dfrac{(Ax+B)(x^2+1) + (Cx+D)}{(x^2+1)^2}$.

Then $x^3-4x = (Ax+B)(x^2+1) + (Cx+D) = Ax^3 + Bx^2 + (A+C)x + (B+D)$.

Then $A=1$, $B=0$, $A+C = -4$ so $C=-5$ [since $A=1$], $B+D=0$ so $D=0$ [since $B=0$].

$\displaystyle\int \dfrac{x^3-4x}{(x^2+1)^2}\,dx = \int\left(\dfrac{1x+0}{x^2+1} + \dfrac{-5x+0}{(x^2+1)^2}\right)dx = \int\left(\dfrac{1}{2}\dfrac{2x}{x^2+1} - \dfrac{5}{2}\dfrac{2x}{(x^2+1)^2}\right)dx$

$= \dfrac{1}{2}\ln(x^2+1) + \dfrac{5}{2}\dfrac{1}{x^2+1} + C$.   [May use $u = x^2+1$ substitution.]

24. $\dfrac{3x+13}{x^2+4x+3} = \dfrac{3x+13}{(x+3)(x+1)} = \dfrac{A}{x+3} + \dfrac{B}{x+1} = \dfrac{A(x+1) + B(x+3)}{(x+3)(x+1)}$.

Then $3x+13 = A(x+1) + B(x+3)$; $x=-1 \Rightarrow B=5$; $x=-3 \Rightarrow A=-2$.

$\displaystyle\int_1^5 \dfrac{3x+13}{x^2+4x+3}\,dx = \int_1^5\left(\dfrac{-2}{x+3} + \dfrac{5}{x+1}\right)dx = \left[-2\ln(x+3) + 5\ln(x+1)\right]_1^5$

$= (-2\ln 8 + 5\ln 6) - (-2\ln 4 + 5\ln 2) = \ln\dfrac{6^5\,4^2}{8^2\,2^5} = \ln\left(\dfrac{243}{4}\right) \approx 4.1068$.

Problem Set 8.5

27. (a) $\frac{dy}{dt} = ky(10-y)$; $\frac{1}{y(10-y)}dy = kdt$; $\int\frac{1}{y(10-y)}dy = \int kdt$

$\frac{1}{y(10-y)} = \frac{1/10}{y} + \frac{1/10}{10-y}$ [partial fraction decomposition]

Then $(1/10)[\ln(y) - \ln(10-y)] = kt + C$ or $\frac{y}{10-y} = e^{10kt}\bar{C}$

When t=0 (1925), y=2, so $\bar{C} = 1/4$. Then $\frac{4y}{10-y} = e^{10kt}$, and solving

for y obtain $y = \frac{10}{1 + 4e^{-10kt}}$

When t=50 (1975), y=4, so $k = \frac{\ln(8/3)}{500}$ and then $y = \frac{10}{1 + 4(3/8)^{0.02t}}$

Then when y=9, $t = \frac{-50\ln 36}{\ln(3/8)} \approx 182.68$; i.e. in about the year 2108.

Problem Set 8.6 Chapter Review Problems

True-False Quiz

3. False. It will be more efficient to use the substitution $u = 9+x^4$.

6. True. Final answer will involve $\sin^{-1}(\sqrt{5}x/2)$.

9. True. See Section 8.2, Type 2 with m and n both even.

12. True. See "Integrands Involving $\sqrt[q]{ax+b}$" type of Section 8.3.

15. True. Let $u = \ln x$ and $dv = x^2 dx$.

18. True. See Section 8.5. [A=-2, B=3/2, C=3/2 work.]

21. True. See the first few sentences of Section 8.4.

24. True. Let $P(x)$ and $Q(x)$ be the polynomials. If any corresponding coefficients of $P(x)$ and $Q(x)$ are not the same, then the polynomial equation $P(x) - Q(x) = 0$ has positive or zero degree and would have at most a finite number of solutions, so $P(x) = Q(x)$ would be true for at most a finite number of values of x.

Miscellaneous Problems

3. $\int_0^{\pi/2} e^{\cos x} \sin x \, dx = \int_1^0 -e^u du = \left[-e^u\right]_1^0 = (-1 + e) \approx 1.7183$

$\boxed{u = \cos x; \; du = -\sin x \, dx; \; x=\pi/2 \Rightarrow u=0; \; x=0 \Rightarrow u=1}$

6. $\int \sin^3 2t \, dt = \int (1-\cos^2 2t) \sin 2t \, dt = \int \frac{-1}{2}(1-u^2) du = \frac{-1}{2}\left(u - \frac{u^3}{3}\right) + C$

$= \frac{-\cos 2t}{2} + \frac{\cos^3 2t}{6} + C \qquad \boxed{u = \cos 2t; \; du = -2\sin 2t \, dt}$

9. $\int \frac{e^{2t}}{e^t - 2} dt = \int \left(e^t + \frac{2e^t}{e^t - 2}\right) dt = e^t + \int \frac{2e^t}{e^t - 2} dt = e^t + \int \frac{2}{u} du = e^t + 2\ln|u| + C$

$= e^t + 2\ln|e^t - 2| + C \qquad \boxed{u = e^t - 2; \; du = e^t dt}$

12. $\int x^2 e^x dx = x^2 e^x - \int 2xe^x dx = x^2 e^x - (2xe^x - \int 2e^x dx) = x^2 e^x - 2xe^x + 2e^x + C$

$\boxed{\text{1st Step: } u = x^2; \; dv = e^x dx \\ \phantom{\text{1st Step: }} du = 2x dx; \; v = e^x} \qquad \boxed{\text{2nd Step: } u = 2x; \; dv = e^x dx \\ \phantom{\text{2nd Step: }} du = 2dx; \; v = e^x}$

15. $\int \frac{\tan x}{\ln|\cos x|} dx = \int \frac{-1}{u} du = -\ln|u| + C = -\ln|\ln|\cos x|| + C$

$\boxed{u = \ln|\cos x|; \; du = -\tan x \, dx}$

18. $\int \frac{(\ln y)^5}{y} dy = \int u^5 du = \frac{u^6}{6} + C = \frac{\ln^6 y}{6} + C \qquad \boxed{u = \ln y; \; du = \frac{1}{y} dy}$

Problem Set 8.6

21. $\int \frac{\ln t^2}{t} dt = \int \frac{2\ln|t|}{t} dt = \int 2u\, du = u^2 + C = \ln^2|t| + C$ $\boxed{u = \ln|t|;\ du = \frac{1}{t} dt}$

24. $\frac{t+9}{t^3+9t} = \frac{A}{t} + \frac{Bt+C}{t^2+9}$; $t+9 = A(t^2+9) + t(Bt+C)$; $A=1,\ B=-1,\ C=1$

$\int \frac{t+9}{t^3+9t} dt = \int \left(\frac{1}{t} + \frac{-t+1}{t^2+9}\right) dt = \int \left(\frac{1}{t} - \frac{t}{t^2+9} + \frac{1}{t^2+9}\right) dt$ [May use $u = t^2+9$ on second term.]

$= \ln|t| - \frac{1}{2}\ln(t^2+9) + \frac{1}{3}\tan^{-1}\left(\frac{t}{3}\right) + C$

27. $\int \tan^3 2x\, \sec 2x\, dx = \int (\sec^2 2x - 1)\sec 2x\, \tan 2x\, dx = \int \frac{1}{2}(u^2 - 1)\, du$

$= \frac{1}{2}\left(\frac{u^3}{3} - u\right) + C = \frac{\sec^3 2x}{6} - \frac{\sec 2x}{2} + C$ $\boxed{u = \sec 2x \\ du = 2\sec 2x\, \tan 2x\, dx}$

30. $\int \frac{1}{t(t^{1/6}+1)} dt = \int \frac{1}{u^6(u+1)} 6u^5 du = \int \frac{6}{u(u+1)} du$ $\boxed{t = u^6,\ u > 0 \\ dt = 6u^5 du}$

$= \int \left(\frac{6}{u} - \frac{6}{u+1}\right) du = 6\ln|u| - 6\ln|u+1| + C$ [used partial fractions]

$= 6\ln|t^{1/6}| - 6\ln|t^{1/6}+1| + C = \ln|t| - 6\ln|t^{1/6}+1| + C$

33. $\int e^{\ln(3\cos x)} dx = \int 3\cos x\, dx = 3\sin x + C$

36. $\int \frac{\sqrt{x^2+a^2}}{x^4} dx = \int \frac{a\sec u}{a^4 \tan^4 u} a\sec^2 u\, du$ $\boxed{x = a\tan u;\ dx = a\sec^2 u\, du}$

$= a^{-2} \int (\sin u)^{-4} \cos u\, du$ $\boxed{v = \sin u;\ dv = \cos u\, du}$

$= a^{-2} \int v^{-4} dv = a^{-2} \frac{v^{-3}}{-3} + C$

$= \frac{\csc^3 u}{-3a^2} + C = \frac{(a^2+x^2)^{3/2}}{-3a^2 x^2} + C$

Problem Set 8.6

39. $\int \frac{\sin y \cos y}{9+\cos^4 y} dy = \int \frac{1}{9+(\cos^2 y)^2} (\sin y \cos y) dy$ $\boxed{\begin{array}{l} u = \cos^2 y \\ du = -2\sin y \cos y \, dy \end{array}}$

$= \int \frac{-1}{2} \frac{1}{9+u^2} du = \frac{-1}{2} \frac{1}{3} \tan^{-1}\left(\frac{u}{3}\right) + C = \frac{-1}{6} \tan^{-1}\left(\frac{\cos^2 y}{3}\right) + C$

42. $\int \frac{1}{(16+x^2)^{3/2}} dx = \int \frac{1}{(4 \sec u)^3} 4 \sec^2 u \, du$ $\boxed{\begin{array}{l} x = 4 \tan u \\ dx = 4 \sec^2 u \, du \end{array}}$

$= \int \frac{\cos u}{16} du = \frac{\sin u}{16} + C = \frac{x}{16\sqrt{16+x^2}} + C$

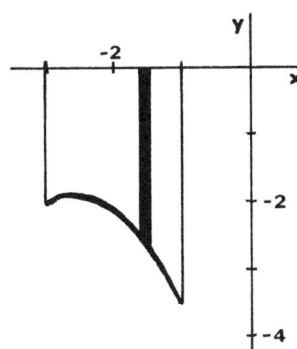

45.

[figure]

Volume $= \int_{-3}^{-1} \pi \left(\frac{6}{x\sqrt{x+4}}\right)^2 dx = 36\pi \int_{-3}^{-1} \frac{1}{x^2(x+4)} dx$

$= 36\pi \int_{-3}^{-1} \left(\frac{-1/16}{x} + \frac{1/4}{x^2} + \frac{1/16}{x+4}\right) dx$ [using partial fractions]

$= 36\pi \left[\frac{-\ln|x|}{16} - \frac{1}{4x} + \frac{\ln|x+4|}{16}\right]_{-3}^{-1}$

$= 36\pi \left[\left(\frac{-\ln(1)}{16} + \frac{1}{4} + \frac{\ln(3)}{16}\right) - \left(\frac{-\ln(3)}{16} + \frac{1}{12} + \frac{\ln(1)}{16}\right)\right]$

$= \frac{3\pi(4 + 3\ln 3)}{2} \approx 34.3808$

Problem Set 8.6

CHAPTER 9 INDETERMINATE FORMS AND IMPROPER INTEGRALS

Problem Set 9.1 Indeterminate Forms of Type 0/0

Each step where l'Hôpital's Rule is used will be indicated by Ⓛ.

3. $\lim_{x\to 0} \dfrac{x-2\sin x}{\tan x} \overset{Ⓛ}{=} \lim_{x\to 0} \dfrac{1-2\cos x}{\sec^2 x} = \dfrac{1-2}{1} = -1$

6. $\lim_{x\to 0} \dfrac{x^3-3x^2+5x}{x^3-x} \overset{Ⓛ}{=} \lim_{x\to 0} \dfrac{3x^2-6x+5}{3x^2-1} = \dfrac{5}{-1} = -5$

9. $\lim_{x\to \pi/2} \dfrac{\ln(\sin x)}{(\pi/2)-x} \overset{Ⓛ}{=} \lim_{x\to \pi/2} \dfrac{\cot x}{-1} = \dfrac{0}{-1} = 0$

12. $\lim_{x\to 0^+} \dfrac{8^{\sqrt{x}}-1}{3^{\sqrt{x}}-1} \overset{Ⓛ}{=} \lim_{x\to 0^+} \dfrac{8^{\sqrt{x}} \cdot \frac{1}{2\sqrt{x}} \ln 8}{3^{\sqrt{x}} \cdot \frac{1}{2\sqrt{x}} \ln 3} = \lim_{x\to 0^+} \dfrac{8^{\sqrt{x}} \ln 8}{3^{\sqrt{x}} \ln 3} = \dfrac{\ln 8}{\ln 3} \approx 1.8928$

15. $\lim_{x\to 0} \dfrac{\tan x - x}{\sin x - x} \overset{Ⓛ}{=} \lim_{x\to 0} \dfrac{\sec^2 x - 1}{\cos x - 1} \overset{Ⓛ}{=} \lim_{x\to 0} \dfrac{2\sec^2 x \tan x}{-\sin x} = \lim_{x\to 0}(-2\sec^3 x) = -2$

18. $\lim_{x\to 0} \dfrac{e^x-\ln(1+x)-1}{x^2} \overset{Ⓛ}{=} \lim_{x\to 0} \dfrac{e^x-(1+x)^{-1}}{2x} \overset{Ⓛ}{=} \lim_{x\to 0} \dfrac{e^x+(1+x)^{-2}}{2} = \dfrac{1+1}{2} = 1$

21. $\lim_{x\to 0^+} \dfrac{1-\cos x-x\sin x}{2-2\cos x-\sin^2 x} \overset{Ⓛ}{=} \lim_{x\to 0^+} \dfrac{\sin x-x\cos x-\sin x}{2\sin x-2\sin x\cos x} = \lim_{x\to 0^+} \dfrac{-x\cos x}{2\sin x-\sin 2x}$

$\overset{Ⓛ}{=} \lim_{x\to 0^+} \dfrac{x\sin x-\cos x}{2\cos x-2\cos 2x} = -\infty$ since $\begin{bmatrix} \to -1 \\ \to 0^+ \end{bmatrix}$

Note: $2\cos x-2\cos 2x > 0$ as $x\to 0^+$ since $x < 2x$ and cos is a decreasing function on $[0,\pi]$.

24. $\lim_{x\to 0^+} \dfrac{\int_0^x (\sqrt{t}\cos t)dt}{x^2} \overset{Ⓛ}{=} \lim_{x\to 0^+} \dfrac{\sqrt{x}\cos x}{2x} = \lim_{x\to 0^+} \dfrac{\cos x}{2\sqrt{x}} = \infty \qquad \begin{bmatrix} \to 1 \\ \to 0^+ \end{bmatrix}$

Problem Set 9.2 Other Indeterminate Forms

3. $\lim\limits_{x\to\infty} \dfrac{x^{10}}{e^x} \overset{L}{=} \lim\limits_{x\to\infty} \dfrac{10x^9}{e^x} \overset{L}{=} \lim\limits_{x\to\infty} \dfrac{90x^8}{e^x} \overset{L}{=} \lim\limits_{x\to\infty} \dfrac{720x^7}{e^x} \overset{L}{=} \lim\limits_{x\to\infty} \dfrac{5040x^6}{e^x}$
$\overset{L}{=} \lim\limits_{x\to\infty} \dfrac{30240x^5}{e^x} \overset{L}{=} \lim\limits_{x\to\infty} \dfrac{151200x^4}{e^x} \overset{L}{=} \lim\limits_{x\to\infty} \dfrac{604800x^3}{e^x} \overset{L}{=} \lim\limits_{x\to\infty} \dfrac{1814400x^2}{e^x}$
$\overset{L}{=} \lim\limits_{x\to\infty} \dfrac{3628800x}{e^x} \overset{L}{=} \lim\limits_{x\to\infty} \dfrac{10!}{e^x} = 0.$

6. $\lim\limits_{x\to 0^+} \dfrac{\ln \sin x}{\ln \tan x} \overset{L}{=} \lim\limits_{x\to 0^+} \dfrac{(\cos x)/(\sin x)}{(\sec^2 x)/(\tan x)} = \lim\limits_{x\to 0^+} (\cos^2 x) = 1.$

9. $\lim\limits_{x\to 0^+} \dfrac{\cot x}{\ln x} \overset{L}{=} \lim\limits_{x\to 0^+} \dfrac{-\csc^2 x}{1/x} = \lim\limits_{x\to 0^+} \dfrac{-x}{\sin^2 x} \overset{L}{=} \lim\limits_{x\to 0^+} \dfrac{-1}{2\sin x \cos x} = -\infty. \quad \begin{bmatrix} \to -1 \\ \to 0^+ \end{bmatrix}$

12. $\lim\limits_{x\to 0} x^2 \csc x = \lim\limits_{x\to 0} \dfrac{x^2}{\sin x} \overset{L}{=} \lim\limits_{x\to 0} \dfrac{2x}{\cos x} = \dfrac{0}{1} = 0.$

15. $\lim\limits_{x\to 0^+} (2x)^{x^2} = \lim\limits_{x\to 0^+} \exp(x^2 \ln 2x) = \exp\left(\lim\limits_{x\to 0^+} \dfrac{\ln 2 + \ln x}{x^{-2}}\right)$ [since exp is a cont. function]
$\overset{L}{=} \exp\left(\lim\limits_{x\to 0^+} \dfrac{1/x}{-2x^{-3}}\right) = \exp\left(\lim\limits_{x\to 0^+} \dfrac{x^2}{-2}\right) = \exp(0) = e^0 = 1.$

18. $\lim\limits_{x\to 0}\left(\csc^2 x - \dfrac{1}{x^2}\right) = \lim\limits_{x\to 0} \dfrac{x^2 \csc^2 x - 1}{x^2}.$ [Numerator approaches 0 since $x^2 \csc^2 x = (x/\sin x)^2$ approaches 1]
$\overset{L}{=} \lim\limits_{x\to 0} \dfrac{-2x^2 \csc^2 x \cot x + 2x \csc^2 x}{2x} = \lim\limits_{x\to 0}(-x \csc^2 x \cot x + \csc^2 x)$
$= \lim\limits_{x\to 0} \dfrac{\sin x - x \cos x}{\sin^3 x} \overset{L}{=} \lim\limits_{x\to 0} \dfrac{\cos x + x \sin x - \cos x}{3 \sin^2 x \cos x} = \lim\limits_{x\to 0} \dfrac{x}{3 \sin x \cos x}$
$= \lim\limits_{x\to 0} \dfrac{2x}{3 \sin 2x} \overset{L}{=} \lim\limits_{x\to 0} \dfrac{2}{6 \cos 2x} = \dfrac{2}{6} = \dfrac{1}{3}.$

21. 1 (a determinate form).

24. $\lim_{x\to 0}(\cos x)^{1/x^2} = \lim_{x\to 0}[\exp(1/x^2)\ln\cos x] = \exp\left(\lim_{x\to 0}\frac{\ln\cos x}{x^2}\right)$

$\stackrel{L}{=} \exp\left(\lim_{x\to 0}\frac{-\tan x}{2x}\right) \stackrel{L}{=} \exp\left(\lim_{x\to 0}\frac{-\sec^2 x}{2}\right) = \exp(-1/2) = e^{-1/2} \approx 0.6065$

27. $\lim_{x\to 0^+}(\sin x)^x = \lim_{x\to 0^+}\exp[x\ln(\sin x)] = \exp\left(\lim_{x\to 0^+}\frac{\ln\sin x}{x^{-1}}\right) \stackrel{L}{=} \exp\left(\lim_{x\to 0^+}\frac{\cot x}{-x^{-2}}\right)$

$= \exp\left(\lim_{x\to 0^+}\frac{x^2}{-\tan x}\right) \stackrel{L}{=} \exp\left(\lim_{x\to 0^+}\frac{2x}{-\sec^2 x}\right) = \exp\left(\frac{0}{-1}\right) = e^0 = 1$

30. $\lim_{x\to\infty}\left(1+\frac{1}{x}\right)^x = \lim_{y\to 0^+}(1+y)^{1/y}$ [letting $y = 1/x$]

$= \lim_{y\to 0^+}\exp\left(\frac{\ln(1+y)}{y}\right) = \exp\left(\lim_{y\to 0^+}\frac{\ln(1+y)}{y}\right) \stackrel{L}{=} \exp\left(\lim_{y\to 0^+}\frac{1}{1+y}\right) = e^1 \approx 2.7183$

33. $\lim_{x\to 0}(\cos x)^{1/x} = \lim_{x\to 0}\exp\left(\frac{\ln\cos x}{x}\right) = \exp\left(\lim_{x\to 0}\frac{\ln\cos x}{x}\right) \stackrel{L}{=} \exp\left(\lim_{x\to 0}\frac{-\tan x}{1}\right)$

$= \exp(0) = e^0 = 1$

36. $\lim_{x\to\infty}[\ln(x+1) - \ln(x-1)] = \lim_{x\to\infty}\ln\left(\frac{x+1}{x-1}\right) = \ln\left(\lim_{x\to\infty}\frac{x+1}{x-1}\right) \stackrel{L}{=} \ln\left(\lim_{x\to\infty}\frac{1}{1}\right)$

$= \ln(1) = 0$

39. $\lim_{x\to\infty}\dfrac{\int_1^x \sqrt{1+e^{-t}}\,dt}{x} \stackrel{L}{=} \lim_{x\to\infty}\dfrac{\sqrt{1+e^{-x}}}{1} = 1$

Problem Set 9.3   Improper Integrals: Infinite Limits

3. $\int xe^{-x^2}dx = \int\frac{e^u}{-2}du = \frac{e^u}{-2} + C = \frac{e^{-x^2}}{-2} + C$ $\boxed{u = -x^2;\ du = -2x\,dx}$

3. (continued)

Then $\int_4^\infty xe^{-x^2}dx = \lim_{b\to\infty} \int_4^b xe^{-x^2}dx = \lim_{b\to\infty}\left[\frac{e^{-x^2}}{-2}\right]_4^b = \frac{-1}{2}\lim_{b\to\infty}\left(e^{-b^2} - e^{-16}\right)$

$= \frac{-1}{2}(0 - e^{-16}) = \frac{1}{2e^{16}} \approx 5.6268 \times 10^{-8}$

6. $\int_1^b \frac{1}{\sqrt{3x}}dx = \int_1^b \frac{x^{-1/2}}{\sqrt{3}}dx = \left[\frac{2x^{1/2}}{\sqrt{3}}\right]_1^b = \frac{2\sqrt{b} - 2}{\sqrt{3}} \to \infty$ as $b\to\infty$

so the improper integral diverges.

9. The improper integral diverges. See Example 6 ($p = 0.99 < 1$).

12. $\int_1^b \frac{\ln x}{x}dx = \left[\frac{\ln^2 x}{2}\right]_1^b = \frac{\ln^2 b}{2} \to \infty$ as $b\to\infty$, so the integral diverges.

15. $\int_a^0 (2x-1)^{-3}dx = \left[\frac{(2x-1)^{-2}}{-4}\right]_a^0 = \frac{1 - (2a-1)^{-2}}{-4} = \frac{1}{4(2a-1)^2} - \frac{1}{4}$

[Use a substitution of $u = 2x-1$ if necessary to see the above.]

Then, $\int_{-\infty}^0 (2x-1)^{-3}dx = \lim_{a\to-\infty}\left(\frac{1}{4(2a-1)^2} - \frac{1}{4}\right) = \frac{-1}{4}$

18. $\int_0^b x(x^2+4)^{-2}dx = \left[\frac{(x^2+4)^{-1}}{-2}\right]_0^b = \frac{-1}{2(b^2+4)} + \frac{1}{8}$ [Use $u = x^2+4$ if needed to see it.]

Then, $\int_0^\infty x(x^2+4)^{-2}dx = \lim_{b\to\infty}\left(\frac{-1}{2(b^2+4)} + \frac{1}{8}\right) = \frac{1}{8}$

Therefore, $\int_{-\infty}^\infty x(x^2+4)^{-2}dx = 0$, since the integrand defines an odd function.

21. $\int_0^b \text{sech } x \, dx = \left[\tan^{-1}(\sinh x)\right]_0^b = \tan^{-1}(\sinh b) - 0$

Then, $\int_0^\infty \text{sech } x \, dx = \lim_{b \to \infty} \tan^{-1}(\sinh b) = \pi/2$

Thus, $\int_{-\infty}^\infty \text{sech } x \, dx = \pi$ since the integrand defines an even function.

24. $\int_0^b e^{-x}\sin x \, dx = \left[(1/2)e^{-x}(-\sin x - \cos x)\right]_0^b = \frac{1}{2} - \frac{\sin b + \cos b}{2e^b}$

[Use integration by parts twice.]

Then, $\int_0^\infty e^{-x}\sin x \, dx = \lim_{b \to \infty}\left(\frac{1}{2} - \frac{\sin b + \cos b}{2e^b}\right) = 1/2$

27. $F = k/x$ would be the force required to counteract the force due to gravity. At $x = 3960$ miles, $F = m$ lbs (weight of object at surface of earth is m lbs), so $k = 3960m$, and the force formula for the object is $F = 3960m/x$. Therefore, the work required to send the object out of earth's gravitational field would be

$\int_{3960}^\infty (3960m/x) dx = \lim_{b \to \infty} \int_{3960}^b (3960m/x) dx = \lim_{b \to \infty}\left[3960m \, \ell n x\right]_{3960}^b$

$= \lim_{b \to \infty} (3960m \, \ell n b - 3960m \, \ell n 3960) = \infty$

30. $FP = \int_0^\infty e^{-0.08t}(100000 + 1000t) dt = 1250000 + 1000 \int_0^\infty t e^{-0.08t} dt$

[The 1250000 is for the first term and is the answer for Problem 29.]

Use integration by parts to obtain

$\int_0^b t e^{-0.08t} dt = \left[\frac{-(1+0.08t)}{0.0064 e^{0.08t}}\right]_0^b = \frac{1}{0.0064}\left(1 - \frac{(1+0.08b)}{e^{0.08b}}\right)$

Problem Set 9.3

30. (continued)

Then, $\int_0^\infty te^{-0.08t}dt = \frac{1}{0.0064} \lim_{b\to\infty}\left(1 - \frac{(1+0.08b)}{e^{0.08b}}\right) = \frac{1}{0.0064}$

Therefore, FP = 1250000 + 1000(1/0.0064) = 1406250 [dollars].

33. (a) $\int_0^b \sin x\, dx = \left[-\cos x\right]_0^b = -\cos b + 1$; $\int_0^\infty \sin x\, dx$ and hence $\int_{-\infty}^\infty \sin x\, dx$

diverge since $\lim_{b\to\infty}(-\cos b + 1)$ doesn't exist.

(b) $\lim_{a\to\infty}\int_{-a}^a \sin x\, dx = \lim_{a\to\infty}(0) = 0$ [since sinx defines an odd function]

Problem Set 9.4   Improper Integrals: Infinite Integrands

3. The asymptotic discontinuity is at x=1.

$\int_0^1 \frac{1}{\sqrt{1-x^2}}\, dx = \lim_{t\to 1^-}\int_0^t \frac{1}{\sqrt{1-x^2}}\, dx = \lim_{t\to 1^-}\left[\sin^{-1}x\right]_0^t = \lim_{t\to 1^-}(\sin^{-1}t - 0) = \pi/2$

6. The asymptotic discontinuity is at x=0.

$\int_{-1}^{27} x^{-2/3}dx = \int_{-1}^0 x^{-2/3}dx + \int_0^{27} x^{-2/3}dx = \lim_{t\to 0^-}\int_{-1}^t x^{-2/3}dx + \lim_{s\to 0^+}\int_s^{27} x^{-2/3}dx$

$= \lim_{t\to 0^-}\left[3x^{1/3}\right]_{-1}^t + \lim_{s\to 0^+}\left[3x^{1/3}\right]_s^{27} = \lim_{t\to 0^-}(3t^{1/3} + 3) + \lim_{s\to 0^+}(9 - 3s^{1/3})$

$= 3 + 9 = 12$

9. The asymptotic discontinuity is at x=-2. Use a u = (x^2-4) substitution to find that an antiderivative of $x(x^2-4)^{-2/3}$ is $(3/2)(x^2-4)^{1/3}$.

$\int_{-3}^0 x(x^2-4)^{-2/3}dx = \int_{-3}^{-2} x(x^2-4)^{-2/3}dx + \int_{-2}^0 x(x^2-4)^{-2/3}dx$

9. (continued)

$$= \lim_{t \to -2^-}\left[(3/2)(x^2-4)^{1/3}\right]_{-3}^{t} + \lim_{s \to -2^+}\left[(3/2)(x^2-4)^{1/3}\right]_{s}^{0}$$

$$= \lim_{t \to -2^-}\frac{3[(t^2-4)^{1/3} - (5)^{1/3}]}{2} + \lim_{s \to -2^+}\frac{3[(-4)^{1/3} - (s^2-4)^{1/3}]}{2}$$

$$= \frac{-3[(5)^{1/3} + (4)^{1/3}]}{2} \approx -4.9461$$

12. The asymptotic discontinuity is at x=1.

$$\int\frac{3}{(x+2)(x-1)}dx = \int\left(\frac{-1}{x+2} + \frac{1}{x-1}\right)dx = -\ell n|x+2| + \ell n|x-1| + C = \ell n\left|\frac{x-1}{x+2}\right| + C$$

$$\int_0^1 \frac{3}{(x+2)(x-1)}dx = \lim_{t \to 1^-}\left[\ell n\left|\frac{x-1}{x+2}\right|\right]_0^t = \lim_{t \to 1^-}\left(\ell n\left|\frac{t-1}{t+2}\right| - \ell n(1/2)\right) = -\infty$$

Therefore, $\int_0^1 \frac{3}{(x+2)(x-1)}dx$ diverges; hence, $\int_0^2 \frac{3}{(x+2)(x-1)}dx$ diverges.

15. The asymptotic discontinuity is at x=0.

$$\int_0^1 \frac{\ell nx}{x} dx = \lim_{s \to 0^+}\int_s^1 \frac{\ell nx}{x} dx = \lim_{s \to 0^+}\left[\frac{\ell n^2 x}{2}\right]_s^1 = \lim_{s \to 0^+}\left(0 - \frac{\ell n^2 s}{2}\right) = -\infty$$

so the improper integral diverges. [Use u = ℓnx substitution at the second step if necessary to see the result.]

18. The asymptotic discontinuities are at x=-2 and x=2.

$$\int_0^2 \frac{1}{4-x^2} dx = \lim_{t \to 2^-}\int_0^t \frac{1}{4-x^2} dx = \lim_{t \to 2^-}\int_0^t \left(\frac{1/4}{2+x} + \frac{1/4}{2-x}\right)dx$$

$$= \lim_{t \to 2^-}\left[(1/4)[\ell n(2+x) - \ell n(2-x)]\right]_0^t = \lim_{t \to 2^-}\left[\frac{1}{4}\ell n\left(\frac{2+t}{2-t}\right) - 0\right] = \infty$$

Therefore, $\int_{-2}^2 \frac{1}{4-x^2} dx$ diverges.

Problem Set 9.4

21. Area $= \int_0^8 (x-8)^{-2/3}dx = \lim_{t \to 8^-} \int_0^t (x-8)^{-2/3}dx$

$= \lim_{t \to 8^-} \left[3(x-8)^{1/3} \right]_0^t = \lim_{t \to 8^-} [3(t-8)^{1/3} + 6]$

$= 6$ square units.

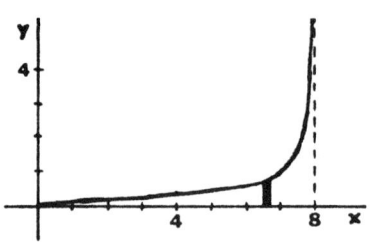

24. There is an asymptotic discontinuity at $x=0$.

$\int_0^b \ell nx \, dx = \lim_{s \to 0^+} \int_s^b \ell nx \, dx = \lim_{s \to 0^+} \left[x(\ell nx - 1) \right]_s^b$

$= \lim_{s \to 0^+} [b(\ell nb - 1) - s(\ell ns - 1)] = b(\ell nb - 1) - \lim_{s \to 0^+} \frac{\ell ns - 1}{s^{-1}}$

$\overset{\text{\textcircled{L}}}{=} b(\ell nb - 1) - \lim_{s \to 0^+} \frac{s^{-1}}{-s^{-2}} = b(\ell nb - 1) - \lim_{s \to 0^+}(-s) = b(\ell nb - 1)$.

Thus, the integral equals zero if $\ell nb = 1$; i.e., $b = e$. [Note that b cannot be 0.]

27. $\int_1^\infty e^{-x}dx = \lim_{b \to \infty} \int_1^b e^{-x}dx = \lim_{b \to \infty} \left[-e^{-x} \right]_1^b = \lim_{b \to \infty} \left(\frac{-1}{e^b} + \frac{1}{e} \right) = 1/e$.

The fact that this improper integral converges, along with the hint and the convergence test, implies that the integral in question converges.

30. If $n \geq 1$, the integral is not an improper integral; and the integral exists since the integrand defines a continuous function for $x \in [0,1]$.

If $0 < n < 1$, (1) $0 < e^{-x} \leq 1$ for x in $[0,1]$, so $0 < x^{n-1}e^{-x} \leq x^{n-1}$,

(2) $\int_0^1 x^{n-1}dx = \lim_{s \to 0^+} \left[\frac{x^n}{n} \right]_s^1 = \lim_{s \to 0^+} \left(\frac{1}{n} - \frac{s^n}{n} \right) = \frac{1}{n}$.

(1) and (2), along with the convergence test (Problem 28), implies that the integral in question converges.

Problem Set 9.5 Chapter Review Problems

True-False Quiz

3. False. It equals 1,000,000. (Use l'Hôpital's Rule four times.)

6. False. C^{ex}: For $a=0$, $f(x) = (1+x)$, $g(x) = 1/x$, the limit is e.

9. True. Let $-M < 0$ be given. Then $g(x) > 2M$, and $-3/2 < f(x) < -1/2$ for x in some deleted neighborhood of a, so $f(x)g(x) < -M$ for x in the deleted neighborhood.

12. True. Determinate form.

15. True. Use l'Hôpital's Rule n times, where n is the degree of p(x).

18. False. See Example 4 of Section 9.4. [p = 1.001 > 1]

21. True. $\int_{-a}^{0} f(x)dx = \int_{0}^{a} f(x)dx$ for all $a \geq 0$, so $\int_{-\infty}^{0} f(x)dx$ converges.

24. True. See Problems 26 and 27, Section 9.4.

Miscellaneous Problems

3. $\lim\limits_{x \to 0} \dfrac{\sin x - \tan x}{x^2/3} \overset{\text{L}}{=} \lim\limits_{x \to 0} \dfrac{\cos x - \sec^2 x}{2x/3} \overset{\text{L}}{=} \lim\limits_{x \to 0} \dfrac{-\sin x - 2\sec^2 x \tan x}{2/3} = 0$

6. $\lim\limits_{x \to 1^-} \dfrac{\ln(1-x)}{\cot \pi x} \overset{\text{L}}{=} \lim\limits_{x \to 1^-} \dfrac{-1/(1-x)}{-\pi \csc^2 \pi x} = \lim\limits_{x \to 1^-} \dfrac{\sin^2 \pi x}{\pi(1-x)} \overset{\text{L}}{=} \lim\limits_{x \to 1^-} \dfrac{2\pi \sin \pi x \cos \pi x}{-\pi} = 0$

9. $\lim\limits_{x \to 0^+} (\sin x)^{1/x} = \lim\limits_{x \to 0^+} e^{(1/x)\ln(\sin x)} = \lim\limits_{x \to 0^+} \exp\left(\dfrac{\ln(\sin x)}{x}\right) = 0$

since $\lim\limits_{x \to 0^+} \dfrac{\ln(\sin x)}{x} = -\infty$

12. $\lim\limits_{x \to 0} (1+\sin x)^{2/x} = \lim\limits_{x \to 0} e^{(2/x)\ln(1+\sin x)} = \lim\limits_{x \to 0} \exp \dfrac{2\ln(1+\sin x)}{x}$

$= \exp\left(\lim\limits_{x \to 0} \dfrac{2\ln(1+\sin x)}{x}\right) \overset{\text{L}}{=} \exp\left(\lim\limits_{x \to 0} \dfrac{2\cos x}{1+\sin x}\right) = \exp(2) = e^2 \approx 7.3891$

15. $\lim\limits_{x \to 0^+} \left(\dfrac{1}{\sin x} - \dfrac{1}{x}\right) = \lim\limits_{x \to 0^+} \dfrac{x - \sin x}{x \sin x} \overset{\text{L}}{=} \lim\limits_{x \to 0^+} \dfrac{1 - \cos x}{x \cos x + \sin x}$

$\overset{\text{L}}{=} \lim\limits_{x \to 0^+} \dfrac{\sin x}{-x \sin x + \cos x + \cos x} = \dfrac{0}{2} = 0$

18. $\lim\limits_{x \to \pi/2} [x \tan x - (\pi/2)\sec x] = \lim\limits_{x \to \pi/2} \left(\dfrac{x \sin x}{\cos x} - \dfrac{\pi}{2\cos x}\right) = \lim\limits_{x \to \pi/2} \dfrac{2x \sin x - \pi}{2 \cos x}$

$\overset{\text{L}}{=} \lim\limits_{x \to \pi/2} \dfrac{2x \cos x + 2 \sin x}{-2 \sin x} = \dfrac{0 + 2}{-2} = -1$

21. $\int_{-\infty}^{1} e^{2x} dx = \lim\limits_{a \to -\infty} \int_{a}^{1} e^{2x} dx = \lim\limits_{a \to -\infty} \dfrac{e^{2x}}{2} \Big|_a^1 = \lim\limits_{a \to -\infty} \left[\dfrac{e^2}{2} - \dfrac{e^{2a}}{2}\right] = \dfrac{e^2 - 0}{2} = \dfrac{e^2}{2}$

≈ 3.6945

24. The integrand has an asymptotic discontinuity at x=1.

$$\int \frac{1}{x(\ln x)^{1/5}} dx = \int u^{-1/5} du = \frac{5u^{4/5}}{4} + C = \frac{5(\ln x)^{4/5}}{4} + C \qquad \boxed{\begin{array}{l} u = \ln x \\ du = (1/x)dx \end{array}}$$

Therefore, $\int_{1/2}^{2} \frac{1}{x(\ln x)^{1/5}} dx = \int_{1/2}^{1} \frac{1}{x(\ln x)^{1/5}} dx + \int_{1}^{2} \frac{1}{x(\ln x)^{1/5}} dx$

$$= \lim_{t \to 1^-} \int_{1/2}^{t} \frac{1}{x(\ln x)^{1/5}} dx + \lim_{s \to 1^+} \int_{s}^{2} \frac{1}{x(\ln x)^{1/5}} dx$$

$$= \lim_{t \to 1^-} \left[\frac{5(\ln x)^{4/5}}{4} \right]_{1/2}^{t} + \lim_{s \to 1^+} \left[\frac{5(\ln x)^{4/5}}{4} \right]_{s}^{2}$$

$$= \lim_{t \to 1^-} \frac{5(\ln t)^{4/5} - 5(\ln[1/2])^{4/5}}{4} + \lim_{s \to 1^+} \frac{5(\ln 2)^{4/5} - 5(\ln s)^{4/5}}{4}$$

$$= \frac{-5(\ln[1/2])^{4/5}}{4} + \frac{5(\ln 2)^{4/5}}{4} = \frac{-5(-\ln 2)^{4/5} + 5(\ln 2)^{4/5}}{4}$$

$$= \frac{-5(\ln 2)^{4/5} + 5(\ln 2)^{4/5}}{4} = 0$$

27. There is an asymptotic discontinuity at x=-3/2.

$$\int_{-3/2}^{0} \frac{1}{2x+3} dx = \lim_{s \to -3/2^+} \int_{s}^{0} \frac{1}{2x+3} dx \qquad \boxed{\begin{array}{l} \text{Let} \quad u = 2x+3 \\ \text{Then } du = 2dx \\ x=0 \Rightarrow u=3 \\ x=s \Rightarrow u=2s+3 \end{array}}$$

$$= \lim_{s \to -3/2^+} \int_{2s+3}^{3} \frac{1}{u} du = \lim_{v \to 0^+} \int_{v}^{3} \frac{1}{u} du \qquad [\text{letting } v = 2s+3]$$

which diverges [See Example 3 of Section 9.4]. Therefore, the improper integral in question diverges.

30. $\int_{0}^{\infty} e^{-x/2} dx = \lim_{b \to \infty} \int_{0}^{b} e^{-x/2} dx = \lim_{b \to \infty} \left[-2e^{-x/2} \right]_{0}^{b} = \lim_{b \to \infty} \left(\frac{-2}{e^{b/2}} + 2 \right) = 0 + 2 = 0$

Problem Set 9.5

33. $\int \frac{x}{x^2+1} dx = \int \frac{1}{2} \frac{1}{u} du = \frac{\ln|u|}{2} + C = \frac{\ln(x^2+1)}{2} + C$ $\boxed{\text{Let } u = x^2+1 \\ \text{Then } du = 2xdx}$

Then, $\int_0^\infty \frac{x}{x^2+1} dx = \lim_{b\to\infty} \int_0^b \frac{x}{x^2+1} dx = \lim_{b\to\infty} \left[\frac{\ln(x^2+1)}{2}\right]_0^b$

$= \lim_{b\to\infty} \left(\frac{\ln(b^2+1)}{2} - 0\right) = \infty$

Therefore, $\int_0^\infty \frac{x}{x^2+1} dx$ diverges, so $\int_{-\infty}^\infty \frac{x}{x^2+1} dx$ diverges.

36. For $x > 1$, $\ln x < x < e^x$, so $\ln x < e^x$.

Therefore, $0 < \frac{\ln x}{e^{2x}} < \frac{e^x}{e^{2x}} = e^{-x}$

$\int_1^\infty e^{-x} dx = \lim_{b\to\infty} \int_1^b e^{-x} dx = \lim_{b\to\infty} \left[-e^{-x}\right]_1^b = \lim_{b\to\infty} (-e^{-b} + e^{-1}) = e^{-1} \approx 0.3679$

Therefore, $\int_1^\infty \frac{\ln x}{e^{2x}} dx$ converges, by a comparison test.

CHAPTER 10 NUMERICAL METHODS, APPROXIMATIONS

Problem Set 10.1 Taylor's Approximation to Functions

3. 0.1193318 (both ways)

6. 5204.8596

9. $f(x) = \sin 2x$ $f(0) = 0$
    $f'(x) = 2\cos 2x$ $f'(0) = 2$
    $f''(x) = -4\sin 2x$ $f''(0) = 0$
    $f^{(3)}(x) = -8\cos 2x$ $f^{(3)}(0) = -8$
    $f^{(4)}(x) = 16\sin 2x$ $f^{(4)}(0) = 0$

$$P_4(x) = 0 + 2x + \frac{0x^2}{2!} + \frac{-8x^3}{3!} + \frac{0x^4}{4!} = 2x - \frac{4x^3}{3}$$

$$f(0.23) \approx P_4(0.23) \approx 0.443777333$$

12. $f(x) = (1+x)^{1/2}$ $f(0) = 1$
    $f'(x) = (1/2)(1+x)^{-1/2}$ $f'(0) = 1/2$
    $f''(x) = (-1/4)(1+x)^{-3/2}$ $f''(0) = -1/4$
    $f^{(3)}(x) = (3/8)(1+x)^{-5/2}$ $f^{(3)}(0) = 3/8$
    $f^{(4)}(x) = (-15/16)(1+x)^{-7/2}$ $f^{(4)}(0) = -15/16$

$$P_4(x) = 1 + (1/2)x + \frac{(-1/4)x^2}{2!} + \frac{(3/8)x^3}{3!} + \frac{(-15/16)x^4}{4!}$$

$$= 1 + \frac{x}{2} - \frac{x^2}{8} + \frac{x^3}{16} - \frac{5x^4}{128}$$

$$f(0.23) \approx P_4(0.23) \approx 1.109038625$$

15. Let $f(x) = e^x$. $f^{(n)}(x) = e^x$ and $f^{(n)}(2) = e^2$ for each $n = 0, 1, 2, \cdots$.

$$P_3(x) = e^2 + e^2(x-2) + \frac{e^2(x-2)^2}{2!} + \frac{e^2(x-2)^3}{3!}$$

18. Let $f(x) = \sec x$

$f'(x) = \sec x \tan x$

$f''(x) = (\sec x)(\sec^2 x) + (\tan x)(\sec x \tan x) = \sec x(\sec^2 x + \tan^2 x)$
$= \sec x(1 + 2\tan^2 x)$ [since $\sec^2 x = 1 + \tan^2 x$]

$f^{(3)}(x) = (\sec x)(4\tan x \sec^2 x) + (1 + 2\tan^2 x)(\sec x \tan x)$
$= \sec x \tan x(4[1+\tan^2 x] + [1+2\tan^2 x]) = \sec x \tan x(5 + 6\tan^2 x)$

$f(\pi/6) = 2/\sqrt{3}$

$f'(\pi/6) = (2/\sqrt{3})(1/\sqrt{3}) = 2/3$

$f''(\pi/6) = (2/\sqrt{3})[1 + 2(1/3)] = 10/3\sqrt{3}$

$f^{(3)}(\pi/6) = (2/\sqrt{3})(1/\sqrt{3})[5 + 6(1/3)] = 14/3$

$P_3(x) = (2/\sqrt{3}) + (2/3)(x - \pi/6) + \dfrac{(10/3\sqrt{3})(x-\pi/6)^2}{2!} + \dfrac{(14/3)(x-\pi/6)^3}{3!}$

$= \dfrac{2}{\sqrt{3}} + \dfrac{2(x-\pi/6)}{3} + \dfrac{5(x-\pi/6)^2}{3\sqrt{3}} + \dfrac{7(x-\pi/6)^3}{9}$

21. Let $f(x) = x^3 - 2x^2 + 3x + 5$ $\quad f(1) = 7$

$f'(x) = 3x^2 - 4x + 3$ $\quad f'(1) = 2$

$f''(x) = 6x - 4$ $\quad f''(1) = 2$

$f^{(3)}(x) = 6$ $\quad f^{(3)}(1) = 6$

$P_3(x) = 7 + 2(x-1) + \dfrac{2(x-1)^2}{2!} + \dfrac{6(x-1)^3}{3!}$

$= 7 + (2x-2) + (x^2 - 2x + 1) + (x^3 - 3x^2 + 3x - 1) = x^3 - 2x^2 + 3x + 5 = f(x)$

24. Let $f(x) = \sin x$ $\qquad$ cyclic $\qquad$ $f(0) = 0$

$f'(x) = \cos x$ $\qquad$ pattern $\qquad$ $f'(0) = 1$

$f''(x) = -\sin x$ $\qquad\qquad\qquad$ $f''(0) = 0$

$f^{(3)}(x) = -\cos x$ $\qquad\qquad\qquad$ $f^{(3)}(0) = -1$

$f^{(4)}(x) = \sin x$ $\qquad\qquad\qquad$ $f^{(4)}(0) = 0$

$P_n(x) = x - \dfrac{x^3}{3!} + \dfrac{x^5}{5!} - \dfrac{x^7}{7!} + \cdots + \dfrac{(-1)^n x^n}{n!}$ (n odd)

$P_5(x) = x - \dfrac{x^3}{3!} + \dfrac{x^5}{5!}$

24. (continued)

(a) $\sin(0.1) \approx P_5(0.1) \approx 0.099833416$ $\sin(0.1) \approx 0.099833416$
(b) $\sin(0.5) \approx P_5(0.5) \approx 0.479427083$ $\sin(0.5) \approx 0.479425538$
(c) $\sin(1) \approx P_5(1) \approx 0.841666667$ $\sin(1) \approx 0.841470984$
(d) $\sin(10) \approx P_5(10) \approx 676.6666667$ $\sin(10) \approx -0.544021111$

The values to the right were determined with a calculator. Notice that $P_5(x)$ gives a good approximation for x near 0, and that it worsens as x gets farther from 0. It is still fairly good if $|x| \leq 1$.

Problem Set 10.2 Estimating the Errors

3. $\left|\dfrac{2}{\cos c}\right| = \dfrac{2}{|\cos c|} \leq \dfrac{2}{0.5} = 4$ (For c in $[2\pi/3, \pi]$, the smallest $|\cos c|$ can be is 0.5.)

6. $\left|\dfrac{\sin c}{c+1}\right| = \dfrac{\sin c}{c+1} \leq \dfrac{\sin 1}{1} < 0.8415$ for c in [0,1]

9. Let $f(x) = \ln(1+x)$. $f'(x) = (1+x)^{-1}$, $f''(x) = -(1+x)^{-2}$,
$f^{(3)}(x) = 2(1+x)^{-3}$, $f^{(4)}(x) = -3!(1+x)^{-4}$, $f^{(5)}(x) = 4!(1+x)^{-5}$,
$f^{(6)}(x) = -5!(1+x)^{-6}$, $f^{(7)}(x) = 6!(1+x)^{-7}$.

$R_6(x) = \dfrac{f^{(7)}(c) \, x^7}{7!} = \dfrac{6! x^7}{7!(1+c)^7} = \dfrac{x^7}{7(1+c)^7}$ for some c between 0 and x.

$|R_6(0.5)| = \dfrac{(0.5)^7}{7(1+c)^7} < \dfrac{(0.5)^7}{7} < 0.001117$ for the c between 0 and 0.5.

12. Let $f(x) = (x-2)^{-1}$. $f'(x) = -(x-2)^{-2}$, $f''(x) = 2(x-2)^{-3}$,
$f^{(3)}(x) = -3!(x-2)^{-4}, \ldots, f^{(7)}(x) = -7!(x-2)^{-8}$.

$R_6(x) = \dfrac{f^{(7)}(c)(x-1)^7}{7!} = \dfrac{-7!(x-1)^7}{7!(c-2)^8} = \dfrac{-(x-1)^7}{(c-2)^8}$ for some c between x and 1.

$|R_6(0.5)| = \dfrac{(0.5)^7}{(c-2)^8} < (0.5)^7 < 0.007813$ for the c between 0.5 and 1.

15. Let $f(x) = (1+x)^{-1/2}$ $f(0) = 1$

$f'(x) = (-1/2)(1+x)^{-3/2}$ $f'(0) = -1/2$

$f''(x) = (3/4)(1+x)^{-5/2}$ $f''(0) = 3/4$

$f^{(3)}(x) = (-15/8)(1+x)^{-7/2}$ $f^{(3)}(0) = -15/8$

$f^{(4)}(x) = (105/16)(1+x)^{-9/2}$

$$P_3(x) = 1 + (-1/2)x + \frac{(3/4)x^2}{2!} + \frac{(-15/8)x^3}{3!} = 1 - \frac{x}{2} + \frac{3x^2}{8} - \frac{5x^3}{16}$$

$|R_3(x)| = \dfrac{f^{(4)}(c) x^4}{4!}$ for some c between 0 and x, hence in $(-0.05, 0.05)$,

$$= \frac{105 x^4}{16(1+c)^{9/2}(4!)} \leq \frac{35(0.05)^4}{128(0.95)^{9/2}} < 0.0000022.$$

18. Let $f(x) = \cos x$. $f'(x) = -\sin x$, $f''(x) = -\cos x$, ..., $f^{(6)}(x) = -\cos x$.

$|R_5(x)| = \dfrac{|f^{(6)}(c)x^6|}{6!} = \dfrac{(\cos c)x^6}{6!} \leq \dfrac{(1)(1)}{6!} = \dfrac{1}{720}$ for some c between 0 and x, x in $[0,1]$.

Then, $1 - \dfrac{x^2}{2} + \dfrac{x^4}{24} - \dfrac{1}{720} \leq \cos x \leq 1 - \dfrac{x^2}{2} + \dfrac{x^4}{24} + \dfrac{1}{720}$ for x in $[0,1]$.

$$\int_0^1 \left(1 - \frac{x^2}{2} + \frac{x^4}{24} - \frac{1}{720}\right) dx \leq \int_0^1 \cos x \, dx \leq \int_0^1 \left(1 - \frac{x^2}{2} + \frac{x^4}{24} + \frac{1}{720}\right) dx$$

$$\left[x - \frac{x^3}{6} + \frac{x^5}{120} - \frac{x}{720}\right]_0^1 \leq \int_0^1 \cos x \, dx \leq \left[x - \frac{x^3}{6} + \frac{x^5}{120} + \frac{x}{720}\right]_0^1$$

Thus, $\int_0^1 \cos x \, dx \approx 1 - \dfrac{1}{6} + \dfrac{1}{120} \approx 0.84167$; Error $\leq \dfrac{1}{720} \approx 0.00138$.

21. Let $f(x) = \sin x$ $f(\pi/4) = \sqrt{2}/2$
$f'(x) = \cos x$ $f'(\pi/4) = \sqrt{2}/2$
$f''(x) = -\sin x$ $f''(\pi/4) = -\sqrt{2}/2$
$f^{(3)}(x) = -\cos x$ $f^{(3)}(\pi/4) = -\sqrt{2}/2$
$f^{(4)}(x) = \sin x$

$\sin 43° = \sin(43\pi/180)$ and $(43\pi/180 - \pi/4) = -\pi/90$

$$P_3(x) = (\sqrt{2}/2) + (\sqrt{2}/2)(x-\pi/4) + \frac{(-\sqrt{2}/2)(x-\pi/4)^2}{2} + \frac{(-\sqrt{2}/2)(x-\pi/4)^3}{3!}$$

$$\sin(43\pi/180) \approx \frac{\sqrt{2}}{2} + \frac{\sqrt{2}(-\pi/90)}{2} - \frac{\sqrt{2}(-\pi/90)^2}{4} - \frac{\sqrt{2}(-\pi/90)^3}{12} \approx 0.681998316$$

$$R_3(x) = \frac{f^{(4)}(c)(x-\pi/4)^4}{4!} = \frac{(\sin c)(x-\pi/4)^4}{4!}$$ for some c between x and $\pi/4$.

$$|R_3(43\pi/180)| \leq \frac{(\sqrt{2}/2)(-\pi/90)^4}{24} < 0.0000000438$$

Problem Set 10.3 Numerical Integration

3. $f(x) = \sqrt{x}$

$h = \frac{4-0}{8} = 0.5$

i	x_i	$f(x_i)$	$2f(x_i)$	$4f(x_i)$
0	0	0		
1	0.5	$\sqrt{0.5}$	$2\sqrt{0.5}$	$4\sqrt{0.5}$
2	1	1	2	
3	1.5	$\sqrt{1.5}$	$2\sqrt{1.5}$	$4\sqrt{1.5}$
4	2	$\sqrt{2}$	$2\sqrt{2}$	
5	2.5	$\sqrt{2.5}$	$2\sqrt{2.5}$	$4\sqrt{2.5}$
6	3	$\sqrt{3}$	$2\sqrt{3}$	
7	3.5	$\sqrt{3.5}$	$2\sqrt{3.5}$	$4\sqrt{3.5}$
8	4	2		

Trapezoidal: $\frac{0.5}{2}[0 + 2(\sqrt{0.5} + 1 + \sqrt{1.5} + \sqrt{2} + \sqrt{2.5} + \sqrt{3} + \sqrt{3.5}) + 2]$
≈ 5.2650

Parabolic: $\frac{0.5}{3}[0 + 4\sqrt{0.5} + 2 + 4\sqrt{1.5} + 2\sqrt{2} + 4\sqrt{2.5} + 2\sqrt{3} + 4\sqrt{3.5} + 2]$
≈ 5.3046

Exact: $\int_0^4 x^{1/2} dx = \left[\frac{2x^{3/2}}{3}\right]_0^4 = \frac{16}{3} = 5.3333\ldots$

6. n=2: $h = \frac{\pi-0}{2} = \pi/2$

```
├────┼────┤
0   π/2   π
```

$\frac{\pi/2}{3}[\sin 0 + 4\sin(\pi/2) + \sin\pi] = 2\pi/3 \approx 2.0943951$

n=6: $h = \frac{\pi-0}{6} = \pi/6$

```
├────┼────┼────┼────┼────┼────┤
0   π/6  2π/6  3π/6  4π/6  5π/6   π
```

$\frac{\pi/6}{3}[\sin 0 + 4\sin(\frac{\pi}{6}) + 2\sin(\frac{\pi}{3}) + 4\sin(\frac{\pi}{2}) + 2\sin(\frac{2\pi}{3}) + 4\sin(\frac{5\pi}{6}) + \sin(\pi)]$

$= \frac{\pi[0 + 2 + \sqrt{3} + 4 + \sqrt{3} + 2 + 0]}{18} = \frac{\pi(4 + \sqrt{3})}{9} \approx 2.0008632$

n=12: $h = \frac{\pi-0}{12}$

```
├──┼──┼──┼──┼──┼──┼──┼──┼──┼──┼──┼──┤
0  π  2π 3π 4π 5π 6π 7π 8π 9π 10π 11π 12π
   12 12 12 12 12 12 12 12 12 12  12  12
```

$\frac{\pi/12}{3}[\sin 0 + 4\sin(\pi/12) + 2\sin(\pi/6) + 4\sin(\pi/4) + 2\sin(\pi/3) +$

$+ 4\sin(5\pi/12) + 2\sin(\pi/2) + 4\sin(7\pi/12) + 2\sin(2\pi/3) +$

$+ 4\sin(3\pi/4) + 2\sin(5\pi/6) + 4\sin(11\pi/12) + \sin\pi] \approx 2.000526$

9. Let $f(x) = e^{-x^2}$. $f'(x) = -2xe^{-x^2}$, $f''(x) = 2e^{-x^2}(2x^2-1)$.

$|E_n| = \frac{(1.2-0)^3|f''(c)|}{12n^2} = \frac{(1.728)2e^{-c^2}|2c^2-1|}{12n^2}$ for some c in (0,1.2)

$\leq \frac{(1.728)(2)(1)(1.88)}{12n^2} = \frac{0.54144}{n^2} \leq 0.01$ if $n^2 \geq 54.144$, or $n \geq 8$.

$h = \frac{1.2-0}{8} = 0.15$

```
├────┼────┼────┼────┼────┼────┼────┼────┤
0  0.15  0.3  0.45  0.6  0.75  0.9  1.05  1.2
```

$\frac{0.15}{2}[e^0 + 2(e^{-0.0225} + e^{-0.09} + e^{-0.2025} + e^{-0.36} + e^{-0.5625} + e^{-0.81} +$

$+ e^{-1.1025}) + e^{-1.44}] \approx 0.8057$

12. Let $f(x) = \sin\sqrt{x}$. $f'(x) = \frac{\cos\sqrt{x}}{2\sqrt{x}}$, $f''(x) = \frac{\sin\sqrt{x}}{-4x} + \frac{\cos\sqrt{x}}{-4x^{3/2}}$

$$|E_n| = \frac{(2-1)^3}{12n^2}\left(\frac{\sin\sqrt{c}}{4c} + \frac{\cos\sqrt{c}}{4c^{3/2}}\right) \text{ for some } c \text{ in } (1,2)$$

$$< \frac{1}{12n^2}\left(\frac{\sin\sqrt{2}}{4} + \frac{\cos\sqrt{1}}{4}\right) < \frac{0.04}{n^2} \le 0.01 \text{ if } n^2 \ge 4, \text{ or } n \ge 2.$$

$h = \frac{2-1}{2} = 1/2$

|---|---|---|
| 1 | 1.5 | 2 |

$\frac{1/2}{2}[\sin\sqrt{1} + 2\sin\sqrt{1.5} + \sin\sqrt{2}] \approx 0.9277$

15. $\int_{m-h}^{m+h} (ax^2+bx+cx)dx = \left[\frac{ax^3}{3} + \frac{bx^2}{2} + cx\right]_{m-h}^{m+h}$

$$= \frac{a(m+h)^3}{3} + \frac{b(m+h)^2}{2} + c(m+h) - \frac{a(m-h)^3}{3} - \frac{b(m-h)^2}{2} - c(m-h)$$

$$= \frac{6am^2h + 2ah^3 + 6bmh + 6ch}{3} = \frac{h}{3}[a(6m^2+2h^2) + b(6m) + 6c]$$

$\frac{h}{3}[f(m-h) + 4f(m) + f(m+h)]$

$$= \frac{h}{3}\Big([a(m-h)^2+b(m-h)+c] + [4(am^2+bm+c)] + [a(m+h)^2+b(m+h)+c]\Big)$$

$$= \frac{h}{3}(6am^2 + 2ah^2 + 6bm + 6c) = \frac{h}{3}[a(6m^2+2h^2) + b(6m) + 6c]$$

18. Let $f(x) = x^{-1}$. $f'(x) = -x^{-2}$, $f''(x) = 2x^{-3}$, $f^{(3)}(x) = -6x^{-4}$.

$$|E_n| = \frac{(2-1)^5}{180n^4} \frac{24}{c^2} \text{ for some } c \text{ in } (1,2)$$

$$< \frac{24}{180n^4(1)} = \frac{2}{15n^4} \le 10^{-10} \text{ if } n^4 \ge \frac{10^{10}}{7.5}, \text{ or } n \ge 192.$$

Problem Set 10.3

Problem Set 10.4 Solving Equations Numerically

3. Let $f(x) = \cos x - e^{-x}$ on $[1,2]$. $f'(x) = -\sin x + e^{-x}$. Note that f is decreasing on $[1,2]$ since $-\sin x$ is between -0.84 and -1 while e^{-x} is between 0.37 and 0.13, so $f'(x) < 0$ on $[1,2]$.

$f(1) = \cos 1 - 1/e > 0$, and $f(2) = \cos 2 - 1/e^2 < 0$.

n	h_n	m_n	$f(m_n)$
1	0.5	1.5	-0.2
2	0.25	1.25	0.03
3	0.125	1.375	-0.06
4	0.0625	1.3125	-0.01
5	0.03125	1.28125	0.008
6	0.015625	1.296875	-0.003
7	0.0078125	1.2890625	0.002
8	0.00390625	1.29296875	-0.0002
9	0.991953125	1.291015625	0.001

A real root is 1.29 accurate to two decimal places, since $m_9 \pm h_9$ is in the interval $[1.285, 1.295]$.

6. Let $f(x) = 7x^3 + x - 6 = 0$. $f'(x) = 21x^2 + 1$, $f''(x) = 42x$.

x	f(x)
0	-6
1	2

There is a root between 0 and 1, seemingly closer to 1; let $x_1 = 0.9$.

$$x_{n+1} = x_n - \frac{f(x_n)}{f'(x_n)} = x_n - \frac{7x_n^3 + x_n - 6}{21x_n^2 + 1}$$

n	x_n
1	0.9
2	0.899833425
3	0.899833396

The root is approximately 0.89983.

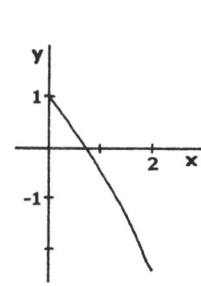

9. Let $f(x) = \cos x - x$. $f'(x) = -\sin x - 1$, $f''(x) = -\cos x$.

x	f(x)
0	1
$\pi/2$	$-\pi/2$

There is a root between 0 and $\pi/2$; let $x_1 = 0.8$.

$$x_{n+1} = x_n - \frac{f(x_n)}{f'(x_n)} = x_n - \frac{x_n - \cos(x_n)}{1 + \sin(x_n)}$$

9. (continued)

n	x_n
1	0.8
2	0.739853306
3	0.739085263
4	0.739085133

The root is approximately 0.73909.

12. Let $f(x) = x^4+x^3-3x^2+4x-28 = (x^2+4)(x^2+x-7)$.

$f(x) = 0$ for real x iff $x^2+x-7=0$.

Let $g(x) = x^2+x-7$. $g'(x) = 2x+1$.

x	g(x)
-4	5
-3	-1
2	-1
3	5

Root between -4 and -3; let $x_1 = -3.2$.

Root between 2 and 3; let $x_1 = 2.2$.

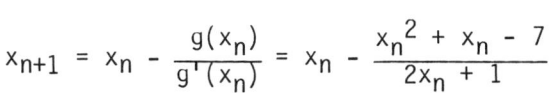

$$x_{n+1} = x_n - \frac{g(x_n)}{g'(x_n)} = x_n - \frac{x_n^2 + x_n - 7}{2x_n + 1}$$

n	x_n
1	-3.2
2	-3.19259259
3	-3.19258240
4	-3.19258240

n	x_n
1	2.2
2	2.19259259
3	2.19258240
4	2.19258240

The roots are approximately -3.19258 and 2.19258.

15. Let $f(x) = x^3-6$. $f'(x) = 3x^2$.

$f(x) = 0$ for some x between 1 and 2 (closer to 2). Let $x_1 = 1.75$.

$$x_{n+1} = x_n - \frac{f(x_n)}{f'(x_n)} = x_n - \frac{x_n^3 - 6}{3x_n^2} = x_n - x_n/3 + 2/x_n^2$$

n	x_n
1	1.75
2	1.819727891
3	1.817124327
4	1.817120593
5	1.817120593

The root is approximately 1.81712.

Therefore, $\sqrt[3]{6} \approx 1.81712$.

Problem Set 10.4

18. Let P_n be the statement $|x_n - r| \le \frac{2m}{M} \left(\frac{M}{2m} |x_1 - r| \right)^{2^{n-1}}$.

P_1 is true since $\frac{2m}{M} \left(\frac{M}{2m} |x_1 - r| \right)^{2^0} = |x_1 - r|$.

Assume P_k is true and show that P_{k+1} is true.

$|x_{k+1} - r| \le \frac{M}{2m} (x_k - r)^2$ [Given]

$= \frac{M}{2m} |x_k - r|^2 \le \frac{M}{2m} \left[\frac{2m}{M} \left(\frac{M}{2m} |x_1 - r| \right)^{2^{k-1}} \right]^2$ [since P_k is true]

$= \frac{2m}{M} \left(\frac{M}{2m} |x_1 - r| \right)^{2^k} = \frac{2m}{M} \left(\frac{M}{2m} |x_1 - r| \right)^{2^{(k+1)-1}}$

so P_{k+1} is true.

21. Let $f(x) = \frac{1 + \ln x}{x}$. $f'(x) = \frac{\ln x}{x^2}$.

$x_{n+1} = x_n - \frac{f(x_n)}{f'(x_n)} = x_n - \frac{[1 + \ln(x_n)]/x_n}{[-\ln(x_n)]/x_n^2} = x_n - \frac{x_n + x_n \ln(x_n)}{-\ln(x_n)}$

$= x_n + x_n/\ln(x_n) + x_n = 2x_n + x_n/\ln(x_n)$

For $x_1 = 1.2$:

n	x_n
1	1.2
2	8.98
3	22.05
4	51.24
5	115.50

For $x_1 = 0.5$:

n	x_n
1	0.5
2	0.278652479
3	0.339231168
4	0.364671303
5	0.367837678
6	0.367879434
7	0.367879411

The root is approximately 0.36788.

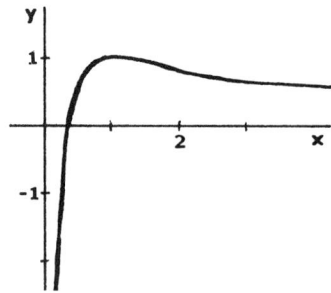

$f'(1.2)$ is small and negative, and $f(1.2)$ is positive, so the tangent at $x=1.2$ intersects the x-axis "far" to the right of $x=1.2$. And the slope remains positive but is decreasing so the tangents at x_n intersect the x-axis farther and farther to the right.

Problem Set 10.5 Fixed-Point Methods

3. $x_{n+1} = \sqrt{2.5 + x_n}$

n	x_n
1	1
2	1.870828693
3	2.090652696
4	2.142580849
5	2.154664904
6	2.157467243
7	2.158116596

n	x_n
8	2.158267035
9	2.158301887
10	2.158309961
11	2.158311831
12	2.158312265
13	2.158312365

Root: 2.15831

6. (a)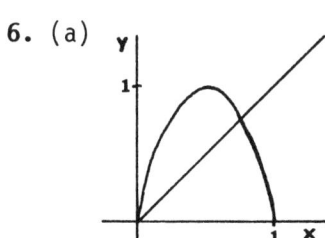

(b) $x_{n+1} = 4(x_n - x_n^2)$

n	x_n
1	0.6
2	0.96
3	0.1536
4	0.52002816
5	0.998395491
6	0.00640773728

(c) $x = 4x - 4x^2$
$4x^2 - 3x = 0$
$x(4x - 3) = 0$
$x = 0$ or $x = 0.75$

(d) $f'(x) = 4(1-2x)$
$f'(0.75) = -2$

9. (a) $x = 2\sin\pi x$
$x + 5x = 5x + 2\sin\pi x$

$x = \dfrac{5x + 2\sin\pi x}{6}$

(b) Let $h(x) = \dfrac{5x + 2\sin\pi x}{6}$.

$x_{n+1} = \dfrac{5x_n + 2\sin(\pi x_n)}{6}$

n	x_n
1	0.9
2	0.853005664
3	0.859357050
4	0.858666231
5	0.858744163
6	0.858735405
7	0.858736389
8	0.858736279

Root: 0.85874

(c) $h'(x) = \dfrac{5 + 2\pi\cos\pi x}{6}$

$h'(x)$ is near 0 in an interval which includes 0.9 and the fixed point. In particular, at the fixed point it is about -0.11242172.

12. Let $x = \sqrt{5+x}$ and $g(x) = \sqrt{5+x}$.

$g'(x) = \dfrac{1}{2\sqrt{5+x}} < 1$ on $[0,\infty)$ so the fixed-point algorithm converges.

12. (continued)

$x_{n+1} = \sqrt{5+x_n}$

n	x_n
1	0
2	2.236067977
3	2.689994048
4	2.773083852
5	2.788025081
6	2.790703331

n	x_n
7	2.791183142
8	2.791269092
9	2.791284488
10	2.791287246
11	2.791287740

Fixed Point: 2.79129

15. (a) $10000 = \dfrac{R}{0.015}[1 - (1+0.015)^{-48}]$

$R = \dfrac{150}{1 - (1.015)^{-48}} \approx 293.75$

(b) $10000 = \dfrac{300}{i}[1 - (1+i)^{-48}]$

$i = 0.03[1 - (1+i)^{-48}]$. Let $g(i) = 0.03[1 - (1+i)^{-48}]$.

$g'(i) = 1.44(1+i)^{-49} < 1$ if $i > .008$

$i_n = 0.03[1 - (1+i_n)^{-48}]$, $i_1 = 0.015$

The convergence is slow. On one calculator it finally stabilized at 0.015990923 beginning at the 36th step. (By starting at 0.016 it was also slow but decreased, reaching 0.015990923 at the 26th step.) It became apparent a bit earlier that 0.01599 would be the correct value accurate to five decimal places. The first few values are:

n	i_n
1	0.015
2	0.015319149
3	0.015539025
4	0.015688550
5	0.015789330
6	0.015856848
7	0.015901897
8	0.015931875
9	0.015951786

Problem Set 10.6 Chapter Review Problems

True-False Quiz

3. True. $P_2(x) = f(0) + f'(0) + \frac{f''(0)x^2}{2} = 0 + 0x + 0x^2 = 0$ is the second-order Maclaurin polynomial of $f(x)$. Note that the second-order polynomial is not a polynomial of degree 2.

6. True. Intuitive: $f'(0)$ exists means there is a tangent line to the graph of f where x is 0. Due to symmetry with respect to the y-axis, the only such tangent line is horizontal, so $f'(0) = 0$.

Analytic: $f'(0) = \lim_{h \to 0} \frac{f(0+h)-f(0)}{h}$ and $f'(0) = \lim_{h \to 0} \frac{f(0-h)-f(0)}{-h}$.

Therefore, $2f'(0) = \lim_{h \to 0}\left(\frac{f(h)-f(0)}{h} + \frac{f(-h)-f(0)}{-h}\right)$

$= \lim_{h \to 0} \frac{f(h)-f(-h)}{h} = 0$ since $f(-h) = f(h)$. [f even]

Therefore, $f'(0) = 0$.

9. False. See the first paragraph of Section 10.3.

12. False. See Section 10.2 dealing with "The Error of Calculation."

15. True. By Intermediate Value Theorem since 0 is between $f(a)$ and $f(b)$.

18. True. Reverse the inequalities in proof of the Fixed-Point Theorem.

Miscellaneous Problems

3. $g(x) = x^3 - 2x^2 + 5x - 7$ $\qquad g(2) = 3$

$\qquad g'(x) = 3x^2 - 4x + 5 \qquad g'(2) = 9$

$\qquad g''(x) = 6x - 4 \qquad\qquad g''(2) = 8$

$\qquad g^{(3)}(x) = 6 \qquad\qquad\quad g^{(3)}(2) = 6$

$P_4(x) = 3 + 9(x-2) + \dfrac{8(x-2)^2}{2} + \dfrac{6(x-2)^3}{3!}$

$\qquad = 3 + (9x-18) + (4x^2-16x+16) + (x^3-6x^2+12x-8) = x^3-2x^2+5x-7$

6. $f^{(5)}(x) = -120(x+1)^{-6}$

$|R_4(x)| = \dfrac{120(c+1)^{-6}|x-1|^5}{5!} = \dfrac{|x-1|^5}{(c+1)^6} \quad$ for some c between 1 and x

$|R_4(1.2)| = \dfrac{(1.2-1)^5}{(c+1)^6} \leq \dfrac{(0.2)^5}{2^6} = 0.000005$

9. $\displaystyle\int_{0.8}^{1.2} P_5(x)\,dx = \int_{0.8}^{1.2}\left((x-1) - \dfrac{(x-1)^2}{2} + \dfrac{(x-1)^3}{3} - \dfrac{(x-1)^4}{4} + \dfrac{(x-1)^5}{5}\right)dx$

$\qquad = \left[\dfrac{(x-1)^2}{2} - \dfrac{(x-1)^3}{6} + \dfrac{(x-1)^4}{12} - \dfrac{(x-1)^5}{20} + \dfrac{(x-1)^6}{30}\right]_{0.8}^{1.2}$

$\qquad = \left(\dfrac{(0.2)^2}{2} - \dfrac{(0.2)^3}{6} + \dfrac{(0.2)^4}{12} - \dfrac{(0.2)^5}{20} + \dfrac{(0.2)^6}{30}\right)$

$\qquad\quad - \left(\dfrac{(-0.2)^2}{2} - \dfrac{(-0.2)^3}{6} + \dfrac{(-0.2)^4}{12} - \dfrac{(-0.2)^5}{20} + \dfrac{(-0.2)^6}{30}\right)$

$\qquad = -2(0.2)^3 - 2(0.2)^5 \approx -0.00269867$

9. (continued)

Therefore, $\int_{0.8}^{1.2} \ln x \, dx \approx -0.00269867$

$|R_5(x)| < \dfrac{(0.2)^6}{6(0.8)^6}$ [from Problem 8]

Therefore, the error of the approximation for the integral is less than

$\int_{0.8}^{1.2} \dfrac{(0.2)^6}{6(0.8)^6} dx = \left[\dfrac{(0.2)^6 x}{6(0.8)^6}\right]_{0.8}^{1.2} = \dfrac{(0.2)^6(1.2 - 0.8)}{6(0.8)^6} < 0.00001628.$

12.

$\int_{0.8}^{1.2} \ln x \, dx = \left[x \ln x - x\right]_{0.8}^{1.2} = (1.2 \ln 1.2 - 1.2) - (0.8 \ln 0.8 - 0.8)$

≈ -0.00269929

Notice that the differences between this value and the approximations obtained in Problems 9 through 11 are within the error estimates found in those problem.

15.
Observe from the graph that the point of intersection $y = x$ and $y = \tan x$ in $(\pi, 2\pi)$ is a little to the left of $3\pi/2$, so we will let $x_1 = 4.6$.

Let $f(x) = x - \tan x$. $f'(x) = 1 - \sec^2 x$

$x_{n+1} = x_n - \dfrac{f(x_n)}{f'(x_n)} = x_n - \dfrac{x_n - \tan(x_n)}{1 - \sec^2(x_n)}$

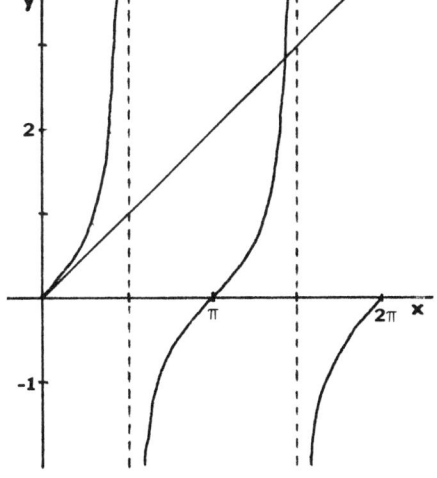

n	x_n
1	4.6
2	4.545732122
3	4.506145588
4	4.494171630
5	4.493412197
6	4.493409458
7	4.493409458

The solution is 4.49341 to five-decimal-place accuracy.

Problem Set 10.6

CHAPTER 11 INFINITE SERIES

Problem Set 11.1 Infinite Sequences

3. 5/2, 17/3, 37/6, 65/11, 101/18. Converges. $a_n = \dfrac{4 + (1/n^2)}{1 -(2/n) +(3/n^2)} \to 4.$
   [2.5, 5.67, 6.17, 5.91, 5.61]

6. 1/2, $\sqrt{2}/3$, $\sqrt{3}/4$, 2/5, $\sqrt{5}/6$. Converges. $a_n = \dfrac{1/\sqrt{n}}{1 + (1/n)} \to 0.$
   [0.5, 0.47, 0.43, 0.40, 0.37]

9. 1, 0, -1/3, 0, 1/5. Converges to 0 by the Squeeze Theorem since $-1/n$ and $1/n$ converge to 0 and $-1/n \leq a_n \leq 1/n$.

12. $e/2$, $e^2/4$, $e^3/8$, $e^4/16$, $e^5/32$. Diverges. $a_n = (e/2)^n$ and $(e/2) > 1$.
    [1.36, 1.85, 2.51, 3.41, 4.64]

15. 1.9, 1.81, 1.729, 1.6561, 1.59049. Converges. $1 + (0.9)^n \to 1 + 0 = 1.$

18. $0, \dfrac{-\ln 2}{\sqrt{2}}, \dfrac{-\ln 3}{\sqrt{3}}, \dfrac{-\ln 4}{\sqrt{4}}, \dfrac{-\ln 5}{\sqrt{5}}.$ $[\ln(1/n) = \ln 1 - \ln(n) = -\ln(n)]$
    [0, -0.49, -0.63, -0.69, -0.72]

    $\lim\limits_{n \to \infty} \dfrac{-\ln(n)}{n} = \lim\limits_{x \to \infty} \dfrac{-\ln x}{x} \stackrel{\text{L}}{=} \lim\limits_{x \to \infty} \dfrac{-1/x}{1/2\sqrt{x}} = \lim\limits_{x \to \infty} \dfrac{-2}{\sqrt{x}} = 0$ [Certainly not clear from 1st five terms]

21. $a_n = \dfrac{n}{n+1} = \dfrac{1}{1 + (1/n)} \to \dfrac{1}{1 + 0} = 1.$

24. $a_n = \dfrac{1}{1 - \frac{n-1}{n}} = n.$ Diverges.

27. $\lim\limits_{n \to \infty} n \sin(1/n) = \lim\limits_{n \to \infty} \dfrac{\sin(1/n)}{1/n} = \lim\limits_{x \to \infty} \dfrac{\sin(1/x)}{1/x} = \lim\limits_{z \to 0^+} \dfrac{\sin z}{z} = 1$ $[z = 1/x]$

30. $a_n = \dfrac{1}{n} - \dfrac{1}{n+1} = \dfrac{1}{n(n+1)} \to 0.$

33. 3/4, 2/3, 5/8, 3/5 [or 0.75, 0.67, 0.62, 0.60]. The sequence is bounded below by 0 and is decreasing since each factor is less than 1. (Multiplying a positive number by a positive number that is less than 1 yields a smaller positive number.)

36. 2, 3/2, 17/12, 577/408 [or 2, 1.5, 1.42, 1.41]. The sequence is bounded below by 0. We will show it is decreasing in two steps.

   (1) $\dfrac{a_{n+1}}{a_n} = \dfrac{a_n/2 + 1/a_n}{a_n} = \dfrac{1}{2} + \dfrac{1}{a_n^2} < 1$ if $a_n^2 > 2$.

   Therefore, $a_{n+1} < a_n$ (so the sequence is decreasing) if $a_n^2 > 2$.

   (2) Use induction. Let P_n be the statement: $a_n^2 > 2$.

   P_1 is true since $a_1^2 = 4$. Now assume that P_n is true.

   $a_{n+1}^2 - 2 = \left(\dfrac{a_n}{2} + \dfrac{1}{a_n}\right)^2 - 2 = \dfrac{a_n^2}{4} + 1 + \dfrac{1}{a_n^2} - 2 = \left(\dfrac{a_n}{2} - \dfrac{1}{a_n}\right)^2 > 0.$

   Therefore, $a_{n+1}^2 > 2$, so P_{n+1} is true.

   Hence $a_n^2 > 2$ for each integer n.

39. Continue where the hint leaves off, and note that u is nonnegative.

   $u^2 = 3+u$, $u^2 - u - 3 = 0$, $u = \dfrac{1 + \sqrt{13}}{2} \approx 2.3028$.

42. $u_1 < 2$, and $u_n < 2$ => $u_{n+1} = 1.1^{u_n} < 1.1^2 = 1.21 < 2$, so the sequence is bounded above by 2. And the sequence is increasing since
   $u_2 > u_1$ since $1 > 0$, and $u_{n+1} > u_n$ => $1.1^{u_n} > u_n$

   => $1.1^{(1.1^{u_n})} > 1.1^{u_n}$

   => $1.1^{u_{n+1}} > u_{n+1}$

   => $u_{n+2} > u_{n+1}$

45. Choose N to be any integer greater than $(1/\varepsilon)$.

   Then $n > N$ => $n+1 > (1/\varepsilon)$ => $\dfrac{1}{n+1} < \varepsilon$ => $\left|\dfrac{n}{n+1} - 1\right| < \varepsilon$.

48. By the Completeness Property $\{a_n\}$ has a least upper bound; call it A. Let $\varepsilon > 0$ be given. Then $A-\varepsilon$ is not an upper bound so there is some term a_N of the sequence such that $A-\varepsilon < a_N \leq A$.

Then $n \geq N \Rightarrow A-\varepsilon < a_N \leq a_n \leq A \Rightarrow |a_n - A| < \varepsilon$, so $a_n \to A$.

Let $c_k = -b_k$. Then $\{c_k\}$ is a nondecreasing sequence that is bounded above by $-L$. Then (by what was proven above) $\{c_k\}$ converges to some number, say C, the least upper bound of $\{c_k\}$.

Then $b_k = -c_k \to -C$, the greatest lower bound of $\{b_k\}$. Let $B = -C$.

51. No. $\{n\}$ and $\{-n\}$ both diverge but $\{n+(-n)\} = \{0\}$ converges to 0.

Problem Set 11.2 Infinite Series

3. $2 \sum_{k=0}^{\infty}(1/3)^k + 3 \sum_{k=0}^{\infty}(1/6)^k = 2 \frac{1}{1-(1/3)} + 3 \frac{1}{1-(1/6)} = 6.6$

6. Diverges since it is geometric series with $r = (4/3) > 1$.

9. Diverges by the nth term test. $a_{n+1} > a_n$ if $n > 9$ since

$$\frac{a_{n+1}}{a_n} = \frac{(n+1)!}{10^{n+1}} \frac{10^n}{n!} = \frac{(n+1)n!}{10^n\, 10^1} \frac{10^n}{n!} = \frac{n+1}{10} > 1 \text{ if } n > 9$$

12. Geometric series with $a = 9$, $r = 3/5$, so the sum is $\frac{9}{1-(3/5)} = 22.5$.

15. $0.2 + 0.02 + 0.002 + \cdots = \frac{0.2}{1-0.1} = \frac{2}{9}$ [Geometric with a=0.2, r=0.1]

18. $0.125 + 0.000125 + 0.000000125 + \cdots = \frac{0.125}{1-0.001} = \frac{125}{999}$
    [Geometric with a=0.125, r=0.001]

21. $r + r(1-r) + r(1-r)^2 + \ldots = \dfrac{r}{1 - (1-r)} = 1$

[Geometric with 1st term = r, common ratio = (1-r), |1-r| < 1]

24. $\ln\left(1 - \dfrac{1}{k^2}\right) = \ln\dfrac{(k+1)(k-1)}{k^2} = \ln(k+1) + \ln(k-1) - 2\ln(k)$

$[\ln 3 + \ln 1 - 2\ln 2] + [\ln 4 + \ln 2 - 2\ln 3] + [\ln 5 + \ln 3 - 2\ln 4] + \ldots$

$S_1 = \ln 3 - 2\ln 2, \quad S_2 = \ln 4 - \ln 3 - \ln 2, \quad S_3 = \ln 5 - \ln 4 - \ln 2, \cdots$

$S_n = \ln(n+2) - \ln(n+1) - \ln 2 = \ln\left(\dfrac{n+2}{n+1}\right) - \ln 2 \to \ln 1 - \ln 2 = -\ln 2$

27. $1 \text{ billion} + (1 \text{ billion})(3/4) + (1 \text{ billion})(3/4)^2 + \ldots = \dfrac{1 \text{ billion}}{1 - (3/4)}$

= 4 billion   [Geometric with a = 1 billion, r = 3/4]

The total increase in spending is 4 billion dollars (including the 1 billion put in by the government).

30.

Step	Fraction Painted	Fraction Unpainted	
1	1/9	8/9	[8/9 of 1]
2	1/9 + (1/9)(8/9)	$(8/9)^2$	[8/9 of 8/9]
3	1/9 + (1/9)(8/9) + (1/9)(8/9)^2	$(8/9)^3$	[8/9 of $(8/9)^2$]

A geometric series with a = 1/9, r = 8/9 is forming. Therefore, the fraction that will be painted is $\dfrac{1/9}{1 - (8/9)} = 1$. [the whole thing]

33. Note that blocks are being added to the pile from the bottom.

(a) The blocks will not topple if the center of mass is above the base. For the six blocks illustrated, with origin at the center of the bottom block, the center of mass would have to have x-coordinate between -1/2 and 1/2. The x-coordinates of the centers of mass of the individual blocks (bottom to top) are:

$$0,\ \dfrac{1}{10},\ \dfrac{1}{10}+\dfrac{1}{8},\ \dfrac{1}{10}+\dfrac{1}{8}+\dfrac{1}{6},\ \dfrac{1}{10}+\dfrac{1}{8}+\dfrac{1}{6}+\dfrac{1}{4},\ \dfrac{1}{10}+\dfrac{1}{8}+\dfrac{1}{6}+\dfrac{1}{4}+\dfrac{1}{2}$$

33. (continued)

Therefore, the x-coordinate of the center of mass of the stack is the average of those numbers:

$$\frac{\frac{5}{10} + \frac{4}{8} + \frac{3}{6} + \frac{2}{4} + \frac{1}{2}}{6} = \frac{5(1/2)}{6} < \frac{1}{2} \quad \text{so the six blocks do not tumble.}$$

For n blocks, one obtains $\frac{(n-1)(1/2)}{n} < \frac{1}{2}$ so the n blocks do not tumble.

(b) Indefinitely since $\frac{1}{2} + \frac{1}{4} + \frac{1}{6} + \frac{1}{8} + \frac{1}{10} + \cdots$ diverges [Problem 32]

36. Σn and $\Sigma(-n)$ both diverge but $\Sigma[n+(-n)] = \Sigma 0$ converges to 0. [Compare with Problem 51 of Section 11.1.]

Problem Set 11.3 Positive Series: The Integral Test

3. $f(x) = \frac{x}{1+x^2}$ satisfies the conditions of the Integral Test since

$$f'(x) = \frac{1-x^2}{(1+x^2)^2} < 0 \text{ for } x > 1.$$

$$\int_1^\infty \frac{x}{1+x^2} dx = \lim_{b \to \infty} \left[(1/2)\ln(1+x^2)\right]_1^b = \lim_{b \to \infty}[(1/2)\ln(1+b^2) - (1/2)\ln 2] = \infty$$

The improper integral diverges, so the series diverges.

6. $f(x) = (x+2)^{-2}$ satisfies the conditions of the Integral Test.

$$\int_1^\infty (x+2)^{-2} dx = \lim_{b \to \infty}\left[\frac{-1}{x+2}\right]_1^b = \lim_{b \to \infty}\left(\frac{-1}{b+2} + \frac{1}{3}\right) = \frac{1}{3}, \text{ so the series converges.}$$

9. $f(x) = (4+3x)^{-3/2}$ satisfies the conditions of the Integral Test. The series converges since

$$\int_1^\infty (4+3x)^{-3/2} dx = \lim_{b \to \infty} \left[\frac{-2}{3(4+3x)^{1/2}}\right]_1^b = \lim_{b \to \infty}\left(\frac{-2}{3(4+3b)^{1/2}} + \frac{2}{3(7)^{1/2}}\right) = \frac{2}{3\sqrt{7}}.$$

12. $f(x) = \dfrac{1}{x(\ell nx)^2}$ satisfies the conditions of the Integral Test.

$$\int_2^\infty \frac{1}{x(\ell nx)^2} dx = \lim_{b \to \infty}\left[\frac{-1}{\ell nx}\right]_2^b = \lim_{b \to \infty}\left(\frac{-1}{\ell nb} + \frac{1}{\ell n2}\right) = \frac{1}{\ell n2} \qquad \text{[Use } u = \ell nx \text{ substitution.]}$$

Therefore, the series converges since the improper integral converges.

15. Diverges since $\left[\left(\frac{1}{2}\right)^k + \frac{k-1}{2k+1}\right] \to 0 + \frac{1}{2} \neq 0.$ [nth-Term Test]

18. Diverges [nth-Term Test] since $k\sin(1/k) \to 1 \neq 0$ (See write-up of Problem 27 of Section 11.1.)

21. $f(x) = \dfrac{\tan^{-1}x}{1+x^2}$ satisfies the conditions of the integral test since

$$f'(x) = \frac{1 - 2x\,\tan^{-1}x}{(1+x^2)^4} < 0 \text{ for } x > 1.$$

$$\int_1^\infty \frac{\tan^{-1}x}{1+x^2} dx = \lim_{b \to \infty}\left[\frac{(\tan^{-1}x)^2}{2}\right]_1^b = \lim_{b \to \infty}\left(\frac{(\tan^{-1}b)^2}{2} - \frac{(\tan^{-1}1)^2}{2}\right)$$

$$= \frac{(\pi/2)^2}{2} - \frac{(\pi/4)^2}{2}, \text{ so the series converges.}$$

24. $f(x) = x^{-3/2}$ satisfies the conditions of the Integral Test.

$$\text{Error} = \sum_{k=6}^\infty k^{-3/2} < \int_5^\infty x^{-3/2} dx = \lim_{b \to \infty}\left[-2x^{-1/2}\right]_5^b = 2/\sqrt{5} < 0.8945$$

27. Case $p < 0$: $(\ln n)^{-p} > 1$ if $n > 3$, so $\dfrac{1}{n(\ln n)^p} = \dfrac{(\ln n)^{-p}}{n} > \dfrac{1}{n}$

Therefore, the partial sums of the given series are larger than the respective partial sums of the harmonic series, which diverges to ∞, so the given series diverges.

$$\left[\text{Note that } f(x) = \dfrac{(\ln x)^{-p}}{x} \text{ satisfies the conditions of the Integral Test on } [2,\infty) \text{ since } f'(x) = \dfrac{(\ln x)^{-p-1}(-xp - \ln x)}{x^2} < 0 \text{ if } p \geq 0.\right]$$

Case $p = 1$: $\displaystyle\int_2^\infty \dfrac{(\ln x)^{-1}}{x}dx = \lim_{b\to\infty}\Big[\ln|\ln x|\Big]_2^b = \infty$. The series diverges.

Other Cases: $\displaystyle\int_2^\infty \dfrac{(\ln x)^{-p}}{x}dx = \lim_{b\to\infty}\left[\dfrac{(\ln x)^{1-p}}{1-p}\right]_2^b = \lim_{b\to\infty}\dfrac{(\ln b)^{1-p} - (\ln 2)^{1-p}}{1-p}$

which converges if $p > 1$, and diverges if $p < 1$.

Summary: The series converges iff $p > 1$.

30. Boundedness: $0 < 1 + \dfrac{1}{2} + \ldots + \dfrac{1}{n} - \ln(n+1) < 1 + \ln(n) - \ln(n+1)$

$$= 1 + \ln\left(\dfrac{n}{n+1}\right) < 1$$

Monotonocity: $B_{n+1} - B_n = \dfrac{1}{n+1} - \ln(n+2) + \ln(n+1) = \dfrac{1}{n+1} + \ln\left(\dfrac{n+1}{n+2}\right)$

(1) If $f(x) = \dfrac{1}{x+1} + \ln\left(\dfrac{x+1}{x+2}\right)$, $f'(x) = \dfrac{-1}{(x+1)^2(x+2)} < 0$

for $x \geq 1$. Hence, f is decreasing for $x \geq 1$.

Thus, $\dfrac{1}{n+1} + \ln\left(\dfrac{n+1}{n+2}\right)$ is decreasing for $n \geq 1$.

(2) $\displaystyle\lim_{n\to\infty}\left[\dfrac{1}{n+1} + \ln\left(\dfrac{n+1}{n+2}\right)\right] = 0 + \ln(1) = 0$

Hence, $\dfrac{1}{n+1} + \ln\left(\dfrac{n+1}{n+2}\right) > 0$ (It is decreasing toward 0.)

Therefore, $B_{n+1} > B_n$, so B_n is increasing.

Problem Set 11.3

Problem Set 11.4 Positive Series: Other Tests

3. Converges. Use $p = 3/2$ series. $\dfrac{1}{n\sqrt{n+1}} \cdot \dfrac{n^{3/2}}{1} = \dfrac{1}{\sqrt{1+(1/n)}} \to 1 > 0$

[Note: Multiplying by $\dfrac{n^{3/2}}{1}$ is equivalent to dividing by $\dfrac{1}{n^{3/2}}$.]

6. Diverges. $\dfrac{5^{n+1}}{(n+1)^5} \cdot \dfrac{n^5}{5^n} = \dfrac{5^1 n^5}{(n+1)^5} = \dfrac{5}{[1+(1/n)]^5} \to 5 > 1$

9. Converges. $\dfrac{(n+1)^3}{[2(n+1)]!} \cdot \dfrac{(2n)!}{n^3} = \dfrac{(n+1)^3 (2n)!}{(2n+2)(2n+1)(2n)! n^3} = \dfrac{(n+1)^3}{(2n+2)(2n+1) n^3}$

$= \dfrac{[1+(1/n)]^3}{(2n+2)(2n+1)(1)} \to 0 < 1$

12. Converges by Ratio Test. $\dfrac{5+(n+1)}{(n+1)!} \cdot \dfrac{n!}{5+n} = \dfrac{n+6}{(n+1)(n+5)}$

$= \dfrac{1+(6/n)}{[1+(1/n)](n+5)} \to 0 < 1$

15. Converges by Ratio Test. $\dfrac{(n+1)^2}{(n+1)!} \cdot \dfrac{n!}{n^2} = \dfrac{(n+1)^2}{(n+1)n^2} = \dfrac{n+1}{n^2} = \dfrac{1}{n} + \dfrac{1}{n^2} \to 0 < 1$

18. Converges by Ratio Test. $\dfrac{(n+1)^2+1}{3^{n+1}} \cdot \dfrac{3^n}{n^2+1} = \dfrac{n^2+2n+2}{3(n^2+1)} \to \dfrac{1}{3} < 1$

21. It is $\sum \dfrac{n+1}{n(n+2)(n+3)}$. It converges by the Limit Comparison Test using the $p=2$ series.

$\dfrac{n+1}{n(n+2)(n+3)} \cdot \dfrac{n^2}{1} = \dfrac{n(n+1)}{(n+2)(n+3)} = \dfrac{[1][1+(1/n)]}{[1+(2/n)][1+(3/n)]} \to 1 > 0$

24. $\sum \dfrac{3^n}{n!}$ converges by the Ratio Test. $\dfrac{3^{n+1}}{(n+1)!} \cdot \dfrac{n!}{3^n} = \dfrac{3}{n+1} \to 0 < 1$

27. Diverges by nth-Term Test. $\dfrac{1}{2+\sin^2 n}$ has values between $1/3$ and $1/2$ so does not converge to 0.

30. Converges by Ratio Test. $\dfrac{5^{2(n+1)}}{(n+1)!} \cdot \dfrac{n!}{5^{2n}} = \dfrac{5^{2n+2}}{(n+1)5^{2n}} = \dfrac{5^2}{n+1} \to 0 < 1$

33. $\sum \dfrac{4^n}{n!}$ converges by Ratio Test. $\dfrac{4^{n+1}}{(n+1)!} \cdot \dfrac{n!}{4^n} = \dfrac{4}{n+1} \to 0 < 1$

$\sum \dfrac{n}{n!}$ converges by the Ratio Test. $\dfrac{n+1}{(n+1)!} \cdot \dfrac{n!}{n} = \dfrac{1}{n} \to 0 < 1$

Therefore, using linearity of Σ, the given series converges.

36. In Example 7 it was proven that the series converges. Therefore, by the nth-Term Test, $\dfrac{n!}{n^n} \to 0$.

39. Diverges by the Limit Comparison Test using the Harmonic Series.

$\dfrac{a_n}{1/n} = na_n \to 1 > 0$

42. (a) Converges. $\left[\left(\dfrac{1}{\ln n}\right)^n\right]^{1/n} = \dfrac{1}{\ln n} \to 0 < 1$

(b) Converges. $\left[\left(\dfrac{n}{3n+2}\right)^n\right]^{1/n} = \dfrac{n}{3n+2} = \dfrac{1}{3+(2/n)} \to \dfrac{1}{3} < 1$

(c) Converges. $\left[\left(\dfrac{1}{2} + \dfrac{1}{n}\right)^n\right]^{1/n} = \left(\dfrac{1}{2} + \dfrac{1}{n}\right) \to \dfrac{1}{2} < 1$

Problem Set 11.5 Alternating Series, Absolute Convergence

3. $\dfrac{1}{\ln(n+1)} \to 0$ and $\dfrac{1}{\ln(n+1)}$ is a dec. sequence so the series converges.

$|\text{Error}| < \dfrac{1}{\ln(10+1)} < 0.4171$

6. $\lim_{n\to\infty} \frac{\ln n}{n^{1/2}} = \lim_{x\to\infty} \frac{\ln x}{x^{1/2}} \stackrel{\text{\large\textcircled{L}}}{=} \lim_{x\to\infty} \frac{x^{-1}}{(1/2)x^{-1/2}} = \lim_{x\to\infty} \frac{2}{x^{1/2}} = 0$

Now let $f(x) = \frac{\ln x}{x^{1/2}}$; then $f'(x) = \frac{2-\ln x}{2x^{3/2}} < 0$ if $x > e^2$.

Then f is dec. for $x > e^2$, so the sequence $\frac{\ln n}{n^{1/2}}$ is dec. for $n \geq 8$.

Therefore, the series converges, by the Alternating-Series Test.

$|\text{Error}| < \frac{\ln 10}{10^{1/2}} < 0.7282$

9. By Absolute Ratio Test. $\frac{n+1}{2^{n+1}} \frac{2^n}{n} = \frac{n+1}{2n} = \frac{1+(1/n)}{2} \to \frac{1}{2} < 1$

12. By Absolute Ratio Test. $\frac{2^{n+1}}{(n+1)!} \frac{n!}{2^n} = \frac{2}{n+1} \to 0 < 1$

15. Divergent by nth-Term Test. $\frac{n}{10n+1} \to \frac{1}{10}$ so the terms do not approach 0.

18. Absolutely convergent. $\sum \frac{1}{n^{3/2}}$ is a p=(3/2) series.

21. Conditionally convergent. Shown to converge in Problem 4,

but $\sum \frac{n}{n^2+1}$ diverges by the Limit Comparison Test, using the Harmonic

Series. $\frac{n}{n^2+1} \frac{n}{1} = \frac{n^2}{n^2+1} = \frac{1}{1+(1/n^2)} \to 1 > 0$

24. Absolutely convergent by Ordinary Comparison Test using the p=2 series.
$\frac{|\sin(n\pi/2)|}{n^2} \leq \frac{1}{n^2}$

27. Conditionally convergent. Converges by the Alternating-Series Test,

but $\sum \frac{1}{[n(n+1)]^{1/2}}$ diverges by the Limit Comparison Test, using the

Harmonic Series. $\frac{1}{[n^2+n]^{1/2}} \frac{n}{1} = \frac{1}{[1+(1/n)]^{1/2}} \to 1 > 0$

30. Conditionally convergent. Converges by the Alternating-Series Test (for $n \geq 2$), but $\Sigma[\sin(\pi/n)]$ diverges by the Limit Comparison Test, using $\Sigma(\pi/n)$. $\quad \dfrac{\sin(\pi/n)}{\pi/n} \to 1 > 0$

33. $1 + \dfrac{1}{3} + \dfrac{1}{5} + \cdots = \sum \dfrac{1}{2n-1}$ which diverges by the Limit Comparison Test, using the Harmonic Series. $\dfrac{1}{2n-1} \dfrac{n}{1} = \dfrac{n}{2n-1} = \dfrac{1}{2-(1/n)} \to \dfrac{1}{2} > 0$

$-\dfrac{1}{2} - \dfrac{1}{4} - \dfrac{1}{6} - \cdots = \sum (-1/2)\dfrac{1}{n}$ diverges since $\Sigma(1/n)$ diverges.

36. 1, 1/3, -1/2, 1/5, 1/7, 1/9, 1/11, -1/4, 1/13, 1/15, 1/17, -1/6, 1/19, 1/21, 1/23, -1/8, 1/25, 1/27, 1/29, 1/31. $\quad S_{20} \approx 1.326464032$.

39. The even-numbered partial sums of the series formed are partial sums of the series $\sum\left(\dfrac{1}{n} - \dfrac{1}{n^2}\right)$ which is the series $\sum \dfrac{n-1}{n^2}$, and this series diverges by the Limit Comparison Test, using the Harmonic Series.

$\dfrac{n-1}{n^2} \dfrac{n}{1} = 1 - \dfrac{1}{n} \to 1 > 0$

Problem Set 11.6 Power Series

3. $\sum \dfrac{(-1)^n x^{2n+1}}{(2n+1)!}$. $\quad \left|\dfrac{x^{2n+3}}{(2n+3)!} \dfrac{(2n+1)!}{x^{2n+1}}\right| = \dfrac{x^2}{(2n+3)(2n+2)} \to 0$ for all x.

Convergence Set: R

6. $\sum (n+1)^2 x^{n+1}$. $\quad \left|\dfrac{(n+2)^2 x^{n+2}}{(n+1)^2 x^{n+1}}\right| = \dfrac{(n+2)^2 |x|}{(n+1)^2} \to |x| < 1$ if x is in $(-1,1)$

At $x = -1$: $\sum (n+1)^2(-1)^{n+1}$ diverges by the nth-Term Test.

At $x = 1$: $\sum (n+1)^2$ diverges by the nth-Term Test.

Convergence Set: $(-1,1)$

9. $1 + \sum_{n=1}^{\infty} \frac{(-1)^n x^n}{n(n+2)}$. $\left|\frac{x^{n+1}}{(n+1)(n+3)} \cdot \frac{n(n+2)}{x^n}\right| = \frac{n(n+2)|x|}{(n+1)(n+3)} \to |x| < 1$, if x is in $(-1,1)$.

At $x=-1$: $\sum \frac{1}{n(n+2)}$ converges by the Ordinary Comparison Test, using the $p=2$ series.

At $x=1$: $\sum \frac{(-1)^n}{n(n+2)}$ converges by the Absolute Convergence Test, using the $x=-1$ case.

Convergence Set: $[-1,1]$.

12. $\sum 2^n x^n = \sum (2x)^n$ is a geometric series. It converges iff $|2x| < 1$.
Convergence Set: $(-1/2, 1/2)$.

15. $\sum \frac{(x-1)^{n+1}}{n+1}$. $\left|\frac{(x-1)^{n+2}}{n+2} \cdot \frac{n+1}{(x-1)^{n+1}}\right| = \frac{(n+1)|x-1|}{n+2} \to |x-1| < 1$, if x is in $(0,2)$.

At $x=0$: Converges since it is an alternating harmonic series.

At $x=2$: Diverges since it is a harmonic series.

Convergence Set: $[0,2)$.

18. $\sum \frac{(x-2)^{n+1}}{(n+1)^2}$. $\left|\frac{(x-2)^{n+2}}{(n+2)^2} \cdot \frac{(n+1)^2}{(x-2)^{n+1}}\right| = \frac{(n+1)^2 |x-2|}{(n+2)^2} \to |x-2| < 1$, if x is in $(1,3)$.

At $x=3$: Converges since it is the $p=2$ series.

At $x=1$: Converges by the Absolute Convergence Test, using $x=3$ case.

Convergence Set: $[1,3]$.

21. By the nth-Term Test.

24. Use Absolute Ratio Test. The radius of convergence is p^p. If the proof that follows is not clear at first, mimic it using $p=2$, $p=3$, etc., until it is clear.

$\frac{[p(n+1)]! x^{n+1}}{[(n+1)!]^p} \cdot \frac{(n!)^p}{(pn)! x^n} = \frac{(pn+p)(pn+p-1) \cdots (pn+1) |x|}{(n+1)^p}$

$= \frac{[p+(p/n)][p+(p/n)-(1/n)] \cdots [p+(1/n)] |x|}{[1+(1/n)]^p} \to p^p |x| < 1$, if x is in $(-p^p, p^p)$.

Problem Set 11.7 Operations on Power Series

3. This is x^2 times the function in Problem 1. The series is then
$$x^2(1-x+x^2-x^3+\cdots) = x^2-x^3+x^4-x^5+\cdots = \sum(-1)^n x^{n+2},$$
and the radius of convergence is 1 (as for Problem 1).

6. $f(x) = \dfrac{1/3}{1-(-2x/3)}$ is the sum of the geometric series with $a = 1/3$ and $r = -2x/3$, which converges if $|-2x/3| < 1$, or x is in $(-3/2, 3/2)$. The radius of convergence is $3/2$ and the series is

$$f(x) = \frac{1}{3} - \frac{2x}{9} + \frac{4x^2}{27} - \frac{8x^3}{81} + \cdots = \sum \frac{(-1)^n (2x)^n}{3^{n+1}} \text{ for } x \text{ in } (-3/2, 3/2).$$

9. The f used here will always be the f of Problem 9. Be careful not to confuse it with the f of Problem 1.

$f'(x) = \ln(1+x)$; $f''(x) = (1+x)^{-1}$, so f is a 2nd antiderivative of the function of Problem 1.

$f''(x) = 1 - x + x^2 - x^3 + \cdots$ [from Problem 1]

$$f'(x) = \int_0^x f''(t)dt = x - \frac{x^2}{2} + \frac{x^3}{3} - \frac{x^4}{4} + \cdots$$

$$f(x) = \int_0^x f'(t)dt = \frac{x^2}{1\cdot 2} - \frac{x^3}{2\cdot 3} + \frac{x^4}{3\cdot 4} - \frac{x^5}{4\cdot 5} + \cdots = \sum \frac{(-1)^n x^{n+2}}{(n+1)(n+2)}$$

for x in $(-1,1)$. The radius of convergence is 1.

12. Solve $M = \dfrac{1+x}{1-x}$ for x and obtain $x = \dfrac{M-1}{M+1}$.

Since $M > 0$, $\dfrac{M-1}{M+1} > -1$ [since $M-1 > -M-1$] and $\dfrac{M-1}{M+1} < 1$ [since $M-1 < M+1$].

Therefore, $x = \dfrac{M-1}{M+1}$ is in $(-1,1)$ for $M > 0$. If $M = 8$, $x = \dfrac{7}{9}$.

Using the series obtain in Problem 11,

$$\ln 8 = 2(7/9) + \frac{2(7/9)^3}{3} + \frac{2(7/9)^5}{5} + \frac{2(7/9)^7}{7} + \cdots$$

The convergence is rather slow. The sum of the first 12 terms to three decimal places is 2.079, which is the correct value.

15. Add the series from Example 3 and the one from Problem 13.

$(1 + x + \frac{x^2}{2!} + \frac{x^3}{3!} + \cdots) + (1 - x + \frac{x^2}{2!} - \frac{x^3}{3!} + \cdots)$

$= 2 + \frac{2x^2}{2!} + \frac{2x^4}{4!} + \cdots = \sum \frac{2x^{2n}}{(2n)!}$ for all x in $\mathbb{R}$

18. See Examples 2 and 3 for the series for e^x and $\tan^{-1}x$. Then for all x in $(-1,1)$, $e^x \tan^{-1}x = (1 + x + \frac{x^2}{2!} + \frac{x^3}{3!} + \cdots)(x - \frac{x^3}{3} + \frac{x^5}{5} - \cdots)$

$= x + x^2 + (\frac{1}{2} - \frac{1}{3})x^3 + (\frac{1}{3!} - \frac{1}{3})x^4 + \cdots = x + x^2 + \frac{x^3}{6} - \frac{x^4}{6} + \cdots$

21. See Example 2 for the series for $\tan^{-1}x$. Then for x in $(-1,1)$,

$(\tan^{-1}x)(1 + x^2 + x^4) = (x - \frac{x^3}{3} + \frac{x^5}{5} - \frac{x^7}{7} + \cdots)(1 + x^2 + x^4)$

$= x + (1 - \frac{1}{3})x^3 + (1 - \frac{1}{3} + \frac{1}{5})x^5 + (-\frac{1}{3} + \frac{1}{5} - \frac{1}{7})x^7 + \cdots$

$= x + 2x^3 + \sum_{n=0}^{\infty}(-1)^n(\frac{1}{2n+1} - \frac{1}{2n+3} + \frac{1}{2n+5})x^{2n+5}$

24. See Example 2 for the series for $\tan^{-1}x$. Then for x in $(-1,1)$,

$\frac{\tan^{-1}t}{t} = \frac{t - (t^3/3) + (t^5/5) - (t^7/7) + \cdots}{t} = 1 - \frac{t^2}{3} + \frac{t^4}{5} - \frac{t^6}{7} + \cdots,$

and $\int_0^x \frac{\tan^{-1}t}{t} dt = x - \frac{x^3}{9} + \frac{x^5}{25} - \frac{x^7}{49} + \cdots = \sum \frac{(-1)^n x^{2n+1}}{(2n+1)^2}$

27. $x + 2x^2 + 3x^3 + 4x^4 + \cdots = x(1 + 2x + 3x^2 + 4x^3 + \cdots)$

$= x D_x(x + x^2 + x^3 + x^4 + \cdots) = x D_x(\frac{x}{1-x}) = x \cdot \frac{1}{(1-x)^2} = \frac{x}{(1-x)^2}$

for x in $(-1,1)$.

30. $a_0 + a_1x + a_2x^2 + \cdots = b_0 + b_1x + b_2x^2 + \cdots$ and $x = 0 \Rightarrow a_0 = b_0$

1st Derivative: $a_1 + 2a_2x + \cdots = b_1 + 2b_2x + \cdots$ and $x=0 \Rightarrow a_1 = b_1$

2nd Derivative: $2a_2 + \cdots = 2b_2 + \cdots$ and $x=0 \Rightarrow a_2 = b_2$ [And so on]

Problem Set 11.8 Taylor and Maclaurin Series

3. $e^x \sin x = (1 + x + \frac{x^2}{2!} + \frac{x^3}{3!} + \frac{x^4}{4!} + \frac{x^5}{5!} + \cdots)(x - \frac{x^3}{3!} + \frac{x^5}{5!} + \cdots)$

$= x + x^2 + (\frac{-1}{6} + \frac{1}{2})x^3 + (\frac{-1}{6} + \frac{1}{6})x^4 + (\frac{1}{120} - \frac{1}{12} + \frac{1}{24})x^5 + \cdots$

$= x + x^2 + \frac{x^3}{3} - \frac{x^5}{30} + \cdots$ for all x in R.

6. $(\sin x)(1+x)^{1/2} = (x - \frac{x^3}{3!} + \frac{x^5}{5!} - \cdots)(1 + \frac{x}{2} - \frac{x^2}{8} + \frac{x^3}{16} - \frac{5x^4}{128} + \cdots)$

$= x + \frac{x^2}{2} + (\frac{-1}{8} - \frac{1}{6})x^3 + (\frac{1}{16} - \frac{1}{12})x^4 + (\frac{-5}{128} + \frac{1}{48} + \frac{1}{120})x^5 + \cdots$

$= x + \frac{x^2}{2} - \frac{7x^3}{24} - \frac{x^4}{48} - \frac{19x^5}{1920} + \cdots$ for all x in (-1,1).

9. $\frac{1}{1-x} \cosh x = (1 + x + x^2 + x^3 + x^4 + x^5 + \cdots)(1 + \frac{x^2}{2!} + \frac{x^4}{4!} + \cdots)$

$= 1 + x + (\frac{1}{2} + 1)x^2 + (\frac{1}{2} + 1)x^3 + (\frac{1}{24} + \frac{1}{2} + 1)x^4 + (\frac{1}{24} + \frac{1}{2} + 1)x^5 + \cdots$

$= 1 + x + \frac{3x^2}{2} + \frac{3x^3}{2} + \frac{37x^4}{24} + \frac{37x^5}{24} + \cdots$ for all x in (-1,1)

12. $\frac{1}{1-\sin x} = \frac{1}{1 - x + (x^3/3!) - (x^5/5!) + \cdots}$ for all x in $(-\pi/2, \pi/2)$

$= 1 + x + x^2 + \frac{5x^3}{6} + \frac{2x^4}{3} + \frac{61x^5}{120} + \cdots$ [by long division]

15. $x \sec(x^2) + \sin x = \frac{x}{\cos(x^2)} + \sin x$

$= \frac{x}{1 - (x^4/2!) + (x^8/4!) - \cdots} + (x - \frac{x^3}{3!} + \frac{x^5}{5!} - \cdots)$

$= (x + \frac{x^5}{2} + \cdots) + (x - \frac{x^3}{5} + \frac{x^5}{120} - \cdots)$ [long division for first part]

$= 2x - \frac{x^3}{6} + \frac{61x^5}{120} + \cdots$ for all x in $(-\sqrt{\pi/2}, \sqrt{\pi/2})$

18. $(1-x^2)^{2/3} = 1 + (2/3)(-x^2) + \frac{(2/3)(-1/3)(-x^2)^2}{2!} + \cdots$

$= 1 - \frac{2x^2}{3} - \frac{x^4}{9} + \cdots$ for all x in (-1,1).

21. $f(x) = \cos x$, $f'(x) = -\sin x$, $f''(x) = -\cos x$, $f^{(3)}(x) = \sin x$.

$f(\pi/3) = 1/2$, $f'(\pi/3) = -\sqrt{3}/2$, $f''(\pi/3) = -1/2$, $f^{(3)}(x) = \sqrt{3}/2$.

$\cos x = \frac{1}{2} + (-\sqrt{3}/2)(x - \pi/3) + \frac{(-1/2)(x - \pi/3)^2}{2!} + \frac{(\sqrt{3}/2)(x - \pi/3)^3}{3!} + \cdots$

$= \frac{1}{2} - \frac{\sqrt{3}(x - \pi/3)}{2} - \frac{(x - \pi/3)^2}{4} + \frac{\sqrt{3}(x - \pi/3)^3}{12} + \cdots$.

24. $f(x) = 2-x+3x^2-x^3$, $f'(x) = -1+6x-3x^2$, $f''(x) = 6-6x$, $f^{(3)}(x) = -6$.

$f(-1) = 7$, $f'(-1) = -10$, $f''(-1) = 12$, $f^{(3)}(-1) = -6$.

$2-x+3x^2-x^3 = 7 + (-10)(x+1) + \frac{12(x+1)^2}{2!} + \frac{-6(x+1)^3}{3!}$

$= 7 - 10(x+1) + 6(x+1)^2 - (x+1)^3$.

27. $(1-t^2)^{-1/2} = 1 + (-1/2)(-t^2) + \frac{(-1/2)(-3/2)(-t^2)^2}{2!} + \frac{(-1/2)(-3/2)(-5/2)(-t^2)^3}{3!} + \cdots$

$= 1 + \frac{t^2}{2} + \frac{3t^4}{8} + \frac{5t^6}{16} + \cdots$.

$\sin^{-1} x = \int_0^x (1 + \frac{t^2}{2} + \frac{3t^4}{8} + \frac{5t^6}{16} + \cdots) dt = x + \frac{x^3}{6} + \frac{3x^5}{40} + \frac{5x^7}{112} + \cdots$.

30. $\sin(x^{1/2}) = x^{1/2} - \frac{x^{3/2}}{3!} + \frac{x^{5/2}}{5!} - \frac{x^{7/2}}{7!} + \cdots$.

$\int_0^{0.5} \sin(x^{1/2}) dx = \left[\frac{2x^{3/2}}{3} - \frac{2x^{5/2}}{3!5} + \frac{2x^{7/2}}{5!7} - \frac{2x^{9/2}}{7!9} + \cdots \right]_0^{0.5}$

$= \frac{2(0.5)^{3/2}}{3} - \frac{2(0.5)^{5/2}}{3!5} + \frac{2(0.5)^{7/2}}{5!7} - \frac{2(0.5)^{9/2}}{7!9} + \cdots$.

This series satisfies the Alternating Series Test criteria so the error is less than the absolute value of the 1st term omitted. We require this error to be less than 0.000005 for five-decimal-place accuracy. The 4th term is the first one with value less than 0.000005, so we use the 1st three terms and obtain a value of 0.22413.

Problem Set 11.9 Chapter Review Problems

True-False Quiz

3. **True.** If a sequence converges to L, every subsequence converges to L.

6. **True.** a_{2n} involves all even-numbered terms and a_{2n+1} involves all odd-numbered terms.

9. **True.** $a_n \to L \Rightarrow L-1 < a_n < L+1$ for all $n > N$ (for some N)

    $\Rightarrow \frac{L}{n} - \frac{1}{n} < \frac{a_n}{n} < \frac{L}{n} + \frac{1}{n}$. Now apply Squeeze Theorem.

12. **True.** $0 < (1/n)^n \leq (1/2)^{n-1}$ for all n, and $\sum_{n=1}^{\infty}(1/2)^{n-1} = \frac{1}{1-(1/2)} = 2$

    Therefore, $\Sigma[(1/n)^n]$ converges by the Ordinary Comparison Tests. Its sum is greater than 1 since its 1st term is 1. Its sum is less than 2 since its partial sums (after the 1st) are less than the corresponding partial sums of the geometric series above.

15. **True.** $\rho = 1$, so the test is inconclusive.

18. **False.** By Limit Comparison Test, using $\sum \frac{1}{\ln n}$.

21. **True.** These conditions imply the convergence of $\sum_{n=101}^{\infty} a_n$ by Ordinary Comparison Test. Then $a_1 + \ldots + a_{100} + \sum_{n=101}^{\infty} a_n$ converges.

24. **False.** c^{ex}: Let $a_n = [(-1)^n]/n$, so the corresponding series [alternating harmonic] converges. However, $(-1)^n a_n = 1/n$, and the corresponding series [harmonic] diverges.

27. **True.** The left-hand side is the error in using the first 99 terms of the alternating harmonic series to approximate its sum. That error is less than the 100th term, 1/100, or 0.01.

30. False. c^{ex}: $\sum \dfrac{x^{n+1}}{(n+1)2^{n+1}}$ has a convergence set of $[-2,2)$.

33. False. c^{ex}: See Problem 32 of Section 11.8 (page 499).

Miscellaneous Problems

3. $\lim\limits_{n\to\infty}\left(1 + \dfrac{4}{n}\right)^n = \lim\limits_{m\to\infty}\left(1 + \dfrac{1}{m}\right)^{4m}$ [letting $n = 4m$]

   $= \lim\limits_{m\to\infty}\left[\left(1 + \dfrac{1}{m}\right)^m\right]^4 = \left[\lim\limits_{m\to\infty}\left(1 + \dfrac{1}{m}\right)^m\right]^4 = e^4 \approx 54.5982$

6. $\lim\limits_{n\to\infty}[n^{-1/3}] + \lim\limits_{n\to\infty}[3^{-1/n}] = 0 + 3^{\left(\lim\limits_{n\to\infty}(-1/n)\right)} = 3^0 = 1$

9. Partial sums, S_n, are collapsing. $S_n = 1 - \dfrac{1}{\sqrt{n+1}} \to 1$, so the sum is 1.

12. Each term is -1 or 1, so the series diverges by the nth-Term Test.

15. Converges to $\dfrac{0.91}{1 - 0.01} = \dfrac{91}{99}$. [Geometric with $a = 0.91$, $r = 0.01$]

18. Converges to $e^{-1} \approx 0.3679$. [Power series for e^x evaluated at $x = -1$]

21. Converges, by Absolute Convergence Test, using the $p=(3/2)$ series.

24. Converges, by Ordinary Comparison Test, using the $p=2$ series, since
    $e^{n^2} > n^3$ for all n so $\dfrac{n}{e^{n^2}} < \dfrac{n}{n^3} = \dfrac{1}{n^2}$. [Can also use Integral Test]

27. Converges by Ratio Test. $\dfrac{(n+1)^2}{(n+1)!}\dfrac{n!}{n^2} = \dfrac{n+1}{n^2} = \dfrac{1}{n} + \dfrac{1}{n^2} \to 0 < 1$

30. Diverges by nth-Term Test. $\lim\limits_{n\to\infty}\left(1 - \dfrac{1}{n}\right)^n = e^{-1} \neq 0$ [See Problem 27 of Section 7.5.]

33. Conditionally convergent. The series converges by the Alternating Series Test, but the corresponding positive-term series diverges by the Limit Comparison Test, using the Harmonic Series.

36. The given series converges by the Alternating Series Test since

(1) for $f(x) = x^{1/x}$, $f'(x) = \dfrac{x^{1/x}(1 - \ln x)}{x^2}$, so f is decreasing

for $x > e$. Therefore, $\dfrac{n^{1/n}}{\ln n}$ is decreasing for $n \geq 3$.

(2) $\dfrac{n^{1/n}}{\ln n} \to 0$ since $n^{1/n} \to 1$.

However, $n^{1/n} > 1$ and $\ln n < n$, so $\dfrac{n^{1/n}}{\ln n} > \dfrac{1}{n}$. Hence, the corresponding positive-term series diverges by the Ordinary Comparison Test, using the Harmonic Series.

Therefore, the given series is conditionally convergent.

39. $\left| \dfrac{(x-4)^{n+1}}{n+2} \cdot \dfrac{n+1}{(x-4)^n} \right| = \dfrac{(n+1)|x-4|}{n+2} \to |x-4| < 1$ if $3 < x < 5$. [Ratio Test]

At $x = 3$: Divergent [Harmonic Series].

At $x = 5$: Convergent [Alternating Harmonic Series].

Convergence Set: $(3,5]$

42. $\left| \dfrac{(n+1)!(x+1)^{n+1}}{3^{n+1}} \cdot \dfrac{3^n}{n!(x+1)^n} \right| = \dfrac{(n+1)|x+1|}{3}$ which converges only for $x = -1$.

Convergence Set: $\{-1\}$

45.
$f(x)$	$= \sin^2 x$	$f(0)$	$= 0$
$f'(x)$	$= 2\sin x \cos x = \sin 2x$	$f'(0)$	$= 0$
$f''(x)$	$= 2\cos 2x$	$f''(0)$	$= 2$
$f^{(3)}(x)$	$= -4\sin 2x$	$f^{(3)}(0)$	$= 0$
$f^{(4)}(x)$	$= -8\cos 2x$	$f^{(4)}(0)$	$= -8$
$f^{(5)}(x)$	$= 16\sin 2x$	$f^{(5)}(0)$	$= 0$
$f^{(6)}(x)$	$= 32\cos 2x$	$f^{(6)}(0)$	$= 32$

Problem Set 11.9

45. (continued)

Therefore, $\sin^2 x = \frac{2x^2}{2!} - \frac{8x^4}{4!} + \frac{32x^6}{6!} - \frac{128x^8}{8!} + \cdots$ for all x in **R**

48. $\cos(x^2) = 1 - \frac{(x^2)^2}{2!} + \frac{(x^2)^4}{4!} - \frac{(x^2)^6}{6!} + \cdots = 1 - \frac{x^4}{2!} + \frac{x^8}{4!} - \frac{x^{12}}{6!} + \cdots$

Therefore, $\int_0^1 \cos(x^2)\,dx = \left[x - \frac{x^5}{2!\cdot 5} + \frac{x^9}{4!\cdot 9} - \frac{x^{13}}{6!\cdot 13} + \frac{x^{17}}{8!\cdot 17} \cdots \right]_0^1$

$= 1 - \frac{1}{2!\cdot 5} + \frac{1}{4!\cdot 9} - \frac{1}{6!\cdot 13} + \frac{1}{8!\cdot 17} - \cdots$

This series satisfies the Alternating Series Test criteria so the error is less than the absolute value of the 1st term omitted. We require this error to be less than 0.00005 for four-decimal-place accuracy. The 5th term is the first one with value less than 0.00005, so we use the 1st four terms and obtain a value of 0.9045.

51. $\cos x = 1 - \frac{x^2}{2!} + \frac{x^4}{4!} - \cdots$

The error will be less than $\frac{x^4}{4!} < \frac{(0.1)^4}{4!} < 0.00000417$.

Problem Set 11.9

CHAPTER 12 CONICS AND POLAR COORDINATES

Problem Set 12.1 The Parabola

3. $x^2 = -4(4)y$, so $p = 4$.

   Focus: $(0,-4)$

   Directrix: $y = 4$

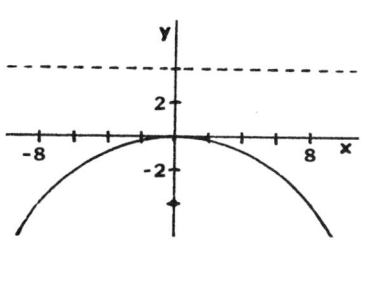

6. $y^2 = 4(7/4)x$, so $p = 7/4$

   Focus: $(7/4, 0)$

   Directrix: $x = -7/4$

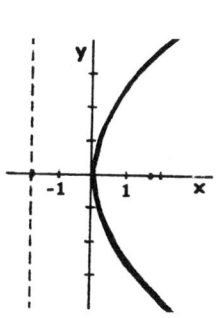

9. $p = 3$. The directrix is $x = -3$. Equation is $y^2 = 4(3)x$ or $y^2 = 12x$.

12. $p = 1/3$. The directrix is $x = 1/3$. Equation is $x^2 = -4(1/3)y$.

15. $y^2 = 4px$

    $(-1)^2 = 4p(3) \implies p = 1/12$

    $y^2 = 4(1/12)x; \; y^2 = (1/3)x$

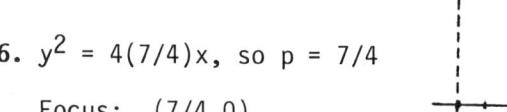

18. $x^2 = 4py$

    $(-3)^2 = 4p(5) \implies p = 9/20$

    $x^2 = 4(9/20)y; \; x^2 = 1.8y$

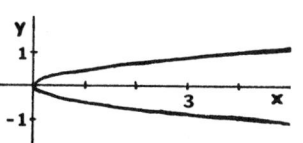

21. $x^2 = 2y \implies 2x = 2y' \implies y' = x$

    At $(4,8)$: (1) $y' = 4$

    (2) Tangent: $y - 8 = 4(x-4)$
    $y = 4x - 8$

    (3) Normal: $y - 8 = (-1/4)(x-4)$
    $y = (-1/4)x + 9$

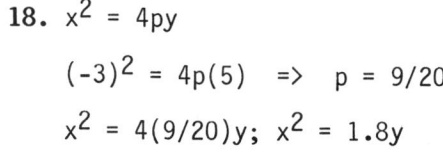

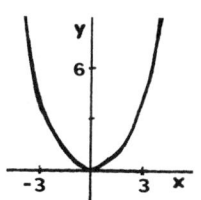

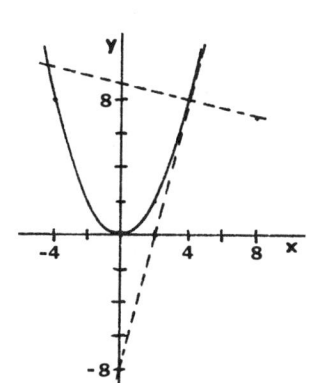

Problem Set 12.1

24. $x^2 = 4y \Rightarrow 2x = 4y' \Rightarrow y' = x/2$

At (4,4): (1) $y' = 4/2 = 2$

(2) Tangent: $y-4 = 2(x-4)$
$y = 2x - 4$

(3) Normal: $y-4 = (-1/2)(x-4)$
$y = (-1/2)x + 6$

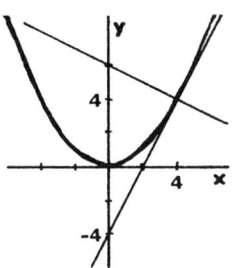

27. $y^2 = 5x \Rightarrow 2yy' = 5 \Rightarrow y' = 5/2y$

The slope, y', is $\sqrt{5}/4$, so $5/2y = \sqrt{5}/4$.

Therefore, $y = 2\sqrt{5}$; then $x = 4$.

The point we seek is $(4, 2\sqrt{5})$.

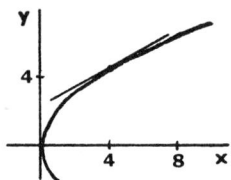

30. The properties are independent of the choice of axes. Using axes as in the figure to the right, we have:

$x^2 = 4py \Rightarrow 2x = 4py' \Rightarrow y' = x/2p$

The slopes of the tangents at (x_i, y_i) are $x_i/2p$ [for $i = 1, 2$].

The equations of the tangents at (x_i, y_i) are: $y - y_i = (x_i/2p)(x - x_i)$.

To find where each intersects the directrix, let $y = -p$, and solve for x:

$x = 2p(y_i - p)/x_i$ (used $x_i^2 = 4py_i$)

But $\dfrac{2p(y_1-p)}{x_1} = \dfrac{2p(y_2-p)}{x_2}$ since $\dfrac{y_1-p}{x_1} = \dfrac{y_2-p}{x_2}$ using the similar triangles in the figure.

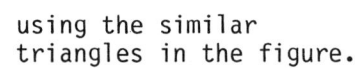

33. Minimize the square of the distance, z, between focus $(0,p)$ and an arbitrary point of the parabola $(x, x^2/4p)$, x any real number.

$z(x) = x^2 + \left(\dfrac{x^2}{4p} - p\right)^2 = \dfrac{x^2}{2} + \dfrac{x^4}{16p^2} + p^2 \Rightarrow z'(x) = x + \dfrac{x^3}{4p^2} = \dfrac{x(4p^2+x^2)}{4p^2}$

$z'(x) = 0$ iff $x = 0$ (and the corresponding value of y is 0).

$z''(x) = 1 + \dfrac{3x^2}{4p^2}$; then $z''(0) = 1 > 0$, so z is minimum at $(0,0)$ [vertex].

Problem Set 12.2 Ellipses and Hyperbolas

3. Vertical hyperbola. The plus sign is associated with the y^2 term.

6. $x^2 = (-4/9)y$. Vertical parabola (since quadratic term is x^2) opening downward (since a negative sign is with the y term).

9. Vertical ellipse since the divisor of y^2 is larger than the divisor of x^2.

   $a = 4$, $b = 3$, $c = \sqrt{16-9} \approx 2.65$

   Vertices: $(0, \pm 4)$

   Foci: $(0, \pm 2.65)$

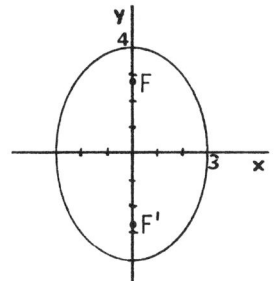

12. Horizontal ellipse since the divisor of x^2 is larger than the divisor of y^2.

    $a = \sqrt{10} \approx 3.16$, $b = 2$, $c = \sqrt{10-4} \approx 2.45$

    Vertices: $(\pm 3.16, 0)$

    Foci: $(\pm 2.45, 0)$

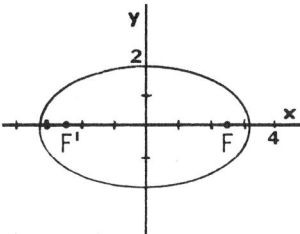

15. $\dfrac{x^2}{25} - \dfrac{y^2}{4} = 1$ Horizontal hyperbola since the plus sign is with the x^2 term.

    $a = 5$, $b = 2$, $c = \sqrt{25+4} \approx 5.39$

    Vertices: $(\pm 5, 0)$

    Foci: $(\pm 5.39, 0)$

    Asymptotes: $y = \pm \dfrac{2}{5} x$

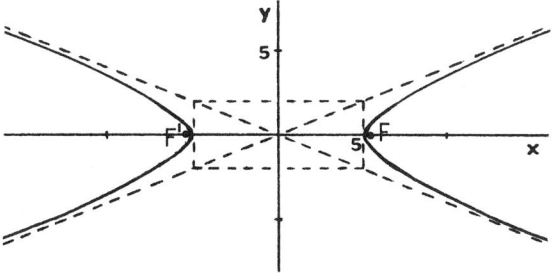

18. Horizontal ellipse since the foci are on the x-axis.
    $c = 6$, $e = 2/3$, $a = c/e = 9$, $b^2 = a^2 - c^2 = 81 - 36 = 45$ $\dfrac{x^2}{81} + \dfrac{y^2}{45} = 1$

21. Horizontal ellipse since the vertices are on the x-axis. $a=5$. Then the equation has the form $\frac{x^2}{25} + \frac{y^2}{b^2} = 1$ so $\frac{4}{25} + \frac{9}{b^2} = 1$. Thus, $b^2 = \frac{225}{21}$.
$\frac{x^2}{25} + \frac{y^2}{225/21} = 1$ or $\frac{x^2}{25} + \frac{21y^2}{225} = 1$.

24. Vertical hyperbola since vertices are on y-axis.
$a = 3$, $e = 3/2$, $c = ae = 9/2$, $b^2 = c^2 - a^2 = (81/4) - 9 = 45/4$.
$\frac{-x^2}{45/4} + \frac{y^2}{9} = 1$ or $\frac{-4x^2}{45} + \frac{y^2}{9} = 1$.

27. Horizontal ellipse since foci are on x-axis. $\frac{x^2}{16} + \frac{y^2}{12} = 1$.
$c = 2$, $a^2 = 8c = 16$, $b^2 = a^2 - c^2 = 16 - 4 = 12$.

30. $\frac{25}{a^2} + \frac{1}{b^2} = 1$ and $\frac{16}{a^2} + \frac{4}{b^2} = 1$.
$\frac{1}{b^2} = 1 - \frac{25}{a^2}$ (1st equation); then $\frac{16}{a^2} + 4 - \frac{100}{a^2} = 1$ (into 2nd equation).
Thus, $a^2 = 28$, $b^2 = 28/3$. Equation is $\frac{x^2}{28} + \frac{y^2}{28/3} = 1$ or $\frac{x^2}{28} + \frac{3y^2}{28} = 1$.

33. $\frac{x^2}{a^2} + \frac{y^2}{b^2} = 1$, so $y^2 = \frac{b^2(a^2 - x^2)}{a^2}$.
At the focus, $x = c$, so $y^2 = \frac{b^2(a^2 - c^2)}{a^2} = \frac{b^2(b^2)}{a^2}$, $y = \frac{b^2}{a}$, $2y = \frac{2b^2}{a}$.

36. $e = 0.999925 = c/a$, so $c = 0.999925a$.

$a - c = 0.13$, so $c = a - 0.13$.

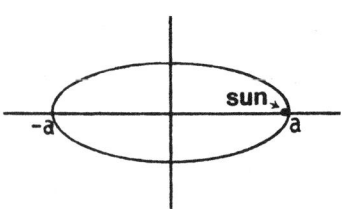

Therefore, $0.999925a = a - 0.13$, so
$a = 130000/75$; $c = 130000/75 - 0.13$.

The maximum distance from the sun is
$c - (-a) = 260000/75 - 0.13 \approx 3466.54$ AU.

Problem Set 12.3 More on Ellipses and Hyperbolas

3. Horizontal hyperbola, $a = 8/2 = 4$, $c = 5$, $b^2 = c^2 - a^2 = 25 - 16 = 9$.

   $\dfrac{x^2}{16} - \dfrac{y^2}{9} = 1$

6. Vertical ellipse [Note that the formula in the table is for horizontal ellipses. Just interchange the roles of a and b.]

   $a^2 = 24$, $b^2 = 16$, $x_0 = -2$, $y_0 = 3\sqrt{2}$. $\dfrac{-2x}{16} + \dfrac{3\sqrt{2}y}{24} = 1$; $y = (\sqrt{2}/2)x + 4\sqrt{2}$.

9. $\dfrac{x^2}{25} + \dfrac{y^2}{25} = 1$. Circle [can be regarded as ellipse with $a = b$]

   $a^2 = b^2 = 25$, $x_0 = 3$, $y_0 = 4$. $\dfrac{3x}{25} + \dfrac{4y}{25} = 1$; $y = \dfrac{-3}{4}x + \dfrac{25}{4}$

12. Since the point is on the x-axis, it must be a vertex, so the tangent is vertical and its equation is $x = 13$. [Or obtain from the formula.]

15. $\dfrac{x^2}{35/2} - \dfrac{y^2}{5} = 1$. Slope is $\dfrac{-2}{3} = \dfrac{5x_0}{(35/2)y_0}$, so $y_0 = (-3/7)x_0$.

    Then substituting into the equation of the hyperbola and solving for x_0, $x_0 = \pm 7$; $y_0 = (-3/7)(\pm 7) = \mp 3$. The points are $(7,-3)$ and $(-7,3)$.

18. By disks: Volume $= 2\displaystyle\int_0^b \pi\left(\dfrac{a}{b}\sqrt{b^2-y^2}\right)^2 dy$

    $= \dfrac{2\pi a^2}{b^2}\displaystyle\int_0^b (b^2 - y^2)\,dy = \dfrac{2\pi a^2}{b^2}\left[b^2 y - \dfrac{y^3}{3}\right]_0^b$

    $= \dfrac{2\pi a^2}{b^2} \cdot \dfrac{2b^2}{3} = \dfrac{4\pi a^2 b}{3}$

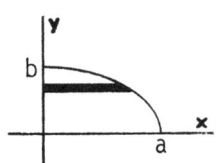

21. Let z denote the square of the area of the rectangle, and (x,y) denote its vertex in the first quadrant. The area will be maximum where z is maximum.

    Maximize $z = [(2x)(2y)]^2 = 16x^2 y^2$.

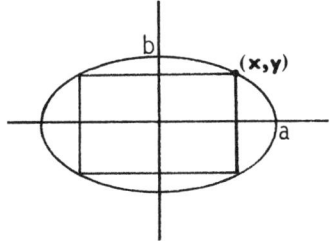

21. (continued)

$z(x) = 16x^2 \dfrac{b^2(a^2-x^2)}{a^2} = \dfrac{16b^2(a^2x^2-x^4)}{a^2}$ where x is in [0,a].

$z'(x) = \dfrac{32b^2x(a^2-2x^2)}{a^2}$ and $z''(x) = \dfrac{32b^2(a^2-6x^2)}{a^2}$

$z'(x) = 0$ if $x = a/\sqrt{2}$ and $z''(a/\sqrt{2}) < 0$, so z is maximum at $x = a/2$.

The corresponding value of y is $b/\sqrt{2}$, so the dimensions of the largest rectangle is $2a/\sqrt{2}$ by $2b/\sqrt{2}$ or $\sqrt{2}a$ by $\sqrt{2}b$.

24. Substitute $x = 6-2y$ into $x^2+4y^2 = 20$, obtaining $(6-2y)^2 + 4y^2 = 20$. Solve for y and obtain $y = 1$ or $y = 2$. The corresponding values of x are 4 and 2, respectively. Intersection points are (4,1) and (2,2).

27. It crosses the major axis between a focus and the neighboring vertex.

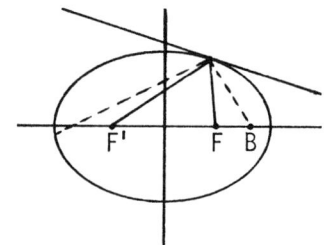

Problem Set 12.4   Translation of Axes

3. $4(x^2-4x+4) + 9(y^2+8y+16) = -124 + 16 + 144$

$4(x-2)^2 + 9(y+4)^2 = 36$ or $\dfrac{(x-2)^2}{9} + \dfrac{(y+4)^2}{4} = 1$ Horizontal ellipse; Center is at (2,-4).

6. $9(x^2+10x+25) + 16(y^2+12y+36) = -1000 + 225 + 576$;

$9(x+5)^2 + 16(y+6)^2 = -199$; empty set.

9. $(x^2-2x+1) + (y^2+4y+4) = -20 + 1 + 4$; $(x-1)^2 + (y+2)^2 = -15$; empty set.

12. $4(x^2-4x+4) = 0$; $4(x-2)^2 = 0$; a single line $x = 2$.

15. $25(x^2+6x+9) - 4(y^2+2y+1) = -129 + 225 - 4$; $25(x+3)^2 - 4(y+1)^2 = 92$;
$\frac{(x+3)^2}{92/25} - \frac{(y+1)^2}{23} = 1$; horizontal hyperbola with center at $(-3,-1)$.

18. Circle with center at $(-3,4)$ and radius 5.

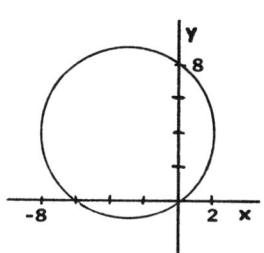

21. Vertical parabola opening upward with vertex at $(-2,1)$.

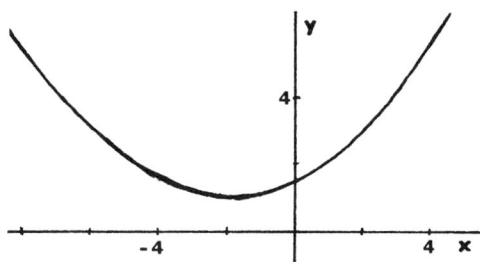

24. Point $(-3,2)$.

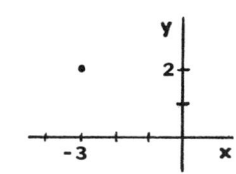

27. $9(x^2+6x+9) - 16(y^2-4y+4) = 127+81-64$

$9(x+3)^2 - 16(y-2)^2 = 144$

$\frac{(x+3)^2}{16} - \frac{(y-2)^2}{9} = 1$

Horizontal hyperbola with center at $(-3,2)$.

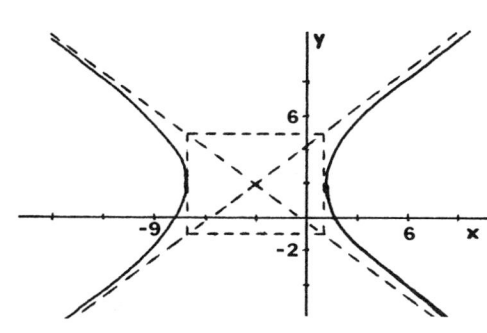

Problem Set 12.4

30. $x^2-4x+4 = -8y+4$

   $(x-2)^2 = -8[y-(1/2)]$

   Vertical parabola with vertex at (2,1/2) and opening downward.

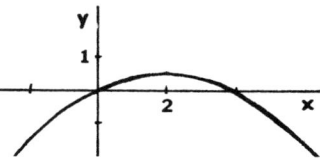

33. $\frac{(x-1)^2}{25} + \frac{(y+2)^2}{16} = 1$ is a horizontal ellipse with center at (1,-2).

   $a = 5$, $b = 4$, $c^2 = a^2-b^2 = 25-16 = 9$, so $c = 3$.
   Then the foci are (1+3,-2) and (1-3,-2); i.e., (4,-2) and (-2,-2).

36. Horizontal hyberbola since the center, vertex, and focus are on the horizontal line $y = -1$.

   $a = 4-2 = 2$ and $c = 5-2 = 3$, so $b^2 = c^2-a^2 = 9-4 = 5$.

   The equation then is $\frac{(x-2)^2}{4} - \frac{(y+1)^2}{5} = 1$

39. Vertical hyperbola since vertices and focus are on vertical line $x = 0$.

   Center is (0,3), $a = 6-3 = 3$, $c = 8-3 = 5$, so $b^2 = c^2-a^2 = 25-9 = 16$.

   $\frac{-x^2}{16} + \frac{(y-3)^2}{9} = 1$

42. Vertical parabola (vertex and focus are on vertical line x=2) opening downward (focus is below vertex) with $p = 6-5 = 1$.

   $(x-2)^2 = -4(1)(y-6)$ or $(x-2)^2 = -4(y-6)$

Problem Set 12.5 Rotation of Axes

3. $\cot 2\theta = \frac{A-C}{B} = \frac{4-0}{-3} = \frac{-4}{3}$

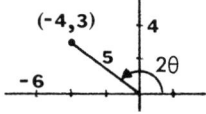

   $\cos 2\theta = \frac{-4}{5} = -0.8$; $\sin 2\theta = \frac{1-\cos 2\theta}{2} = 0.9$; $\cos^2\theta = 1 - \sin^2\theta = 0.1$

3. (continued)

$\tan\theta = \dfrac{\sin\theta}{\cos\theta} = 3$

$x = \sqrt{0.1}u - \sqrt{0.9}v = \sqrt{0.1}(u-3v)$
$y = \sqrt{0.9}u + \sqrt{0.1}v = \sqrt{0.1}(3u+v)$

$4(0.1)(u-3v)^2 - 3(0.1)(u-3v)(3u+v) = 18$

$-5u^2 + 45v^2 = 180$

$\dfrac{-u^2}{36} + \dfrac{v^2}{4} = 1$

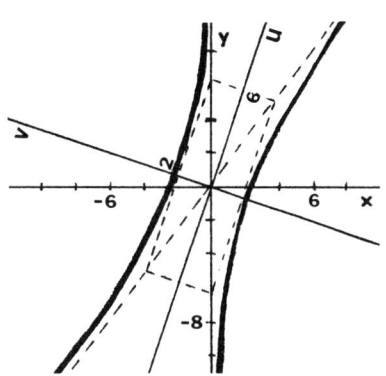

6. $\cot 2\theta = \dfrac{1-3}{2\sqrt{3}} = \dfrac{-\sqrt{3}}{3}$

Choose $2\theta = 2\pi/3$, so $\theta = \pi/3$.

$x = (\cos\theta)u - (\sin\theta)v = \dfrac{u - 3v}{2}$

$y = (\sin\theta)u + (\cos\theta)v = \dfrac{\sqrt{3}u + v}{2}$

Substituting x and y into the given
equation and simplifying yields $u^2 = 4v$.

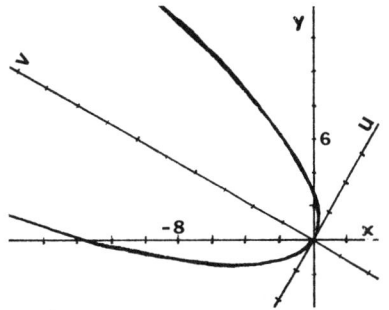

9. $\cot 2\theta = \dfrac{9-16}{-24} = \dfrac{7}{24}$

$\cos 2\theta = 7/25 = 0.28$

$\sin^2\theta = \dfrac{1-\cos 2\theta}{2} = 0.36$, $\cos^2\theta = 1-\sin^2\theta = 0.64$, $\tan\theta = \dfrac{\sin\theta}{\cos\theta} = \dfrac{3}{4}$

$x = 0.8u - 0.6v = 0.2(4u-3v)$
$y = 0.6u + 0.8v = 0.2(3u+4v)$

Substituting x and y into the given
equation and simplifying yields
$v^2 + 4v + 3 = 0$; or $(v+1)(v+3) = 0$, or
$v = -1$, $v = -3$ (parallel lines).

[We could have saved ourselves some
work if we had realized this at the
beginning since it means that the
original equation can be expressed
in a factored form. The factoriza-
tion is $(3x-4y-5)(3x-4y-15) = 0$.]

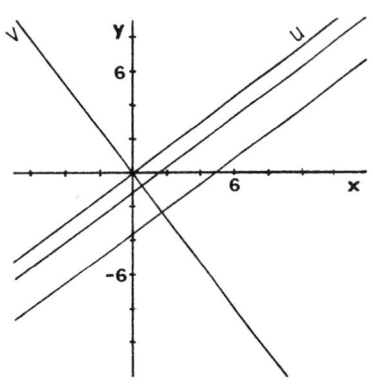

Problem Set 12.5

12. $\cot 2\theta = \frac{6-(-6)}{-5} = \frac{12}{-5}$

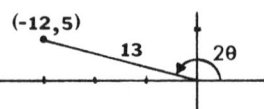

$\cos 2\theta = -12/13$

$\sin^2\theta = \frac{1-\cos 2\theta}{2} = \frac{25}{26}$, $\cos^2\theta = 1-\sin^2\theta = \frac{1}{26}$, $\tan\theta = 5$.

$x = \frac{u-5v}{\sqrt{26}}$, $y = \frac{5u+v}{\sqrt{26}}$

Substituting x and y into the given equation and simplifying yields

$\frac{(u - \sqrt{26})^2}{4} - \frac{(v - \sqrt{26})^2}{4} = 1$

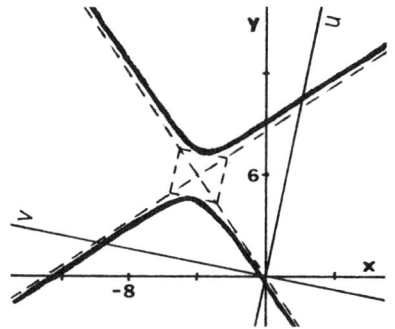

15. $x = (\cos\theta)u - (\sin\theta)v$; $y = (\sin\theta)u + (\cos\theta)v$. Therefore,

   $(\cos\theta)x = (\cos^2\theta)u - (\sin\theta\cos\theta)v$; $(\sin\theta)y = (\sin^2\theta)u + (\sin\theta\cos\theta)v$.

   $(\cos\theta)x + (\sin\theta)y = (\cos^2\theta+\sin^2\theta)u = u$ [adding corresponding sides]

   Similarly, obtain $(-\sin\theta)x + (\cos\theta)y = v$.

   That is, $u = (\cos\theta)x + (\sin\theta)y$, $v = (-\sin\theta)x + (\cos\theta)y$.

18. We substitute $x = (\cos\theta)u - (\sin\theta)v$ and $y = (\sin\theta)u + (\cos\theta)v$ into the equation $Ax^2 + Bxy + Cy^2 + Dx + Ey + F = 0$, and obtain:

   $A[(\cos^2\theta)u^2 - 2(\sin\theta\cos\theta)uv + (\sin^2\theta)v^2]$

   $+ B[(\sin\theta\cos\theta)u^2 + (\cos^2\theta)uv - (\sin^2\theta)uv - (\sin\theta\cos\theta)v^2]$

   $+ C[(\sin^2\theta)u^2 + 2(\sin\theta\cos\theta)uv + (\cos^2\theta)v^2]$

   $+ D[(\cos\theta)u - (\sin\theta)v] + E[(\sin\theta)u + (\cos\theta)v] + F = 0$

   Therefore, $a = A\cos^2\theta + B\sin\theta\cos\theta + C\sin^2\theta$ [the coefficient of u^2]

   $c = A\sin^2\theta - B\sin\theta\cos\theta + C\cos^2\theta$ [the coefficient of v^2]

   Thus, $a + c = a(\cos^2\theta + \sin^2\theta) + C(\sin^2\theta + \cos^2\theta) = A + C$.

Problem Set 12.6 The Polar Coordinate System

3. A(-5,π/4)
    B(5,-3π/4)
    C(2,-π/3)
    D(-2,2π/3)
    E(-6,0)
    F(3,-π)
    G(-4,-2π/3)
    H(-3,π)

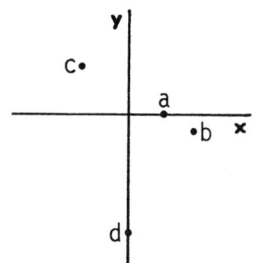

6. (a) (-2,3π), (-2,-π), (2,2π), (2,0) [Points plotted above, right]
   (b) (5,23π/12), (5,-25π/12), (-5,11π/12), (-5,-13π/12)
   (c) (4,11π/4), (4,3π/4), (-4,-π/4), (-4,-9π/4)
   (d) (-7,π/2), (-7,5π/2), (7,-π/2), (7,3π/2)

9. (a) The point is in the 3rd quadrant. $r^2 = 12 + 4 = 16$; $\tan\theta = 1/\sqrt{3}$.
   (-4,π/6) and (4,7π/6) are two of the infinitely many possibilities.

   (b) The point is in the 1st quadrant. $r^2 = 1 + 3 = 4$; $\tan\theta = \sqrt{3}$.
   (2,π/3) or (-2,4π/3) will do.

   (c) The point is in the 4th quadrant. $r^2 = 2 + 2 = 4$; $\tan\theta = -1$.
   (2,-π/4), (-2,3π/4).

   (d) (0,0), or more generally, (0,k) for any k in **R**.

12. θ = π/2

15. r = 4

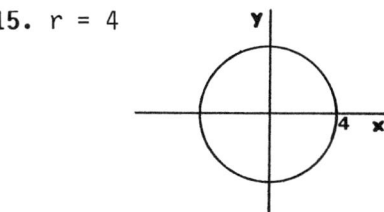

18. r = 2; $r^2 = 4$; $x^2 + y^2 = 4$.

21. $r\sin\theta - 4 = 0$; y - 4 = 0; y = 4.

24. θ = 2π/3 is a line.

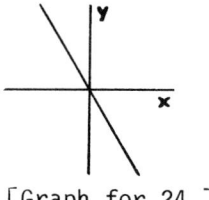

[Graph for 24.]

27. Circle.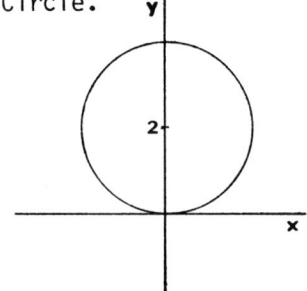

30. Hyperbola, $e = 2$, $\theta_0 = 0$, $d = 2$.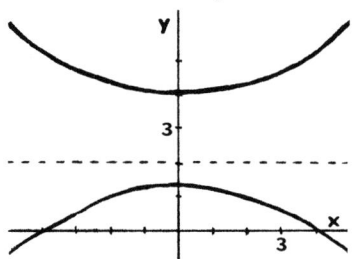

33. Parabola, $e = 1$, $\theta_0 = 0$, $d = 2$.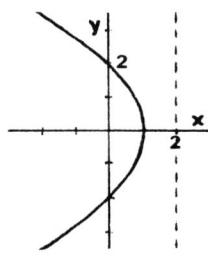

36. Line, $\theta_0 = \pi/3$, $d = 4/3$.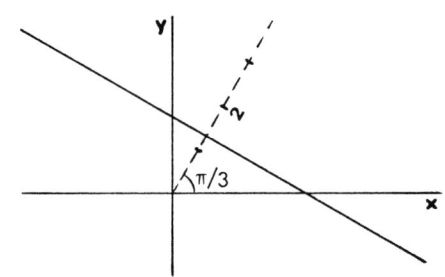

Problem Set 12.7 Graphs of Polar Equations

3. $r\sin\theta + 6 = 0$; $y + 6 = 0$; $y = -6$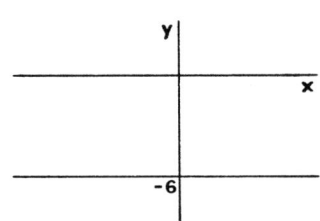

6. $r = 4\cos\theta$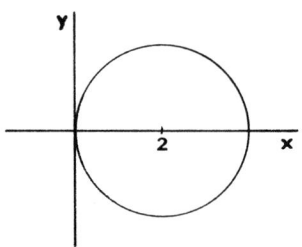

9. $r = 5 - 5\sin\theta$. y-axis symmetry since $r = 5 - 5\sin(\pi - \theta) = 5 - 5\sin\theta$.

r	θ
10	-90°
9.33	-60°
7.50	-30°
6.29	-15°
5	0°
3.71	15°
2.5	30°
0.67	60°
0	90°

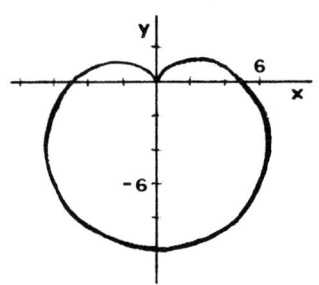

12. $r = 4+4\sin\theta$. y-axis symmetry since $r = 4+4\sin(\pi-\theta) = 4+4\sin\theta$.

r	θ
0	-90°
0.54	-60°
2	-30°
2.96	-15°
4	0°
5.04	15°
6	30°
7.46	60°
8	90°

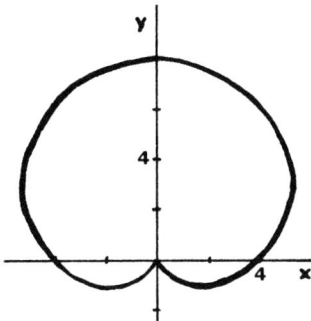

15. $r = 4-3\sin\theta$. y-axis symmetry since $4-3\sin(\pi-\theta) = 4-3\sin\theta$.

r	θ
7	-90°
6.60	-60°
5.50	-30°
4.77	-15°
4	0°
3.22	15°
2.50	30°
1.40	60°
1	90°

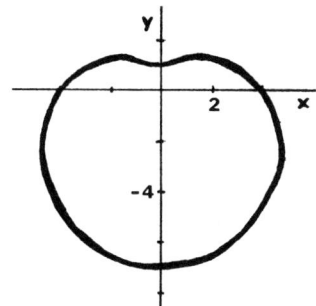

18. $r^2 = 4\cos 2\theta$. Origin symmetry since $(-r)^2 = r^2$; x-axis symmetry since $4\cos 2(-\theta) = 4\cos(-2\theta) = 4\cos 2\theta$; y-axis symmetry follows from these two.

r	θ
2	0°
1.86	15°
1.41	30°
0	45°

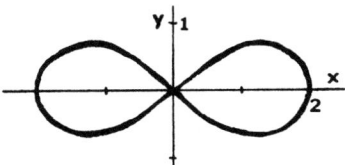

21. $r = 5\cos 3\theta$. x-axis symmetry since $5\cos 3(-\theta) = 5\cos(-3\theta) = 5\cos 3\theta$.

r	θ
5	0°
4.83	5°
4.33	10°
3.54	15°
2.50	20°
1.29	25°
0	30°

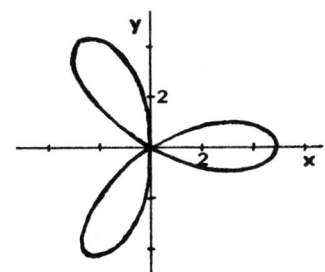

Problem Set 12.7

24. r = 4cos2θ. x-axis symmetry since 4cos2(-θ) = 4cos(-2θ) = 4cos2θ.
    y-axis symmetry since 4cos2(π-θ) = 4cos(2π-2θ) = 4cos(-2θ) = 4cos2θ.
    Origin symmetry follows from the other two.

r	θ
4	0°
3.94	5°
3.76	10°
3.46	15°
3.06	20°
2.57	25°
2	30°
1.37	35°
0.69	40°
0	45°

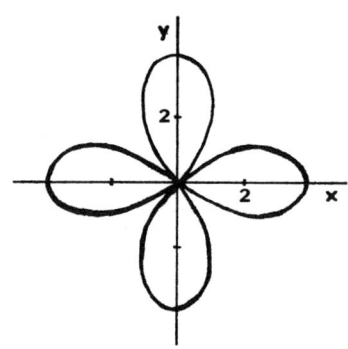

27. r = (1/2)θ, θ ≥ 0.   [Note that θ is in radians.]

r	θ
0	0
π/4	π/2
π/2	π
3π/4	3π/2
π	2π
5π/4	5π/2
3π/2	3π
7π/4	7π/2
2π	4π

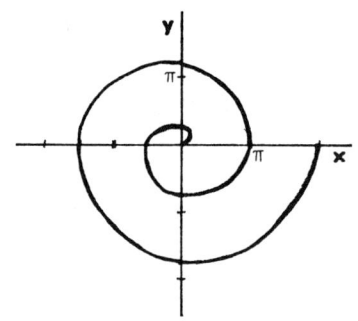

30. r = e^(θ/2), θ ≥ 0.   [Note that θ is in radians.]

r	θ
1	0
1.48	π/4
2.19	π/2
3.25	3π/4
4.81	π
7.12	5π/4
10.55	3π/2
15.63	7π/4
23.14	2π

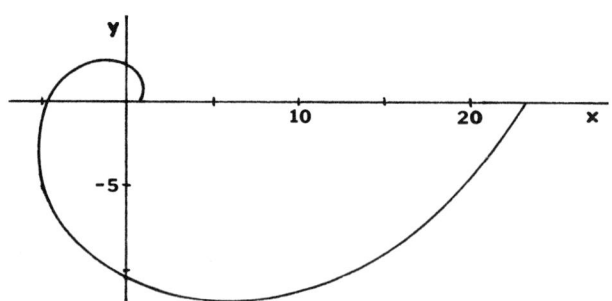

33. $r = 6$ [circle]; $r = 4+4\cos\theta$ [cardioid]

$6 = 4+4\cos\theta$

$2 = 4\cos\theta$

$\cos\theta = 1/2$

$\theta = \pm\pi/3$ and $r = 6$

Points: $(6,\pi/3)$ and $(6,-\pi/3)$

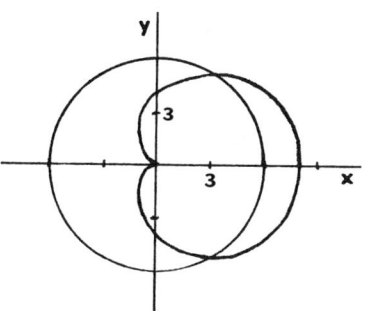

36. $r = \dfrac{5}{1-2\cos\theta} = \dfrac{2(5/2)}{1 + 2\cos(\theta-\pi)}$ [hyperbola]; $r = 5$ [circle]

$5 = 5/(1-2\cos\theta)$

$1-2\cos\theta = 1$

$\cos\theta = 0$, so $\theta = \pm\pi/2$; $r = 5$

Points: $(5,\pi/2)$ and $(5,-\pi/2)$
$(5,\pi)$ is a 3rd point.

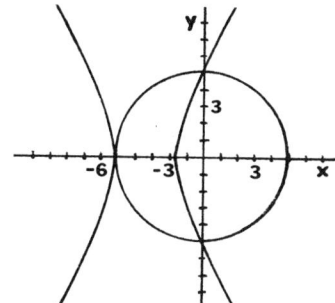

Problem Set 12.8 Calculus in Polar Coordinates

3. $r = 3+\cos\theta$ [limacon, $a > b$ case]

Making use of symmetry,

$\text{Area} = 2\displaystyle\int_0^{\pi} (1/2)(3+\cos\theta)^2 d\theta$

$= \displaystyle\int_0^{\pi} \left(9 + 6\cos\theta + \dfrac{1+\cos 2\theta}{2}\right) d\theta$

$= \left[9\theta + 6\sin\theta + \dfrac{\theta}{2} + \dfrac{\sin 2\theta}{4}\right]_0^{\pi} = 9\pi + \dfrac{\pi}{2} = \dfrac{19\pi}{2} \approx 29.8451$

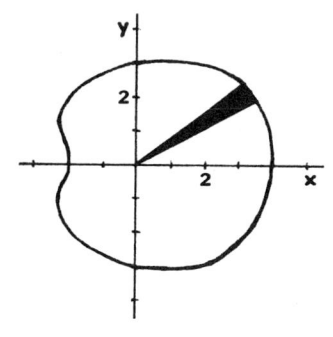

6. r = 7-7sinθ [cardioid]

$$\text{Area} = \int_0^{2\pi} (1/2)(7-7\sin\theta)^2 d\theta$$

$$= \frac{49}{2}\int_0^{2\pi} \left(1 - 2\sin\theta + \frac{1-\cos 2\theta}{2}\right) d\theta$$

$$= \frac{49}{2}\left[\theta + 2\cos\theta + \frac{\theta}{2} - \frac{\sin 2\theta}{4}\right]_0^{2\pi} = \frac{49}{2}[(2\pi+2+\pi) - (2)] = \frac{147\pi}{2} \approx 230.907$$

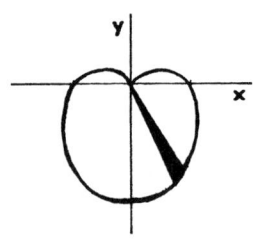

9. r^2 = 4sin2θ [lemniscate]

Making use of symmetry,

$$\text{Area} = 4\int_0^{\pi/4} (1/2)4\sin 2\theta \, d\theta$$

$$= 4\left[-\cos 2\theta\right]_0^{\pi/4} = 4[0 - (-1)] = 4$$

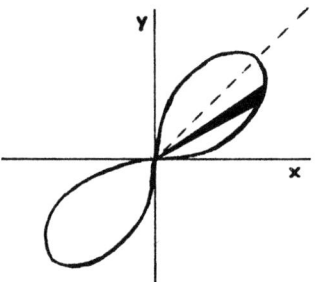

12. r = 3-6sinθ

Tangents at pole: Let r=0. Then sinθ = 1/2
θ = ±π/6

Half of the small loop is traced out as θ goes from π/6 to π/2.

$$\text{Area} = 2\int_{\pi/6}^{\pi/2} (1/2)(3-6\sin\theta)^2 d\theta$$

$$= \int_{\pi/6}^{\pi/2} [9 - 36\sin\theta + (18 - 18\cos 2\theta)] d\theta$$

$$= \left[9\theta + 36\cos\theta + 18\theta - 9\sin 2\theta\right]_{\pi/6}^{\pi/2} = 9\pi - \frac{27\sqrt{3}}{2} \approx 4.8916$$

15. $r = 4\cos 3\theta$

Tangents at pole: Let $r=0$. Then $\cos 3\theta = 0$.
$3\theta = \pm\pi/2, \pm 3\pi/2, \ldots$
$\theta = \pm\pi/6, \pm\pi/2, \ldots$

Area $= 6\int_0^{\pi/6} (1/2)(4\cos 3\theta)^2 d\theta$

$= 3\int_0^{\pi/6} (8 + 8\cos 6\theta) d\theta$

$= 3\left[8\theta + \dfrac{8\sin 6\theta}{6}\right]_0^{\pi/6} = 4\pi \approx 12.5664$

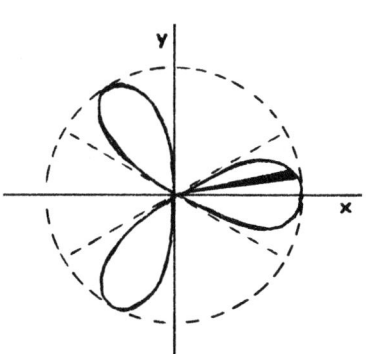

18. Points of intersection are $(1.5, \pi/6)$ and $(1.5, 5\pi/6)$.

Area $= 2\int_{\pi/6}^{\pi/2} (1/2)[(3\sin\theta)^2 - (1+\sin\theta)^2] d\theta$

$= \int_{\pi/6}^{\pi/2} (8\sin^2\theta - 1 - 2\sin\theta) d\theta$

$= \int_{\pi/6}^{\pi/2} (3 - 4\cos 2\theta - 2\sin\theta) d\theta = \left[3\theta - 2\sin 2\theta + 2\cos\theta\right]_{\pi/6}^{\pi/2} = \pi$

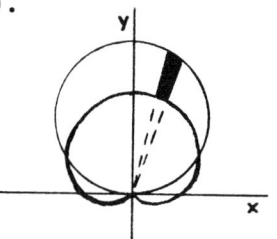

21. Point of intersection is where $\theta = \pi/4$.

Area $= \int_0^{\pi/4} (1/2)[(3+3\cos\theta)^2 - (3+3\sin\theta)^2] d\theta$

$= \int_0^{\pi/4} \left[9\cos\theta - 9\sin\theta + \dfrac{9(1+\cos 2\theta)}{4} - \dfrac{9(1-\cos 2\theta)}{4}\right] d\theta$

$= \int_0^{\pi/4} \left(9\cos\theta - 9\sin\theta + \dfrac{9\cos 2\theta}{2}\right) d\theta = \left[9\sin\theta + 9\cos\theta + \dfrac{9\sin 2\theta}{4}\right]_0^{\pi/4}$

$= \left(9\sqrt{2} + \dfrac{9}{4}\right) - 9 \approx 5.9779$

Problem Set 12.8

24. Let $r = f(\theta) = a(1+\cos\theta)$. Then $f'(\theta) = -a\sin\theta$.

Slope at θ is $\dfrac{a(1+\cos\theta)\cos\theta + (-a\sin\theta)\sin\theta}{-a(1+\cos\theta)\sin\theta + (-a\sin\theta)\cos\theta} = \dfrac{(2\cos\theta - 1)(\cos\theta + 1)}{-(1+2\cos\theta)\sin\theta}$

(a) A tangent line is horizontal if the numerator (in slope formula) is 0 but the denominator is not 0. In the interval $[-\pi,\pi]$, this occurs at $\pm\pi/3$. The points are $(3a/2,\pi/3)$ and $(3a/2,-\pi/3)$.

(b) A tangent line is vertical if the denominator is 0 but the numerator is not 0. In the interval $[-\pi,\pi]$, this occurs at 0 and at $\pm 2\pi/3$. The points are $(2,0)$, $(a/2,2\pi/3)$, and $(a/2,-2\pi/3)$.

[Note: At π, both the numerator and the denominator are 0. Using l'Hopital's Rule to help evaluate the limit of the slope as $\theta \to \pi$ shows that the slope is approaching 0. There is a cusp at the point $(0,\pi)$.]

27. Let $f(\theta) = a(1+\cos\theta)$. Then $f'(\theta) = -a\sin\theta$. Making use of symmetry,

$$\text{Perimeter} = 2\int_0^\pi [a^2(1+\cos\theta)^2 + a^2\sin^2\theta]^{1/2} d\theta$$

$$= 2a\int_0^\pi (1 + 2\cos\theta + \cos^2\theta + \sin^2\theta)^{1/2} d\theta = 2a\int_0^\pi [2(1+\cos\theta)]^{1/2} d\theta$$

$$= 2a\int_0^\pi [4\cos^2(\theta/2)]^{1/2} d\theta = 4a\int_0^\pi \cos(\theta/2) d\theta = 4a\Big[2\sin(\theta/2)\Big]_0^\pi = 8a$$

Problem Set 12.9 Chapter Review Problems

True-False Quiz

3. False. $e = \dfrac{\text{distance to focus}}{\text{distance to directrix}} < 1$ so focus is closer.

6. True. $b < a$. C is between the circumference of a circle of radius b $(2\pi b)$ and the circumference of a circle of radius a $(2\pi a)$.

9. False. It represents the two intersecting lines, $y = \pm x$.

12. False. If $k < 0$, the graph is the empty set.

15. True. See write-up of Problem 27 of Section 12.3.

18. True. $a=4$, $c=1$, so $b^2 = 16-1 = 15$, minor diameter $= 2b = 2\sqrt{15} = \sqrt{60}$.

21. False. c^{ex}: For $B = C = E = 0$, $A = 1$, $D = -3$, $F = 2$ $[x^2-3x+2 = 0]$, the graph is two parallel lines, which can not be obtained from such an intersection.

24. False. There are also ellipses and hyperbolas whose axes are the lines $y = \pm x$ that do so.

27. False. See Example 5 on page 540 of the text.

30. True. See the graph in Problem 15 of Section 12.8. The graph is inscribed in the circle of radius 4 centered at the origin. The leaves are in three of the six regions the circle is divided into by the tangents at the pole. Hence, the area is less than half that of the circle.

Miscellaneous Problems

3. $\dfrac{x^2}{4} + \dfrac{y^2}{9} = 1$

Vertical ellipse with vertices $(0,\pm 3)$.

$c^2 = a^2 - b^2 = 9-4 = 5$; foci $(0,\pm\sqrt{5})$.

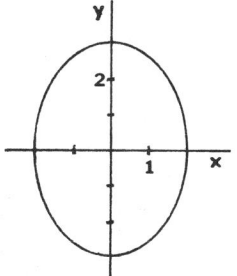

6. $\dfrac{x^2}{16} - \dfrac{y^2}{4} = 1$

Horizontal hyperbola with vertices $(\pm 4, 0)$.

$c^2 = a^2 + b^2 = 16+4 = 20$ so foci are $(\pm\sqrt{20}, 0)$.

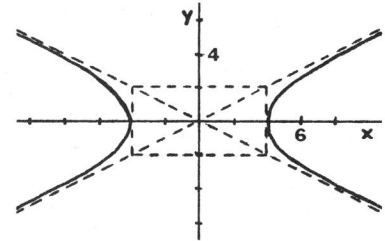

Problem Set 12.9

9. $r = \dfrac{(5/2)(1)}{1+1\sin\theta}$; $e = 1$, $d = \dfrac{5}{2}$.

   Vertical parabola opening downward with vertex $(5/4, \pi/2)$ and focus $(0,0)$.

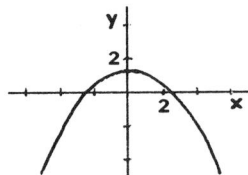

12. Parabola (since $e = 1$) that is vertical (since focus and vertex are on the vertical line $x = 0$) and which opens downward (since the focus is below the vertex); $p = 0-(-3) = 3$. An equation is

    $x^2 = -4(3)y$ or $x^2 = -12y$.

15. Hyperbola (since there are asymptotes) that is horizontal (since the vertices are on the horizontal line $y = 0$); $a = 2$.

    $\dfrac{b}{a} = \dfrac{1}{2}$ (from asymptotes), so $b = 1$. An equation is $\dfrac{x^2}{4} - \dfrac{y^2}{1} = 1$.

18. Hyperbola is vertical (vertices are on vertical line $x = 2$); center is $(2,3)$ (midway between vertices); $2a = 6-0$, so $a=3$; $c=ae = 3(10/3) = 10$. $b^2 = c^2-a^2 = 100-9 = 91$. An equation is $\dfrac{-(x-2)^2}{91} + \dfrac{(y-3)^2}{9} = 1$.

21. $(x^2+8x+16) = -6y-28+16$

    $(x+4)^2 = -6(y+2)$

    Vertical parabola opening downward.

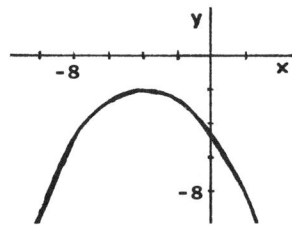

24. $\cot 2\theta = \dfrac{A-C}{B} = \dfrac{7-1}{8} = \dfrac{3}{4}$

    $\cos 2\theta = 3/5 = 0.6$

    $\sin^2\theta = \dfrac{1-\cos 2\theta}{2} = \dfrac{1-0.6}{2} = 0.2$; $\cos^2\theta = 1-\sin^2\theta = 0.8$; $\tan\theta = \dfrac{\sin\theta}{\cos\theta} = \dfrac{1}{2}$.

    $x = \sqrt{0.8}u - \sqrt{0.2}v = \sqrt{0.2}(2u-v)$; $y = \sqrt{0.2}u + \sqrt{0.8}v = \sqrt{0.2}(u+2v)$

    Substituting these expressions for x and y into $7x^2+8xy+y^2 = 9$ yields

    $9u^2-v^2 = 9$ or $\dfrac{u^2}{1} - \dfrac{v^2}{9} = 1$ which represents a hyperbola.

27. $r = \cos 2\theta$ is a four-leaved rose.

For tangents at the pole, let $r = 0$.
Then $2\theta = \pm\pi/2, \pm 3\pi/2, \ldots$, so
$\theta = \pm\pi/4, \pm 3\pi/4, \ldots$

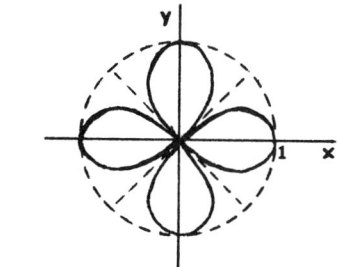

30. $r = 5 - 5\cos\theta$ is a cardioid.
(a=b case)

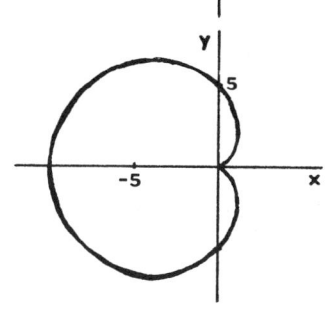

33. $\theta = 2\pi/3$ is a line.

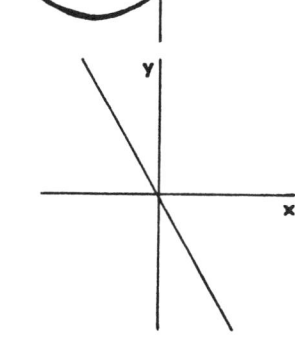

36. $r = -\theta$, $\theta \geq 0$ is a spiral of Archimedes.

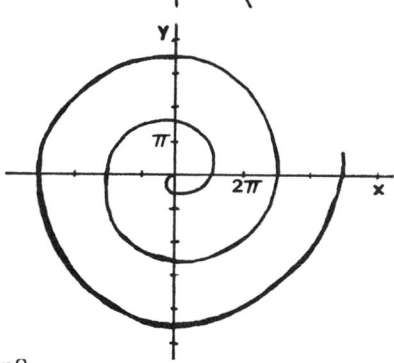

39. Let $r = f(\theta) = 3 + 3\cos\theta$. Then $f'(\theta) = -3\sin\theta$.

$f(\pi/6) = 3 + 3(\sqrt{3}/2)$ and $f'(\pi/6) = -3(1/2) = -3/2$. Then the slope of the tangent where $\theta = \pi/6$ is $\dfrac{(3+3\sqrt{3}/2)(\sqrt{3}/2) + (-3/2)(1/2)}{-(3+3\sqrt{3}/2)(1/2) + (-3/2)(\sqrt{3}/2)}$

$= \dfrac{6\sqrt{3} + 9 - 3}{-6 - 3\sqrt{3} - 3\sqrt{3}} = -1.$

Problem Set 12.9

42. Points of intersection are $(2.5, \pi/6)$ and $(2.5, 5\pi/6)$.

$$\text{Area} = 2\int_{\pi/6}^{\pi/2} (1/2)[(5\sin\theta)^2 - (2+\sin\theta)^2]d\theta$$

$$= \int_{\pi/6}^{\pi/2} (24\sin^2\theta - 4 - 4\sin\theta)d\theta$$

$$= \int_{\pi/6}^{\pi/2} (8 - 12\cos 2\theta - 4\sin\theta)d\theta$$

$$= \left[8\theta - 6\sin 2\theta + 4\cos\theta\right]_{\pi/6}^{\pi/2}$$

$$= 4\pi - \left(\frac{4\pi}{3} - 3\sqrt{3} + 2\sqrt{3}\right)$$

$$= \frac{8\pi}{3} + \sqrt{3}$$

$$\approx 10.1096.$$

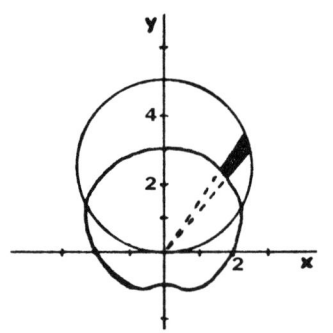

CHAPTER 13 GEOMETRY IN THE PLANE, VECTORS

Problem Set 13.1 Plane Curves: Parametric Representation

3. $x = t-4$, $y = t^{1/2}$, t in $[0,4]$

(a)
t	x	y
0	-4	0
1	-3	1
2	-2	1.41
3	-1	1.73
4	0	2

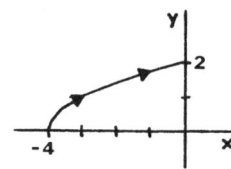

(b) Simple; not closed

(c) $t = y^2$
$x = y^2 - 4$, y in $[0,2]$

6. $x = t^2-1$, $y = t^3-t$, t in $[-3,3]$

(a)
t	x	y
-3	8	-24
-2	3	-6
-1	0	0
-0.5	-0.75	0.38
0	-1	0
0.5	-0.75	0.38
1	0	0
2	3	6
3	8	24

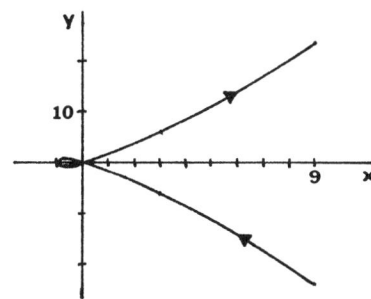

(b) Not simple; not closed

(c) $t^2 = x+1$, and
$y^2 = t^2(t^2-1)^2$
Therefore,
$y^2 = (x+1)^2 x^2$

9. $x = 3\sin t$, $y = 5\cos t$, t in $[0,2\pi]$

(a)
t	x	y
0	0	5
π/6	1.50	4.33
π/4	2.12	3.54
π/3	2.60	2.50
π/2	3	0

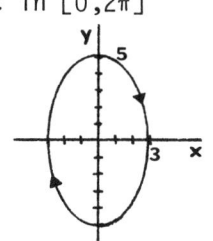

(b) Simple; closed

(c) $x/3 = \sin t$, $y/3 = \cos t$
Therefore,
$(x/3)^2 + (y/5)^2 = 1$

12. $x = 2\cos\theta$, $y = 2\cos(\theta/2)$, θ in R

(a)
θ	x	y
0	2	2
π/3	1	1.73
π/2	0	1.41
2π/3	-1	1
π	-2	0
4π/3	-1	-1
3π/2	0	-1.41
5π/3	1	-1.73
2π	2	-2

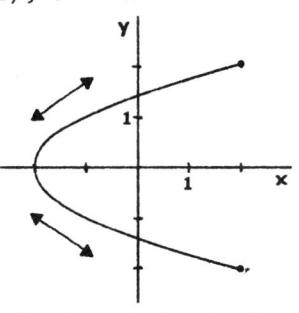

(b) Not simple; not closed

(c) $y^2 = 4\cos^2(\theta/2)$
$= 2 + 2\cos\theta$

Then $y^2 = 2 + x$,
y in $[-2,2]$

The point (x,y) travels back and forth along part of a parabola.

15. $\dfrac{dy}{dx} = \dfrac{dy/dt}{dx/dt} = \dfrac{2-3t^{-2}}{2+3t^{-2}} = \dfrac{2t^2-3}{2t^2+3}.$

$\dfrac{d^2y}{dx^2} = \dfrac{dy'}{dx} = \dfrac{dy'/dt}{dx/dt} = \dfrac{[(2t^2+3)4t - (2t^2-3)4t]}{(2t^2+3)^2} \cdot \dfrac{t^2}{2t^2+3} = \dfrac{24t^3}{(2t^2+3)^3}.$

18. $dy/dt = -2(t+t^3)^{-2}(1+3t^2) = -2t^{-2}(1+t^2)^{-2}(1+3t^2); \; dx/dt = -2(1+t^2)^{-2}2t.$

Thus, $\dfrac{dy}{dx} = \dfrac{dy/dt}{dx/dt} = \dfrac{t^{-2}(1+3t^2)}{2t} = \dfrac{1+3t^2}{2t^3}.$

$\dfrac{d^2y}{dx^2} = \dfrac{dy'}{dx} = \dfrac{(-3/2)t^{-4} - (3/2)t^{-2}}{-2(1+t^2)^{-2}2t} = \dfrac{-3t^{-4}(1+t^2)}{-8t(1+t^2)^{-2}} = \dfrac{3(1+t^2)^3}{8t^5}.$

21. $\dfrac{dy}{dx} = \dfrac{dy/dt}{dx/dt} = \dfrac{2\sec^2 t}{2\sec t \tan t} = \csc t.$ At $t = \dfrac{-\pi}{6},\; x = \dfrac{4\sqrt{3}}{3},\; y = \dfrac{-2\sqrt{3}}{3},\; y' = -2.$

Equation of Tangent: $y + \dfrac{2\sqrt{3}}{3} = -2(x - \dfrac{4\sqrt{3}}{3})$ or $y = -2x + 2\sqrt{3}.$

24. $\displaystyle\int_1^{\sqrt{3}} xy\, dy$

$\quad\quad dy = \sec^2 t\, dt$
$\quad\quad y = \sqrt{3} \Rightarrow t = \tan^{-1}\sqrt{3} = \pi/3$
$\quad\quad y = 1 \Rightarrow t = \tan^{-1}(1) = \pi/4$

$= \displaystyle\int_{\pi/4}^{\pi/3} \sec t \tan t \sec^2 t\, dt$

[Use $u = \sec t$ substitution if next step is not clear.]

$= \left[\dfrac{\sec^3 t}{3}\right]_{\pi/4}^{\pi/3} = \dfrac{8}{3} - \dfrac{2\sqrt{2}}{3} \approx 1.7239.$

27. Add the points Q and T (as indicated to the right) to the Figure 6 on page 554 of the text. The coordinates of Q are:

$x = |ON| - |QT| = at - b\sin t,$
$y = |NC| + |CT| = a - b\cos t.$

For $x = 8t - 4\sin t,\; y = 8 - 4\cos t.$

t	x	y
0	0	4
π/2	4π−4	8
π	8π	12
3π/2	12π+4	8
2π	16π	4

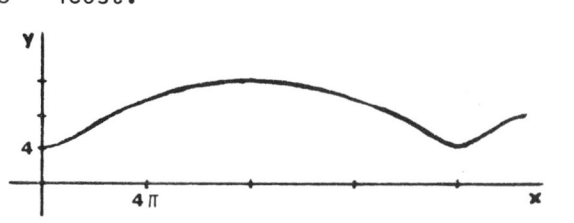

30. Refer to Figure 6 on page 554 of the text.

For the cycloid: $\dfrac{dy}{dx} = \dfrac{dy/dt}{dx/dt} = \dfrac{a\sin t}{a(1-\cos t)} = \dfrac{\sin t}{1-\cos t}$.

At $t=t_1$: $x = a(t_1 - \sin t_1)$, $y = a(1 - \cos t_1)$, $y' = \dfrac{\sin t_1}{1-\cos t_1}$.

Tangent is $y - a(1-\cos t_1) = \dfrac{\sin t_1}{1-\cos t_1}[x - a(t_1 - \sin t_1)]$. $[\cos t_1 \neq 1]$

At $t = t_1$, the highest point of the circle is at $(at_1, 2a)$. Substituting those coordinates into the equation of the tangent line above and simplifying each side, we obtain an identity, so the point is on the tangent line.

It can be similarly shown that the normal goes through $(at_1, 0)$, the lowest point of the circle.

33. Use the figure below to help obtain the coordinates of P. Note that:
  (1) $b\beta$ (length of arc BP) = at (length of arc AB), so $\beta = at/b$,
  (2) Therefore, $\beta + t = \dfrac{at}{b} + t = \left(\dfrac{a+b}{b}\right)t$, and
  (3) the hypotenuse of triangle ONC is $a+b$.

$\begin{aligned}
x &= |OM| = |ON| + |TP| \\
&= (a+b)\cos t + b\sin[\beta - (\tfrac{\pi}{2} - t)] \\
&= (a+b)\cos t - b\cos(\beta + t) \\
&= (a+b)\cos t - b\cos\left(\tfrac{a+b}{b}\right)t.
\end{aligned}$

$\begin{aligned}
y &= |MP| = |NC| - |CT| \\
&= (a+b)\sin t - b\cos[\beta - (\tfrac{\pi}{2} - t)] \\
&= (a+b)\sin t - b\sin(\beta + t) \\
&= (a+b)\sin t - b\cos\left(\tfrac{a+b}{b}\right)t.
\end{aligned}$

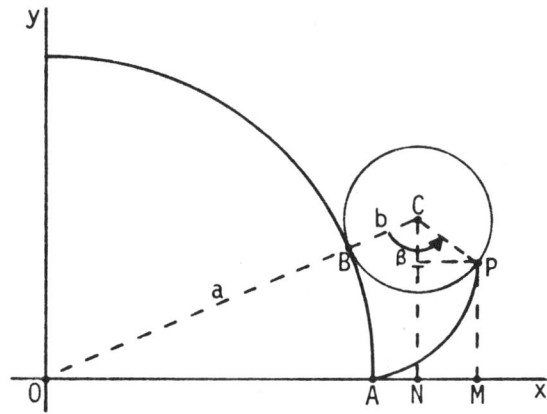

Problem Set 13.1

Problem Set 13.2 Vectors in the Plane: Geometric Approach

3.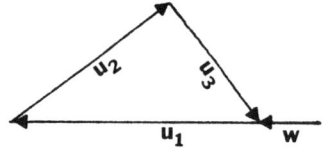

6. $m = (1/2)(u + v)$ and $n = (1/2)(v - u)$.

9. u and v together have a N-S force of
$|v|\sin(45°) + |u|\sin(120°)$
$= 10(\sqrt{2}/2) + 10(\sqrt{3}/2) \approx 15.73$;
E-W force of $|v|\cos(45°) + |u|\cos(120°)$
$= 10(\sqrt{2}/2) + 10(-1/2) \approx 2.07$.

$\alpha = \tan^{-1}(2.07/15.73) \approx 7.50°$.
$[(2.07)^2 + (15.73)^2]^{1/2} \approx 15.87$

Thus, w needs to have direction S7.50°W and a magnitude of 15.87 lbs.

12. The vertical component of the resultant force exerted by the ropes is 237.5 lbs. so $|v|\sin(62.66°) + |w|\sin(129.22°) = 237.5$.

The horizontal component of the resultant force exerted by the ropes is 0 lbs. so $|v|\cos(62.66°) + |w|\cos(129.22°) = 0$.

Solving these two equations simultaneously, we obtain $|v| \approx 163.68$ and $|w| \approx 118.89$.

15. Magnitude of vertical components are equal. Therefore, $|p|\sin(30°) = |w| = 80$, so $|p| = 80/\sin(30°) = 160$ mph.

18. ABC and ACD are triangles. Using the result of Problem 17, line segment EF is parallel to GH (since each is parallel to AC). Similarly, obtain the result for EG and FH using triangles ABD and BDC. Therefore, EFHG is a parallelogram since opposite sides are parallel.

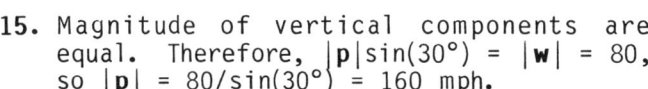

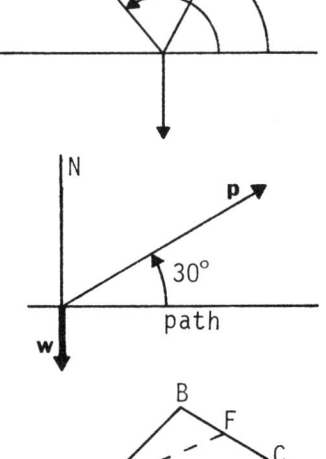

Problem Set 13.3 Vectors in the Plane: Algebraic Approach

3. (a) $\cos\theta = \dfrac{\langle 2,-3\rangle \cdot \langle -1,4\rangle}{|\langle 2,-3\rangle||\langle -1,4\rangle|} = \dfrac{-2-12}{\sqrt{4+9}\sqrt{1+16}} = \dfrac{-14}{\sqrt{221}} \approx -0.9417$

   (b) $\cos\theta = \dfrac{\langle -5,-2\rangle \cdot \langle 6,0\rangle}{|\langle -5,-2\rangle||\langle 6,0\rangle|} = \dfrac{-30+0}{\sqrt{25+4}\sqrt{36+0}} = \dfrac{-5}{\sqrt{29}} \approx -0.9285$

   (c) $\cos\theta = \dfrac{\langle -3,-1\rangle \cdot \langle -2,-4\rangle}{|\langle -3,-1\rangle||\langle -2,-4\rangle|} = \dfrac{6+4}{\sqrt{9+1}\sqrt{4+16}} = \dfrac{10}{\sqrt{200}} \approx 0.7071$

   (d) $\cos\theta = \dfrac{\langle 4,-5\rangle \cdot \langle -8,10\rangle}{|\langle 4,-5\rangle||\langle -8,10\rangle|} = \dfrac{-32-50}{\sqrt{16+25}\sqrt{64+100}} = -1$

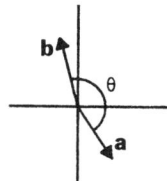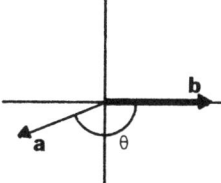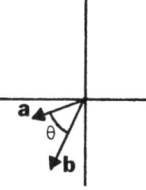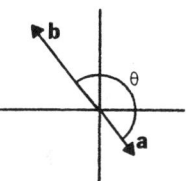

6. (a) $\mathbf{u} = \dfrac{\mathbf{a}}{|\mathbf{a}|} = \dfrac{\langle -3,4\rangle}{\sqrt{9+16}} = \langle -3/5, 4/5\rangle = -0.6\mathbf{i} + 0.8\mathbf{j}$

   (b) $\mathbf{u} = \dfrac{\langle 1,-7\rangle}{\sqrt{1+49}} = \langle 1/\sqrt{50}, -7/\sqrt{50}\rangle = (1/\sqrt{50})\mathbf{i} + (-7/\sqrt{50})\mathbf{j}$

   (c) $\mathbf{a} = -2\mathbf{j}$, so $\mathbf{u} = -\mathbf{j}$

   (d) $\mathbf{u} = \dfrac{\langle -5,-12\rangle}{\sqrt{25+144}} = \langle -5/13, -12/13\rangle = (-5/13)\mathbf{i} + (-12/13)\mathbf{j}$

9. $a\mathbf{u} + b\mathbf{u} = a\langle u_1,u_2\rangle + b\langle u_1,u_2\rangle = \langle au_1, au_2\rangle + \langle bu_1, bu_2\rangle$
   $= \langle au_1+bu_1, au_2+bu_2\rangle = \langle (a+b)u_1, (a+b)u_2\rangle = (a+b)\langle u_1,u_2\rangle = (a+b)\mathbf{u}$

12. $\mathbf{u}+\mathbf{v} = \mathbf{u}$ $\Rightarrow$ $(-\mathbf{u}) + (\mathbf{u}+\mathbf{v}) = (-\mathbf{u}) + \mathbf{u}$
    $\Rightarrow$ $[(-\mathbf{u}) + \mathbf{u}] + \mathbf{v} = \mathbf{u} + (-\mathbf{u})$ [Theorem A, (1) and (2)]
    $\Rightarrow$ $[\mathbf{u} + (-\mathbf{u})] + \mathbf{v} = \mathbf{0}$ [Theorem A, (1) and (4)]
    $\Rightarrow$ $\mathbf{0} + \mathbf{v} = \mathbf{0}$ [Theorem A, (4)]
    $\Rightarrow$ $\mathbf{v} = \mathbf{0}$ [Theorem A, (3)]

15. $\langle 6,3\rangle \cdot \langle -1,2\rangle = -6+6 = 0$, so they are perpendicular.   [Theorem C]

18. $2c\mathbf{i}-4\mathbf{j}$ and $3\mathbf{i}+c\mathbf{j}$ are orthogonal $\Leftrightarrow$ $(2c\mathbf{i}-4\mathbf{j})\cdot(3\mathbf{i}+c\mathbf{j}) = 0$
    $\Leftrightarrow$ $6c-4c = 0$
    $\Leftrightarrow$ $c = 0$

21. $r_1\mathbf{i} + r_2\mathbf{j} = k(a_1\mathbf{i} + a_2\mathbf{j}) + m(b_1\mathbf{i} + b_2\mathbf{j}) = (ka_1 + mb_1)\mathbf{i} + (ka_2 + mb_2)\mathbf{j}$

$\Leftrightarrow \quad r_1 = ka_1 + mb_1 \quad \text{and} \quad r_2 = ka_2 + mb_2.$

Solve these two equations simultaneously [noting that $a_1b_2 - a_2b_1 \neq 0$ since **a** and **b** are noncollinear] and obtain

$$k = \frac{b_2 r_1 - b_1 r_2}{a_1 b_2 - a_2 b_1} \quad \text{and} \quad m = \frac{a_1 r_2 - a_2 r_1}{a_1 b_2 - a_2 b_1}$$

24. Let $\mathbf{D} = \langle 50, 0 \rangle$, and
$\mathbf{F} = \langle 100\cos(-20°), 100\sin(-20°) \rangle$.

Work $= \mathbf{F} \cdot \mathbf{D} = 500\cos(20°) + 0 \approx 4698.46$ dyne-cm.
$\qquad = 4698.46$ ergs

27. $|\mathbf{u} \cdot \mathbf{v}| = |\cos\theta||\mathbf{u}||\mathbf{v}| \leq |\mathbf{u}||\mathbf{v}|$ since $|\cos\theta| \leq 1$.

Problem Set 13.4 Vector-Valued Functions and Curvilinear Motion

3. $\left(\lim\limits_{t \to 2} \dfrac{t-2}{t^2-4}\right)\mathbf{i} + \left(\lim\limits_{t \to 2} \dfrac{t^2+t-6}{t-2}\right)\mathbf{j} \overset{\text{\textcircled{L}}}{=} \left(\lim\limits_{t \to 2} \dfrac{1}{2t}\right)\mathbf{i} + \left(\lim\limits_{t \to 2} \dfrac{2t+1}{1}\right)\mathbf{j} = (1/4)\mathbf{i} + 5\mathbf{j}$

6. $\left(\lim\limits_{t \to \infty} \dfrac{\sin t}{t}\right)\mathbf{i} + \left(\lim\limits_{t \to \infty} \dfrac{t^2+1}{2t^2-3t}\right)\mathbf{j} \overset{\text{\textcircled{L}}}{=} 0\mathbf{i} + \left(\lim\limits_{t \to \infty} \dfrac{2t}{4t-3}\right)\mathbf{j} \overset{\text{\textcircled{L}}}{=} \left(\lim\limits_{t \to \infty} \dfrac{2}{4}\right)\mathbf{j} = (0.5)\mathbf{j}$

[In the first step the Squeeze Theorem was used: $\dfrac{-1}{t} \leq \dfrac{\sin t}{t} \leq \dfrac{1}{t}$ and both $\dfrac{-1}{t}$ and $\dfrac{1}{t}$ approach 0 as $t \to \infty$.]

9. (a) Require that $t - 2 \neq 0$ and $4 + t \geq 0$, so $t \neq 2$ and $t \geq -4$.
   The domain is $[-4, 2) \cup (2, \infty)$.

   (b) Require that $t^2 + 1 \geq 0$. The domain is **R**.

12. (a) $\ln(t^2+1)$ is continuous for all t; $\tan^{-1}t$ is continuous for all t. Then $\mathbf{r}$ is continuous for all t.

(b) $\ln(2t^{-1})$ is continuous if $t > 0$; $-(6-t)^{1/2}$ is continuous if $t < 6$. Then $\mathbf{r}$ is continuous for all t in the interval $(0,6)$.

15. $\mathbf{r}(t) = \langle e^{2t}, 3\ln t\rangle$; $\mathbf{r}'(t) = \langle 2e^{2t}, 3t^{-1}\rangle$; $\mathbf{r}''(t) = \langle 4e^{2t}, -3t^{-2}\rangle$.

$$D_t[\mathbf{r}(t)\cdot\mathbf{r}'(t)] = \mathbf{r}(t)\cdot\mathbf{r}''(t) + \mathbf{r}'(t)\cdot\mathbf{r}'(t)$$
$$= [(e^{2t})(4e^{2t}) + (3\ln t)(-3t^{-2})] + [(2e^{2t})^2 + (3t^{-1})^2]$$
$$= 8e^{4t} + 9t^{-2}(1 - \ln t).$$

18. $h(t)\mathbf{r}(t) = [\ln(3t-2)]\langle\sin 2t, \cosh t\rangle$.

$$D_t[h(t)\mathbf{r}(t)] = \frac{3}{3t-2}\langle\sin 2t, \cosh t\rangle + \ln(3t-2)\langle 2\cos 2t, \sinh t\rangle$$
$$= \langle\frac{3\sin 2t}{3t-2} + 2(\cos 2t)\ln(3t-2)), \frac{3\cosh t}{3t-2} + (\sinh t)\ln(3t-2)\rangle.$$

21. $\int_0^1 \langle e^t, e^{-t}\rangle dt = \left[\langle e^t, -e^{-t}\rangle\right]_0^1 = \langle e, -e^{-1}\rangle - \langle 1, -1\rangle = \langle e-1, 1-e^{-1}\rangle$
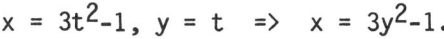
$\approx \langle 1.7182, 0.6321\rangle$.

24. $\mathbf{r}(t) = \langle 3t^2-1, t\rangle$; $\mathbf{v}(t) = \mathbf{r}'(t) = \langle 6t, 1\rangle$; $\mathbf{a}(t) = \mathbf{r}''(t) = \langle 6, 0\rangle$.

At $t_1 = 0.5$:
Position is $\mathbf{r}(0.5) = \langle -0.25, 0.5\rangle$.
Velocity is $\mathbf{v}(0.5) = \langle 3, 1\rangle$.
Speed is $|\mathbf{v}(0.5)| = \sqrt{9+1} \approx 3.1623$.
Acceleration is $\mathbf{a}(0.5) = \langle 6, 0\rangle$.

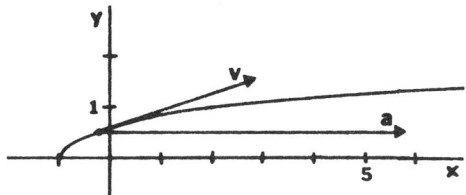

$x = 3t^2-1$, $y = t$ $\Rightarrow$ $x = 3y^2-1$.

27. $\mathbf{r}(t) = \langle 3t^2, t^3\rangle$; $\mathbf{v}(t) = \langle 6t, 3t^2\rangle$; $\mathbf{a}(t) = \langle 6, 6t\rangle$.

$\mathbf{r}(2) = \langle 12, 8\rangle$.
$\mathbf{v}(2) = \langle 12, 12\rangle$.
$|\mathbf{v}(2)| = \sqrt{144 + 144} \approx 16.97$.
$\mathbf{a}(2) = \langle 6, 12\rangle$.

$x = 3t^2$, $y = t^3$
$\Rightarrow x^3 = 27t^6$, $y^2 = t^6$
$\Rightarrow x^3 = 27y^2$.

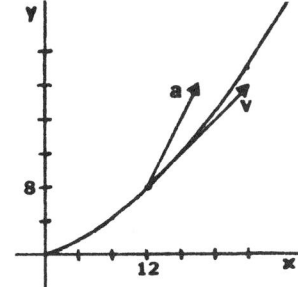

Problem Set 13.4

30. $r(t) = \langle e^{t/2}, e^{-t} \rangle$; $v(t) = \langle 0.5e^{t/2}, -e^{-t} \rangle$; $a(t) = \langle 0.25e^{t/2}, e^{-t} \rangle$.

$r(2) = \langle e, e^{-2} \rangle \approx \langle 2.7183, 0.1353 \rangle$

$v(2) = \langle 0.5e, -e^{-2} \rangle \approx \langle 1.3591, -0.1353 \rangle$

$|v(2)| = \sqrt{0.25e^2 + e^{-4}} \approx 1.3659$

$a(2) = \langle 0.25e, e^{-2} \rangle \approx \langle 0.6796, 0.1353 \rangle$

$x = e^{t/2}$, $y = e^{-t}$ $\Rightarrow$ $y = x^{-2}$

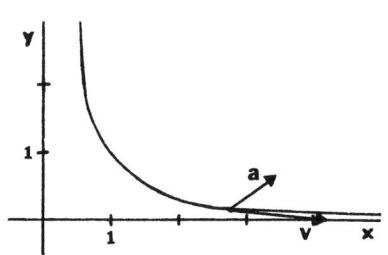

33. $a(t) = \langle 1, e^{-t} \rangle$.

$v(t) = \langle t + C_1, -e^{-t} + C_2 \rangle$. $v(0) = \langle 2, 1 \rangle$, so $C_1 = C_2 = 2$.

$v(t) = \langle t + 2, -e^{-t} + 2 \rangle$

$r(t) = \langle \frac{t^2}{2} + 2t + k_1, e^{-t} + 2t + k_2 \rangle$. $r(0) = \langle 1, 1 \rangle$, so $k_1 = 1$, $k_2 = 0$.

$r(t) = \langle \frac{t^2}{2} + 2t + 1, e^{-t} + 2t \rangle$

36. $D_t[v(t) \cdot v(t)] = D_t[c]$. Therefore, $v(t) \cdot a(t) + v(t) \cdot a(t) = 0$; $2v(t) \cdot a(t) = 0$; $v(t) \cdot a(t) = 0$. Then $v(t)$ is perdendicular to $a(t)$.

39. $r(t) = \langle \cosh \omega t, \sinh \omega t \rangle$; $v(t) = \langle \omega \sinh \omega t, \omega \cosh \omega t \rangle$;

$a(t) = \langle \omega^2 \cosh \omega t, \omega^2 \sinh \omega t \rangle = \omega^2 \langle \cosh \omega t, \sinh \omega t \rangle = \omega^2 r(t)$. $c = \omega^2$.

Problem Set 13.5 Curvature and Acceleration

3. $r(t) = \langle 4\cos t, 3\sin t \rangle$; $v(t) = \langle -4\sin t, 3\cos t \rangle$; $v(\pi/4) = \langle -2\sqrt{2}, 3\sqrt{2}/2 \rangle$.

Therefore, $T(\pi/4) = \frac{v(\pi/4)}{|v(\pi/4)|} = \langle -0.8, 0.6 \rangle$.

3. (continued)

Use Theorem A, noting: $\mathbf{v}(t) = \langle x', y' \rangle$ so $x' = -4\sin t$, $y' = 3\cos t$, $x'' = -4\cos t$, $y'' = -3\sin t$.

$$\kappa(t) = \frac{|12\sin^2 t + 12\cos^2 t|}{(16\sin^2 t + 9\cos^2 t)^{3/2}}, \quad \text{so } \kappa(\pi/4) = \frac{12}{(8 + 9/2)^{3/2}} \approx 0.2715.$$

6. $y = x(x-2)^2 = x^3 - 4x^2 + 4x$

   $y' = 3x^2 - 8x + 4 = (3x-2)(x-2)$

   $y'' = 6x - 8 = 2(3x-4)$

   $y'(2) = 0$; $y''(2) = 4$

   $\kappa(2) = \dfrac{|4|}{(1+0)^{3/2}} = 4$; $R(2) = \dfrac{1}{4}$

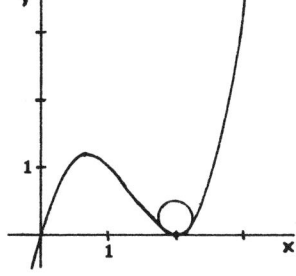

9. $y = e^x - x$; $y' = e^x - 1$; $y'' = e^x$

   $y'(0) = 0$; $y''(0) = 1$

   $\kappa(0) = \dfrac{|1|}{(1+0)^{3/2}} = 1$; $R(0) = 1$

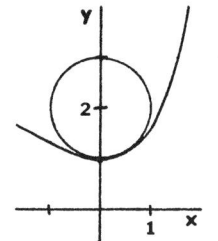

12. $y = \ln \sin x$; $y' = \cot x$; $y'' = -\csc^2 x$

    $y'(\pi/4) = 1$; $y''(\pi/4) = -2$

    $\kappa(\pi/4) = \dfrac{|-2|}{(1+1)^{3/2}} = \dfrac{1}{\sqrt{2}} \approx 0.71$

    $R(\pi/4) = \sqrt{2} \approx 1.41$

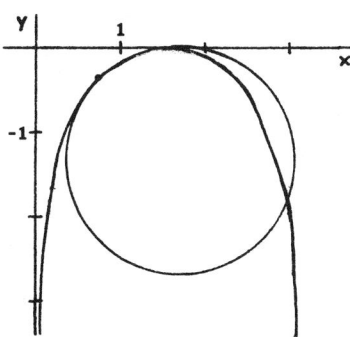

15. $y = \ln x$; $y' = x^{-1}$; $y'' = -x^{-2}$; $\kappa = \dfrac{|-x^{-2}|}{(1+x^{-2})^{3/2}} = \dfrac{x}{(x^2+1)^{3/2}}$

    $\kappa'(x) = \dfrac{1-2x^2}{(x^2+1)^{5/2}}$; Aux. axis for κ':

    ```
         (+)      (0)      (-)
    ←――――――――|――――――――――|―――――――→
    0              √2/2              x
    ```

    κ is maximum at $x = \sqrt{2}/2$, $y = \ln(\sqrt{2}/2) = -\ln\sqrt{2}$; i.e. at $(\sqrt{2}/2, -\ln\sqrt{2})$.

Problem Set 13.5

18. $y = \cosh x$; $y' = \sinh x$; $y'' = \cosh x$.

$$\kappa = \frac{|\cosh x|}{(1 + \sinh^2 x)^{3/2}} = \frac{\cosh x}{(\cosh^2 x)^{3/2}} = \frac{1}{\cosh^2 x} = \mathrm{sech}^2 x$$

$\kappa'(x) = -2\mathrm{sech}^2 x \tanh x = \dfrac{-2 \sinh x}{\cosh^3 x}$; κ': $\underset{}{\underset{0}{\xrightarrow{\hspace{2cm}(+)\hspace{1cm}(0)\hspace{1cm}(-)\hspace{1cm}}}}$ x

κ is maximum where $x = 0$, $y = \cosh(0) = 1$; i.e., at the point $(0,1)$.

21. $\mathbf{r}(t) = \langle -a\sin t, a\cos t\rangle$; $\mathbf{v}(t) = \langle -a\cos t, -a\sin t\rangle$; $\mathbf{a}(t) = \langle a\sin t, -a\cos t\rangle$.

$|\mathbf{v}(t)| = (a^2\cos^2 t + a^2\sin^2 t)^{1/2} = |a|$; similarly, $|\mathbf{a}(t)| = |a|$.

$a_T = \dfrac{d^2 s}{dt^2} = \dfrac{d}{dt}|\mathbf{v}(t)| = \dfrac{d}{dt}|a| = 0$; $a_N^2 = |\mathbf{a}(t)|^2 - a_T^2 = a^2$, so $a_N = |a|$.

24. $\mathbf{r}(t) = \langle a\cos\omega t, b\sin\omega t\rangle$; $\mathbf{v}(t) = \langle -a\omega\sin\omega t, b\omega\cos\omega t\rangle$;

$\mathbf{a}(t) = \langle -a\omega^2\cos\omega t, -b\omega^2\sin\omega t\rangle = -\omega^2 \mathbf{r}(t)$.

$\mathbf{T} = \dfrac{\mathbf{v}}{(\mathbf{v}\cdot\mathbf{v})^{1/2}}$; $\dfrac{d\mathbf{T}}{dt} = \dfrac{(\mathbf{v}\cdot\mathbf{v})\mathbf{a} - (\mathbf{v}\cdot\mathbf{a})\mathbf{v}}{(\mathbf{v}\cdot\mathbf{v})^{3/2}} = \dfrac{-ab\langle b\cos\omega t, a\sin\omega t\rangle}{(a^2\sin^2\omega t + b^2\cos^2\omega t)^{3/2}}$

$|d\mathbf{T}/dt| = \dfrac{ab(b^2\cos^2\omega t + a^2\sin^2\omega t)^{1/2}}{(a^2\sin^2\omega t + b^2\cos^2\omega t)^{3/2}} = \dfrac{ab}{a^2\sin^2\omega t + b^2\cos^2\omega t}$

Then $\dfrac{d\mathbf{T}/dt}{|d\mathbf{T}/dt|} = \dfrac{-\langle b\cos t, a\sin t\rangle}{(a^2\sin^2\omega t + b^2\cos^2\omega t)^{1/2}}$.

[This was done assuming $ab > 0$; if $ab < 0$, drop the $(-)$ in numerator.]

27. $s''(t) = a_T = 0 \Rightarrow$ speed $= s'(t) = c$ (a constant).

$\kappa(ds/dt)^2 = a_N = 0 \Rightarrow \kappa = 0$ or $\dfrac{ds}{dt} = 0$.

$\Rightarrow \kappa = 0$

Problem Set 13.6 Chapter Review Problems

True-False Quiz

3. False. But we can if f^{-1} exists.

6. True. See first curve in Figure 1 on page 552 of the text. There are two tangents at the points where the curve crosses itself.

9. False. $(\mathbf{a} \cdot \mathbf{b}) \cdot \mathbf{c}$ is not defined since $\mathbf{a} \cdot \mathbf{b}$ is a scalar.

12. False. Let $\mathbf{u}$, $\mathbf{v}$, and $\mathbf{u}+\mathbf{v}$ represent the three sides of an equilateral triangle. $\mathbf{u} = \langle 2,0 \rangle$, $\mathbf{v} = \langle -1, \sqrt{3} \rangle$, $\mathbf{u}+\mathbf{v} = \langle 1, \sqrt{3} \rangle$ provide a counterexample.

15. True. See definitions of continuity and limit (page 571).

18. False. The curve is a circle with radius 2 so the curvature is 1/2. (See Example 2 on page 579.)

Miscellaneous Problems

3. $\dfrac{x+2}{4} = \sin t$ and $\dfrac{y+1}{3} = \cos t$

   $\dfrac{(x+2)^2}{16} + \dfrac{(y-1)^2}{9} = 1$

   [since $\sin^2 t + \cos^2 t = 1$]

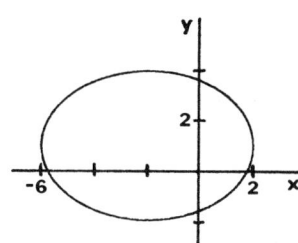

6. $\dfrac{dy}{dx} = \dfrac{dy/dt}{dx/dt} = \dfrac{0.5e^t}{-3e^{-t}} = \dfrac{e^{2t}}{-6}$. At $t = 0$, $x = 3$, $y = \dfrac{1}{2}$, $y' = \dfrac{-1}{6}$.

   Tangent: $y-(1/2) = (-1/6)(x-3)$ or $y = (-1/6)x+1$
   Normal: $y-(1/2) = 6(x-3)$ or $y = 6x-17.5$

9. (a) $3\langle 2,-5 \rangle - 2\langle 1,1 \rangle = \langle 6,-15 \rangle - \langle 2,2 \rangle = \langle 4,-17 \rangle$
   (b) $\langle 2,-5 \rangle \cdot \langle 1,1 \rangle = 2 + (-5) = -3$
   (c) $\langle 2,-5 \rangle \cdot (\langle 1,1 \rangle + \langle -6,0 \rangle) = \langle 2,-5 \rangle \cdot \langle -5,1 \rangle = -10 + (-5) = -15$
   (d) $(4\langle 2,-5 \rangle + 5\langle 1,1 \rangle) \cdot 3\langle -6,0 \rangle = \langle 13,-15 \rangle \cdot \langle -18,0 \rangle = -234 + 0 = -234$
   (e) $\sqrt{36+0} \langle -6,0 \rangle \cdot \langle 1,-1 \rangle = 6(-6 + 0) = -36$
   (f) $\langle -6,0 \rangle \cdot \langle -6,0 \rangle - \sqrt{36+0} = (36+0) - 6 = 30$

12. $y = x^2$; $y' = 2x$, so the slope at $(-1,1)$ is -2 (y changes by -2 as x changes by 1). Therefore, a vector parallel to the tangent line is $\langle 1,-2\rangle$. To obtain a vector of length 3, first divide $\langle 1,-2\rangle$ by its length and then multiply that result by 3.

$$\frac{\langle 1,-2\rangle}{|\langle 1,-2\rangle|}(3) = \frac{3\langle 1,-2\rangle}{\sqrt{1+4}} = \langle 3/\sqrt{5},-6/\sqrt{5}\rangle$$

15. Let the wind vector be $\mathbf{w} = \langle 100\cos 30°, 100\sin 30°\rangle = \langle 50\sqrt{3},50\rangle$
Let $\mathbf{p} = \langle p_1,p_2\rangle$ be the plane's air velocity vector.
We want $\mathbf{w} + \mathbf{p} = 450\mathbf{j} = \langle 0,450\rangle$

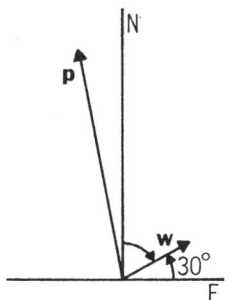

$\langle 50\sqrt{3},50\rangle + \langle p_1,p_2\rangle = \langle 0,450\rangle$

$\Rightarrow$ $50\sqrt{3} + p_1 = 0$, $50 + p_2 = 450$

$\Rightarrow$ $p_1 = -50\sqrt{3}$, $p_2 = 400$

Therefore, $\mathbf{p} = \langle -50\sqrt{3},400\rangle$. The angle β formed with the vertical satisfies

$\cos\beta = \dfrac{\mathbf{p}\cdot\mathbf{j}}{|\mathbf{p}||\mathbf{j}|} = \dfrac{400}{\sqrt{167500}}$; $\beta \approx 12.22°$. Thus, the heading is N12.22°W.

The air speed is $|\mathbf{p}| = \sqrt{167500} \approx 409.27$ mph.

18. Length $= \displaystyle\int_0^2 |\mathbf{r}'(t)|\,dt = \int_0^2 |\langle 6t^{1/2},3\rangle|\,dt = \int_0^2 3(4t+1)^{1/2}\,dt$

$= \left[\dfrac{(4t+1)^{3/2}}{2}\right]_0^2 = \dfrac{27}{2} - \dfrac{1}{2} = 13$

21. (a) $y = x^2-x$; $y' = 2x-1$; $y'' = 2$; $y'(1) = 1$; $y''(1) = 2$;

$\kappa(1) = \dfrac{|2|}{(1+1)^{3/2}} = \dfrac{1}{\sqrt{2}} \approx 0.7071$

(b) Let $x = t+t^3$; $x' = 1+3t^2$; $x'' = 6t$; $y = t+t^2$; $y' = 1+2t$; $y'' = 2$.
At $(2,2)$, $t = 1$, so $x' = 4$, $x'' = 6$, $y' = 3$, $y'' = 2$.
Therefore, $\kappa = \dfrac{|(4)(2)-(3)(6)|}{(16+9)^{3/2}} = 0.08$

(c) $y = a\cosh(x/a)$; $y' = \sinh(x/a)$; $y'' = (1/a)\cosh(x/a)$.
At $(a, a\cosh 1)$, $y' = \sinh 1$, $y'' = (1/a)(\cosh 1)$.
Therefore, $\kappa = \dfrac{(1/a)(\cosh 1)}{(1+\sinh^2 1)^{3/2}} = \dfrac{\cosh 1}{a(\cosh^2 1)^{3/2}} = \dfrac{1}{a\cosh^2 1} \approx \dfrac{0.4120}{a}$

CHAPTER 14 GEOMETRY IN SPACE, VECTORS

Problem Set 14.1 Cartesian Coordinates in Three-Space

3. In the xz-plane, the y-coordinate is 0.
   On the y-axis, the x- and z-coordinates are 0.

6. The squares of the distances between the pairs of points are:
   $(4-1)^2 + (5-7)^2 + (2-3)^2 = 14$, $(4-2)^2 + (5-4)^2 + (2-5)^2 = 14$, and
   $(1-2)^2 + (7-4)^2 + (3-5)^2 = 14$.

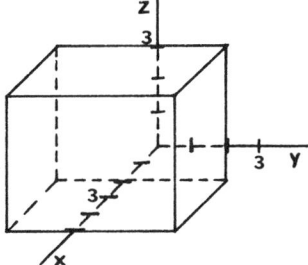

9. Coordinates of lid: $(2,-2,4)$, $(2,3,4)$
   $(5,-2,4)$, $(5,3,4)$

   Coordinates of base: $(2,-2,0)$, $(2,3,0)$
   $(5,-2,0)$, $(5,3,0)$

12. The radius is 5, the distance of the center to the the xy-plane.
    Then an equation of the sphere is $(x-2)^2 + (y-4)^2 + (z-5)^2 = 25$.

15. $4[x^2-x+(1/4)] + 4[y^2+2y+1] + 4[z^2+4z+4] = 13+1+4+16$
    $[x-(1/2)]^2 + [y+1]^2 + [z+2]^2 = 8.5$
    Center: $(1/2,-1,-2)$ Radius: $\sqrt{8.5} \approx 2.92$

18. $3x-4y+2z = 24$

    $x = y = 0 \Rightarrow z = 12$ (z-intercept)

    $x = z = 0 \Rightarrow y = -6$ (y-intercept)

    $y = z = 0 \Rightarrow x = 8$ (x-intercept)

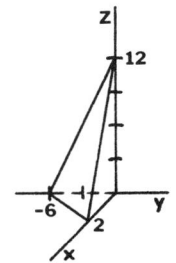

21. $x+3y = 8$. Since no z-term appears, changes in the value of z do not effect changes in x and y, so the graph is parallel to the z-axis. This is analogous to the 2-dimensional situation where, for example, $x = 5$ (no y-term appears) is parallel to the y-axis.

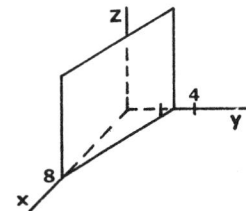

24. Sphere with center at (2,0,0) and radius 2.

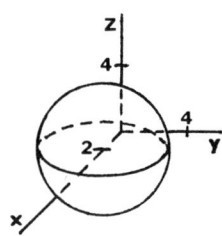

27. The center is in the plane parallel to the xy-plane and 6 units above it, so z = 6. Similarly, x = 6 and y = 6, so the center is (6,6,6).

Equation: $(x-6)^2 + (y-6)^2 + (z-6)^2 = 36$.

Problem Set 14.2 Vectors in Three-Space

3. (a) Length is $\sqrt{21}$; $\cos\alpha = 4/\sqrt{21}$; $\cos\beta = 1/\sqrt{21}$; $\cos\gamma = 2/\sqrt{21}$.

(b) Length is $\sqrt{50}$; $\cos\alpha = -3/\sqrt{50}$; $\cos\beta$ $-4/\sqrt{50}$; $\cos\gamma = 5/\sqrt{50}$.

6. $\dfrac{\langle -2,5,-3\rangle}{\sqrt{4+25+9}}(-10) = \dfrac{-10\langle -2,5,-3\rangle}{\sqrt{38}}$

9. Let $\mathbf{v} = \langle a,b,c\rangle$ be a vector perpendicular to $\langle 4,3,6\rangle$ and $\langle -2,-3,-2\rangle$. Then $\langle a,b,c\rangle \cdot \langle 4,3,6\rangle = 0$ and $\langle a,b,c\rangle \cdot \langle -2,-3,-2\rangle = 0$. Therefore, $4a+3b+6c = 0$ and $-2a-3b-2c = 0$. This pair of equations has infinitely many solutions. Let c be any nonzero constant and solve for a and b.

Let c = 3. Then a = -6 and b = 2. Thus, $\langle -6,2,3\rangle$, which has length 7, is perpendicular to $\langle 4,3,6\rangle$ and $\langle -2,-3,-2\rangle$. Then the vectors required are: $\dfrac{10\langle -6,2,3\rangle}{7}$ and $\dfrac{-10\langle -6,2,3\rangle}{7}$.

12. BA = $\langle 4-1, 3-1, 5-1\rangle$ = $\langle 3,2,4\rangle$ and BC = $\langle -1-1, 10-1, -2-1\rangle$ = $\langle -2,9,-3\rangle$.

BA·BC = $\langle 3,2,4\rangle \cdot \langle -2,9,-3\rangle$ = -6+18-12 = 0, so the vectors, and the sides they represent, are perpendicular.

15. $\mathbf{m}$ = (scaler projection) $\dfrac{\mathbf{v}}{|\mathbf{v}|} = \left(\dfrac{\mathbf{u}\cdot\mathbf{v}}{|\mathbf{v}|}\right)\dfrac{\mathbf{v}}{|\mathbf{v}|} = \left(\dfrac{\mathbf{u}\cdot\mathbf{v}}{\mathbf{v}\cdot\mathbf{v}}\right)\mathbf{v}$

$= \left(\dfrac{-6+4+5}{4+16+5}\right)\langle 2,4,-\sqrt{5}\rangle = \langle 6/25, 12/25, -3\sqrt{5}/25\rangle$

$\mathbf{n} = \mathbf{u}-\mathbf{m} = \langle -3,1,-\sqrt{5}\rangle - \langle 6/25, 12/25, -3\sqrt{5}/25\rangle$

$= \langle -81/25, 13/25, -22\sqrt{5}/25\rangle$

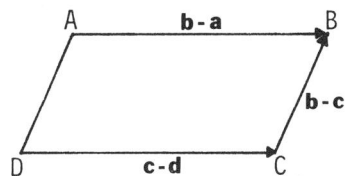

18. $\cos^2(46°) + \cos^2(108°) + \cos^2\gamma = 1 \;\Rightarrow\; \cos\gamma \approx \pm 0.6496$
$\Rightarrow\; \gamma \approx 49.49°$ or $\gamma \approx 130.51°$

21. There are infinitely many such pairs. By inspection, note that $\langle 1,2,0\rangle$ is perpendicular to $\langle -4,2,5\rangle$ (dot product is 0). And $\langle -2,1,c\rangle$ is perpendicular to $\langle 1,2,0\rangle$ for every real number c (dot product is 0). Now find c so that $\langle -2,1,c\rangle$ is perpendicular to $\langle -4,2,5\rangle$. The dot product is $8+2+5c$ which will be 0 if c is -2.

Hence, a pair of perpendicular vectors, each perpendicular to $\langle -4,2,5\rangle$ are $\langle 1,2,0\rangle$ and $\langle -2,1,-2\rangle$.

24. $\mathbf{b}-\mathbf{a} = \mathbf{c}-\mathbf{d}$, and

$\mathbf{b}-\mathbf{a} \neq k(\mathbf{b}-\mathbf{c})$ for each real number k.

27. Let Ω be the smaller angle. Then $\cos\Omega = \dfrac{|\langle 2,-4,3\rangle \cdot \langle 3,-2,-1\rangle|}{\sqrt{4+16+9}\,\sqrt{9+4+1}} = \dfrac{11}{\sqrt{406}}$

Therefore, $\Omega = \cos^{-1}(11/\sqrt{406}) \approx 0.9933$ rad [56.91°]

30. Distance $= \dfrac{|(1)+3(-1)+(2)-7|}{\sqrt{1+9+1}} = \dfrac{7}{\sqrt{11}} \approx 2.1106$

33. $(1,0,0)$ is on $5x-3y-2z = 5$ (by inspection). The distance from $(1,0,0)$ to $-5x+3y+2z = 7$ is $\dfrac{|-5(1)+3(0)+2(0)-7|}{\sqrt{25+9+4}} = \dfrac{12}{\sqrt{38}} \approx 1.9467$

36. $|\mathbf{u}+\mathbf{v}|^2 - |\mathbf{u}-\mathbf{v}|^2 = (\mathbf{u}+\mathbf{v})\cdot(\mathbf{u}+\mathbf{v}) - (\mathbf{u}-\mathbf{v})\cdot(\mathbf{u}-\mathbf{v})$
$= \mathbf{u}\cdot\mathbf{u} + 2\mathbf{u}\cdot\mathbf{v} + \mathbf{v}\cdot\mathbf{v} - (\mathbf{u}\cdot\mathbf{u} - 2\mathbf{u}\cdot\mathbf{v} + \mathbf{v}\cdot\mathbf{v}) = 4\mathbf{u}\cdot\mathbf{v}$.
Now divide through by 4 to finish the proof.

Problem Set 14.2

39. Place the box in the "corner" of the 1st octant so that its vertices are $(0,0,0)$, $(4,0,0)$, $(0,6,0)$, $(4,6,0)$, $(0,0,10)$, $(4,0,10)$, $(0,6,10)$, and $(4,6,10)$.

The main diagonals are $(0,0,0)$ to $(4,6,10)$, $(4,0,0)$ to $(0,6,10)$, $(0,6,0)$ to $(4,0,10)$, $(4,6,0)$ to $(0,0,10)$.

Corresponding vectors are $\langle 4,6,10 \rangle$, $\langle -4,6,10 \rangle$, $\langle 4,-6,10 \rangle$, $\langle -4,-6,-10 \rangle$.

The smallest angle Ω between main diagonals is obtained if the numerator in the formula $\cos\Omega = \dfrac{|u \cdot v|}{|u||v|}$ is largest.

There are six ways of pairing up the four main diagonals. The largest value of $|u \cdot v|$ is 120, using $\langle 4,6,10 \rangle$ and $\langle -4,6,10 \rangle$ [or use $\langle 4,-6,10 \rangle$ and $\langle -4,-6,10 \rangle$].

That is, $\cos\Omega = \dfrac{|\langle 4,6,10 \rangle \cdot \langle -4,6,10 \rangle|}{\sqrt{16+36+100}\sqrt{16+36+100}} = \dfrac{15}{19}$, so $\Omega \approx 0.6608$ rad. [37.86°]

Problem Set 14.3 The Cross Product

3. $\langle -2,1,-4 \rangle \times \langle 3,-4,5 \rangle = \langle 5-16, -12+10, 8-3 \rangle = \langle -11,-2,5 \rangle$ is perpendicular to both. Therefore, every vector perpendicular to both is of the form $k\langle -11,-2,5 \rangle$.

6. Two vectors in the plane are $\langle 2,3,-3 \rangle$ (between 1st two points) and $\langle 5,6,1 \rangle$ (last two points). Then a vector that is normal to the plane is $\langle 2,3,-3 \rangle \times \langle 5,6,1 \rangle = \langle 3+18, -15-2, 12-15 \rangle = \langle 21,-17,-3 \rangle$.

The required vectors are $\dfrac{\pm\langle 21,-17,-3 \rangle}{\sqrt{441+289+9}} = \dfrac{\pm\langle 21,-17,-3 \rangle}{\sqrt{739}}$.

9. The area of the triangle is half the area of the corresponding parallelogram. Adjacent sides of the triangle can be represented by the vectors $\langle -1,2,7 \rangle$ and $\langle -4,0,8 \rangle = 4\langle -1,0,2 \rangle$ using $(3,2,-1)$ as the vertex.

Then Area = $(1/2)|\langle -1,2,7 \rangle \times 4\langle -1,0,2 \rangle| = 2|\langle -1,2,7 \rangle \times \langle -1,0,2 \rangle|$
= $2|\langle 4-0, -7+2, 0+2 \rangle| = 2|\langle 4,-5,2 \rangle| = 2\sqrt{45} \approx 13.4164$.

12. Two vectors in the plane are $\langle 5,0,4\rangle$ and $\langle 1,3,5\rangle$. A vector normal to the plane is then $\langle 5,0,4\rangle \times \langle 1,3,5\rangle = \langle 0-12, 4-25, 15-0\rangle = \langle -12,-21,15\rangle$ $= -3\langle 4,7,-5\rangle$. An equation of the plane is $4(x+1)+7(y+2)-5(x+3) = 0$ or $4x+7y-5z = -3$.

15. Each vector normal to the plane is parallel to the line of intersection of the given planes. Also, the cross product of vectors normal to those planes is parallel to each of those plane, and therefore is parallel to the line of intersection of the planes.

    Thus, $\langle 4,-3,2\rangle \times \langle 3,2,-1\rangle = \langle 3-4, 6+4, 8+9\rangle = \langle -1,10,17\rangle$ is normal to the plane we seek. An equation of the plane is $-1(x-6)+10(y-2)+17(x+1) = 0$ or $x-10y-17z = 3$.

18. Volume $= |\langle 3,-4,2\rangle \cdot \langle -1,2,1\rangle \times \langle 3,-2,5\rangle| = |\langle 3,-4,2\rangle \cdot \langle 10+2, 3+5, 2-6\rangle|$

    $= |\langle 3,-4,2\rangle \cdot \langle 12,8,-4\rangle| = |36-32-8| = 4$.

21. Volume = (1/3)(Area of triangular base)(height)
    = (1/3)[(1/2)(Area of corresponding parallelogram base)](height)
    = (1/6)(Volume of corresponding parallelopiped)
    = $(1/6)|\mathbf{a} \cdot \mathbf{b} \times \mathbf{c}|$

24. $\mathbf{u} \times (\mathbf{v}+\mathbf{w}) = \langle u_1, u_2, u_3\rangle \times \langle v_1+w_1, v_2+w_2, v_3+w_3\rangle$

    $= \langle u_2(v_3+w_3)-u_3(v_2+w_2), u_3(v_1+w_1)-u_1(v_3+w_3), u_1(v_2+w_2)-u_2(v_1+w_1)\rangle$

    $(\mathbf{u} \times \mathbf{v})+(\mathbf{u} \times \mathbf{w}) = (\langle u_1,u_2,u_3\rangle \times \langle v_1,v_2,v_3\rangle) + (\langle u_1,u_2,u_3\rangle \times \langle w_1,w_2,w_3\rangle)$

    $= \langle u_2v_3-u_3v_2, u_3v_1-u_1v_3, u_1v_2-u_2v_1\rangle + \langle u_2w_3-u_3w_2, u_3w_1-u_1w_3, u_1w_2-u_2w_1\rangle$

    $= \langle u_2(v_3+w_3)-u_3(v_2+w_2), u_3(v_1+w_1)-u_1(v_3+w_3), u_1(v_2+w_2)-u_2(v_1+w_1)\rangle$

    Therefore, $\mathbf{u} \times (\mathbf{v} + \mathbf{w}) = (\mathbf{u} \times \mathbf{v}) + (\mathbf{u} \times \mathbf{w})$.

Problem Set 14.4 Lines and Curves in Three-Space

3. $\langle 6-4, 2-2, -1-3\rangle = \langle 2,0,-4\rangle = 2\langle 1,0,-2\rangle$ so $\langle 1,0,-2\rangle$ is a vector in the direction of the line. Therefore, using point $(4,2,3)$, parametric equations of the line are $x = 4+1t$, $y = 2+0t$, $z = 3-2t$, or more simply $x = 4+t$, $y = 2$, $z = 3-2t$.

6. Parametric: $x = -1+4t$, $y = 3+2t$, $z = 2-t$

   Symmetric: $\dfrac{x+1}{4} = \dfrac{y-3}{2} = \dfrac{z-2}{-1}$

9. A vector in the direction of the line is $\langle 5,2,-5\rangle \times \langle 10,6,-5\rangle$ which is $\langle -10+30,-50+25,30-20\rangle = \langle 20,-25,10\rangle = 5\langle 4,-5,2\rangle$.

   To find a point on the intersection, let $x = 0$ and solve $2y-5z = 5$ and $6y-5z=25$ simultaneously, obtaining $y = 5$, $z = 1$. Therefore, $(0,5,1)$ is a point on the line. Then symmetric equations of the line are

   $\dfrac{x-0}{4} = \dfrac{y-5}{-5} = \dfrac{z-1}{2}$.

12. $\langle 1,-2,1\rangle \times \langle 6,-5,4\rangle = \langle -3,2,7\rangle$ is in the direction of the line. Let $y = 0$ and solve $x+z = 1$ and $6x+4z = 10$, obtaining $x = 3$ and $z = -2$.

    Symmetric Equations: $\dfrac{x-3}{-3} = \dfrac{y-0}{2} = \dfrac{z+2}{7}$

15. The point of intersection on the z-axis is $(0,0,4)$. A vector in the direction of the line is $\langle 5-0,-3-0,4-4\rangle = \langle 5,-3,0\rangle$. Parametric equations are $x = 0+5t$, $y = 0-3t$, $z = 4+0t$; i.e., $x = 5t$, $y = -3t$, $z = 4$.

18. Solve $\dfrac{x-1}{-4} = \dfrac{y-2}{3}$ and $\dfrac{x-2}{-1} = \dfrac{y-1}{1}$ simultaneously, obtaining $x = 1$, $y = 2$.

    From $\dfrac{2-2}{3} = \dfrac{z-4}{-2}$ (using y-z equation from 1st line), conclude that $z = 4$,

    Therefore, $(1,2,4)$ is on the first line. The point also satisfies the set of symmetric equations of the 2nd line, so it is the point of intersection of the two lines.

    $\langle -4,3,-2\rangle \times \langle -1,1,6\rangle = \langle 18+2,2+24,-4+3\rangle = \langle 20,26,-1\rangle$ is perpendicular to each line, so is normal to the plane containing the lines. Therefore, an equation of the plane is $20(x-1)+26(y-2)-1(z-4) = 0$ or $20x+26y-z = 68$.

21. (b) Vectors in the direction of the lines are $\langle -1,4,2\rangle$ and $\langle 1,0,2\rangle$, so $\langle -1,4,2\rangle \times \langle 1,0,2\rangle = \langle 8-0,2+2,0-4\rangle = \langle 8,4,-4\rangle = 4\langle 2,1,-1\rangle$ is perpendicular to both lines, so is normal to π. Then an equation of π is $2(x-2)+1(y-3)-1(z-0) = 0$, or $2x+y-z = 7$.

    (c) Let $t = 0$. Then $x = -1$, $y = 2$, $z = -1$, so let $Q = (-1,2-1)$.

    (d) $d(Q,\pi) = \dfrac{|2(-1)+(2)-(-1)-7|}{\sqrt{4+1+1}} = \sqrt{6} \approx 2.4495$

24. Let $r(t) = \langle 2t^2, 4t, t^3 \rangle$. Then $r'(t) = \langle 4t, 4, 3t^2 \rangle$.

$r(1) = \langle 2,4,1 \rangle$ determines the point $(2,4,1)$ of the curve, and a vector in the direction of the tangent at that point is $r'(1) = \langle 4,4,3 \rangle$.

Parametric Equations of tangent: $x = 2+4t$, $y = 4+4t$, $z = 1+3t$.

Problem Set 14.5 Velocity, Acceleration, and Curvature

3. $r(t) = \langle 2t-t^2, 3t, t^3+1 \rangle$; $v(t) = \langle 2-2t, 3, 3t^2 \rangle$; $a(t) = \langle -2, 0, 6t \rangle$.

   $v(1) = \langle 0,3,3 \rangle$; $s(1) = |v(1)| = \sqrt{0+9+9} \approx 4.2426$; $a(1) = \langle -2,0,6 \rangle$.

6. $r(t) = \langle t\sin t, e^t, \cos t \rangle$; $v(t) = \langle t\cos t + \sin t, e^t, -\sin t \rangle$;
   $$a(t) = \langle -t\sin t + 2\cos t, e^t, -\cos t \rangle.$$
   $v(0) = \langle 0,1,0 \rangle$; $s(0) = \sqrt{0+1+0} = 1$; $a(0) = \langle 2,1,-1 \rangle$.

9. $v(t) = \langle t\cos t + \sin t, -t\sin t + \cos t, \sqrt{8} \rangle$

   $|v(t)| = [(t^2\cos^2 t + 2t\sin t\cos t + \sin^2 t) + (t^2\sin^2 t - 2t\sin t\cos t + \cos^2 t) + (8)]^{1/2}$
   $= [t^2(\cos^2 t + \sin^2 t) + (\sin^2 t + \cos^2 t) + 8]^{1/2} = (t^2+9)^{1/2}$

   Then length $= \int_0^4 \sqrt{t^2+9}\, dt = \left[\frac{t}{2}\sqrt{t^2+9} + \frac{9}{2}\ln|t + \sqrt{t^2+9}|\right]_0^4$ [Formula 44]

   $= [2(5) + 4.5\ln|4+5|] - [0 + 4.5\ln|0+3|] \approx 14.9438$

12. $v(t) = \langle -e^{2t}\sin t + 2e^{2t}\cos t, e^{2t}\cos t + 2e^{2t}\sin t, 2e^{2t} \rangle$
    $= e^{2t}\langle -\sin t + 2\cos t, \cos t + 2\sin t, 2 \rangle$

    $|v(t)| = e^{2t}[(\sin^2 t - 4\sin t\cos t + 4\cos^2 t) + (\cos^2 t + 4\sin t\cos t + 4\sin^2 t) + (4)]^{1/2}$
    $= e^{2t}[(\sin^2 t + \cos^2 t) + 4(\cos^2 t + \sin^2 t) + 4]^{1/2} = 3e^{2t}$

    Then length $= \int_0^5 3e^{2t} dt = \left[\frac{3e^{2t}}{2}\right]_0^5 = \frac{3(e^{10}-1)}{2} \approx 33038.20$

15. $r(t) = \langle t^2-1, 2t+3, t^2-4t\rangle$; $v(t) = \langle 2t,2,2t-4\rangle$; $a(t) = \langle 2,0,2\rangle$.

$r(2) = \langle 3,7,-4\rangle$; $v(2) = \langle 4,2,0\rangle$; $|v(2)| = 2\sqrt{5}$; $a(2) = \langle 2,0,2\rangle$.

$$\kappa = \frac{|v \times a|}{|v|^3} = \frac{|2\langle 2,1,0\rangle \times 2\langle 1,0,1\rangle|}{(2\sqrt{5})^3} = \frac{4|\langle 1,-2,-1\rangle|}{40\sqrt{5}} = \frac{\sqrt{6}}{10\sqrt{5}} \approx 0.1095$$

$$T = \frac{v}{|v|} = \frac{2\langle 2,1,0\rangle}{2\sqrt{5}} = \frac{\langle 2,1,0\rangle}{\sqrt{5}}$$

$$a_N = \kappa|v|^2 = \frac{\sqrt{6}}{10\sqrt{5}}(2\sqrt{5})^2 = \frac{2\sqrt{30}}{5}; \quad a_T = \frac{v \cdot a}{|v|} = \frac{2\langle 2,1,0\rangle \cdot 2\langle 1,0,1\rangle}{2\sqrt{5}} = \frac{4}{\sqrt{5}}.$$

$$N = \frac{a - a_T T}{a_N} = \frac{2\langle 1,0,1\rangle - \langle 8/5, 4/5, 0\rangle}{2\sqrt{30}/5} = \frac{\langle 1,-2,5\rangle}{\sqrt{30}}$$

$$B = T \times N = \frac{\langle 2,1,0\rangle \times \langle 1,-2,5\rangle}{\sqrt{5}\sqrt{30}} = \frac{\langle 1,-2,-1\rangle}{\sqrt{6}}$$

18. $v(t) = \langle 8\cos t, -8\sin t, 4\rangle$; $a(t) = \langle -8\sin t, -8\cos t, 0\rangle$.

$v(\pi/6) = 4\langle \sqrt{3},-1,1\rangle$; $a(\pi/6) = -4\langle 1,\sqrt{3},0\rangle$; $|v(\pi/6)| = 4\sqrt{5}$.

$$a_N = \frac{|v \times a|}{|v|} = 8; \quad a_T = \frac{v \cdot a}{|v|} = 0; \quad \kappa = \frac{a_N}{|v|^2} = 0.1.$$

$$T = \frac{v}{|v|} = \frac{\langle \sqrt{3},-1,1\rangle}{\sqrt{5}}; \quad N = \frac{a - a_T T}{a_N} = \frac{\langle 1,\sqrt{3},0\rangle}{-2}; \quad B = T \times N = \frac{\langle \sqrt{3},-1,-4\rangle}{2\sqrt{5}}.$$

21. $v(t) = \langle 1, 2t, 3t^2\rangle$; $|v(t)| = (1+4t^2+9t^4)^{1/2}$; $a(t) = \langle 0,2,6t\rangle$.

$$a_T = \frac{v \cdot a}{|v|} = \frac{4t+18t^3}{(1+4t^2+9t^4)^{1/2}}; \quad a_N = \frac{|v \times a|}{|v|} = \left[\frac{4(9t^4+9t^2+1)}{(9t^4+4t^2+1)}\right]^{1/2}$$

24. $v(t) = \langle \cot t, -\tan t, 1\rangle$; $a(t) = \langle -\csc^2 t, -\sec^2 t, 0\rangle$.

$|v(t)| = (\cot^2 t + \tan^2 t + 1)^{1/2} = (\cot^2 t + \sec^2 t)^{1/2}$.

$$a_T = \frac{v \cdot a}{|v|} = \frac{(-\cot t)\csc^2 t + (\tan t)\sec^2 t}{(\cot^2 t + \sec^2 t)^{1/2}}$$

$$a_N = \frac{|v \times a|}{|v|} = \left[\frac{\sec^4 t + \csc^4 t + 4\csc^2 t \sec^2 t}{\cot^2 t + \sec^2 t}\right]^{1/2}$$

27. (a) $L'(t) = mr(t) \times a(t) = \tau(t)$ [using result of Problem 26]

(b) $\tau(t) = 0$ for all t $\Rightarrow$ $L'(t) = 0$ for all t $\Rightarrow$ $L(t) = k$ for all t
for some constant k

Problem Set 14.6 Surfaces in Three-Space

3. $2x+5z+12 = 0$ is a plane (cylinder) parallel to the y-axis. The xz-trace is a line.

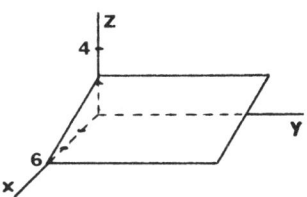

6. $4x^2-9z^2 = 0$, or $(2x+3z)(2x-3z) = 0$ is a pair of planes intersecting on the y-axis. The xz-trace is a pair of lines.

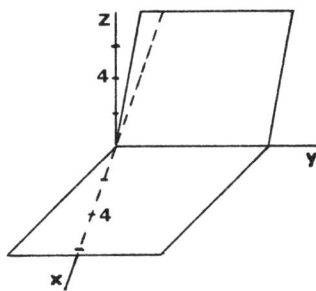

9. $z = \dfrac{x^2}{3} + \dfrac{y^2}{4/3}$ is an elliptic paraboloid with z-axis for the axis of symmetry.

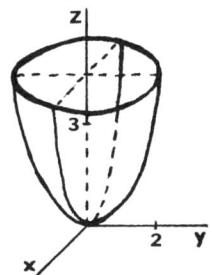

12. $5x + 8y = 0$ is a plane which contains the z-axis. The xy-trace is a line through the origin.

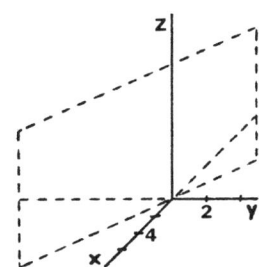

15. $y = \dfrac{x^2}{4} + \dfrac{z^2}{9}$ is an elliptic paraboloid with y-axis for axis of symmetry.

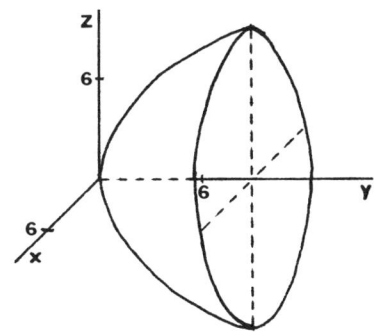

18. $y = \cos x$ is a cylinder parallel to the z-axis. The xy-trace is the graph of the cosine function.

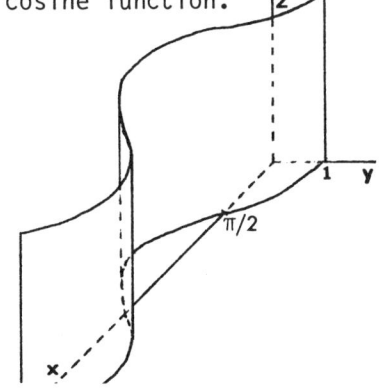

21. (a) Replacing x by -x results in an equivalent equation.
    (b) Replacing x by -x and y by -y results in an equivalent equation.
    (c) Replacing y by -y and z by -z results in an equivalent equation.
    (d) Replacing x by -x, y by -y, and z by -z, results in an equivalent equation.

24. Circular cone $z^2 = 4x^2 + 4y^2$.

27. Substituting $z = 4$ into the given equation results in $\frac{x^2}{16} + \frac{y^2}{36} = 1$.

    $a^2 = 36$, $b^2 = 16$; then $c^2 = a^2 - b^2 = 36 - 16 = 10$, so $c = 2\sqrt{5}$.

    Therefore, the foci are $(0, \pm 2\sqrt{5}, 4)$.

Problem Set 14.7 Cylindrical and Spherical Coordinates

3. (a) $\rho = \sqrt{x^2 + y^2 + z^2} = \sqrt{4 + 12 + 16} = 4\sqrt{2}$.
    $\tan\theta = \frac{y}{x} = \frac{-2\sqrt{3}}{2} = -\sqrt{3}$ and (x,y) is in the 4th quadrant so $\theta = 5\pi/3$.
    $\cos\phi = \frac{z}{\rho} = \frac{4}{4\sqrt{2}} = \frac{\sqrt{2}}{2}$ so $\phi = \pi/4$.   Spherical: $(4\sqrt{2}, 5\pi/3, \pi/4)$.

    (b) $\rho = \sqrt{2 + 2 + 12} = 4$.
    $\tan\theta = \frac{\sqrt{2}}{-\sqrt{2}} = -1$ and (x,y) is in 2nd quadrant so $\theta = \frac{3\pi}{4}$.
    $\cos\phi = \frac{2\sqrt{3}}{4} = \frac{\sqrt{3}}{2}$ so $\phi = \pi/6$.   Spherical: $(4, 3\pi/4, \pi/6)$.

6. $\rho = 5$ is a sphere.

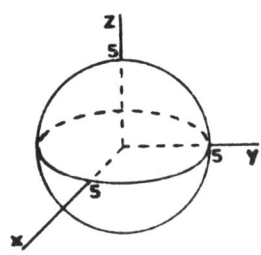

9. $r = 3\cos\theta$ is a circular cylinder.

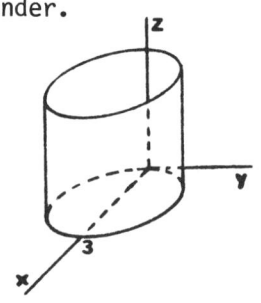

12. $\rho = \sec\phi$
    $\rho\cos\phi = 1$
    $z = 1$ (a plane)

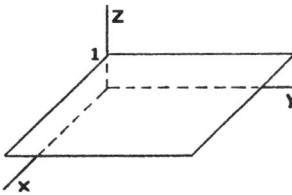

15. $x^2 + y^2 = 9$; $r^2 = 9$; $r = 3$.

18. $(x^2+y^2+z^2) + 3z^2 = 10$; $\rho^2 + 3\rho^2\cos^2\phi = 10$; $\rho^2 = \dfrac{10}{1+3\cos^2\phi}$.

21. $(r^2 + z^2) + z^2 = 4$; $\rho^2 + \rho^2\cos^2\phi = 4$; $\rho^2 = \dfrac{4}{1+\cos^2\phi}$

24. $x+y+z = 1$; $\rho\sin\phi\cos\theta + \rho\sin\phi\sin\theta + \rho\cos\phi = 1$; $\rho = \dfrac{1}{\sin\phi(\sin\theta+\cos\theta)+\cos\phi}$.

27. $r^2\cos 2\theta = z$; $r^2(\cos^2\theta-\sin^2\theta) = z$; $(r\cos\theta)^2-(r\sin\theta)^2 = z$; $x^2-y^2 = z$.

30. $2x^2+2y^2-z^2 = 2$ [Cartesian]; $2r^2-z^2 = 2$ [cylindrical].

33. Using the results of Example 7, the Cartesian coordinates of St. Paul are $(-151.4,-2796,2800)$.

    Turin: $\rho = 3960$, $\theta = 7.4°$, $\phi = 90°-45° = 45°$. Therefore,

    $x = (3960)(\sin 45°)(\cos 7.4°) \approx 2777$
    $y = (3960)(\sin 45°)(\sin 7.4°) \approx 360.6$
    $z = (3960)(\cos 45°) \approx 2800$

    Let Ω be the angle formed by the Earth radius to St. Paul with the Earth radius to Turin.

    Then $\cos\Omega = \dfrac{\langle-151.4,-2796,2800\rangle \cdot \langle 2777,360.6,2800\rangle}{(3960)(3960)} \approx 0.4088$; $\Omega \approx 1.1496$.

    Therefore, the great-circle distance between St. Paul and Turin is

    $3960(1.1496) \approx 4552$ miles.

Problem Set 14.7

Problem Set 14.8 Chapter Review Problems

True-False Quiz

3. True. See "Planes" beginning at the bottom of Page 598.

6. False. It is normal to the plane.

9. True. $||u|u| = |u||u| = |u|^2$. [See page 566.]

12. True. $(kv) \times v = k(v \times v) = k(0) = 0$.

15. True. $\dfrac{|u \times v|}{(u \cdot v)} = \dfrac{|u||v|\sin\theta}{|u||v|\cos\theta} = \tan\theta$.

18. False. c^{ex}: Let $u = v = i$, $w = j$.
    Then $(u \times v) \times w = 0 \times w = 0$; but $u \times (v \times w) = i \times k = -j$.

21. False. c^{ex}: Let $r(t) = \langle 0,1,t \rangle$. Then $|r'(t)| = |\langle 0,0,1 \rangle| = 1$;
    but $D_t |r(t)| = D_t(1+t^2)^{1/2} = \dfrac{t}{(1+t^2)^{1/2}}$ which is never 1.

    In general, $D_t |r(t)| = \dfrac{r(t) \cdot r'(t)}{|r(t)|}$

24. False. It is a parabolic cylinder parallel to the z-axis.

Miscellaneous Problems

3. (a) $|a| = \sqrt{4+1+4} = 3$; $|b| = \sqrt{25+1+9} = \sqrt{35}$.
   (b) 2/3, -1/3, 2/3 for a;
       $5/\sqrt{35}$, $1/\sqrt{35}$, $-3/\sqrt{35}$ for b.
   (c) $\dfrac{a}{3} = \langle 2/3, -1/3, 2/3 \rangle \cdot$
   (d) $\cos\theta = \dfrac{a \cdot b}{|a||b|} = \dfrac{10-1-6}{3\sqrt{35}} = \dfrac{1}{\sqrt{35}}$. $\theta \approx 1.4010$ rad [80.27°]

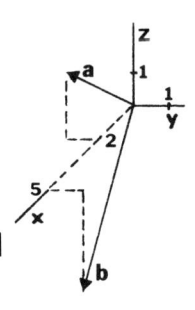

6. Two vectors determined by the points are $\langle -1,7,-3 \rangle$ and $\langle 3,-1,-3 \rangle$. Then $\langle -1,7,-3 \rangle \times \langle 3,-1,-3 \rangle = -4\langle 6,3,5 \rangle$ so $\langle 6,3,5 \rangle$ is normal to the plane. Then $\dfrac{\pm \langle 6,3,5 \rangle}{\sqrt{36+9+25}} = \dfrac{\pm \langle 6,3,5 \rangle}{\sqrt{70}}$ are the unit vectors normal to the plane.

9. Vectors normal to the planes are perpendicular so $\langle 1,5,C \rangle \cdot \langle 4,-1,1 \rangle = 0$. Therefore, $4-5+C = 0$, $C = 1$.

12. In the yz-plane, $x = 0$. Solve $-2y+4z = 14$ and $2y-5z = -30$, obtaining $y = 25$ and $z = 16$.

    In the xz-plane, $y = 0$. Solve $x+4z = 14$ and $-x-5z = -30$, obtaining $x = -50$ and $z = 16$.

    Therefore, the points are $(0,25,16)$ and $(-50,0,16)$.

15. $\langle 5,-4,-3 \rangle$ is a vector in the direction of the line, and $\langle 2,-2,1 \rangle$ is a position vector to the line. Then a vector equation of the line is $\mathbf{r}(t) = \langle 2,-2,1 \rangle + t\langle 5,-4,-3 \rangle$.

18. $\mathbf{r}(t) = \langle t\cos t, t\sin t, 2t \rangle$; $\mathbf{r}'(t) = \langle -t\sin t+\cos t, t\cos t+\sin t, 2 \rangle$; $\mathbf{r}''(t) = \langle -t\cos t-2\sin t, -t\sin t+2\cos t, 0 \rangle$.

    $\mathbf{r}'(\pi/2) = \langle -\pi/2, 1, 2 \rangle$; $\mathbf{r}''(\pi/2) = \langle -2, -\pi/2, 0 \rangle$.

    $|\mathbf{r}'(\pi/2)| = \dfrac{\sqrt{\pi^2+20}}{2}$; $\mathbf{T}(\pi/2) = \dfrac{\mathbf{r}'(\pi/2)}{|\mathbf{r}'(\pi/2)|} = \dfrac{\langle -\pi,2,4 \rangle}{(\pi^2+20)^{1/2}}$.

21. $\mathbf{v}(t) = \langle 1, 2t, 3t^2 \rangle$; $\mathbf{a}(t) = \langle 0, 2, 6t \rangle$.
    $\mathbf{v}(1) = \langle 1,2,3 \rangle$; $|\mathbf{v}(1)| = \sqrt{14}$; $\mathbf{a}(1) = \langle 0,2,6 \rangle$.

    $a_T = \dfrac{\mathbf{v}\cdot\mathbf{a}}{|\mathbf{v}|} = \dfrac{0+4+18}{\sqrt{14}} = \dfrac{22}{\sqrt{14}} \approx 5.880$; $a_N = \dfrac{|\mathbf{v}\times\mathbf{a}|}{|\mathbf{v}|} = \dfrac{|\langle 6,-6,2 \rangle|}{\sqrt{14}} = \dfrac{2\sqrt{19}}{\sqrt{14}} \approx 2.330$

24. $z^2 = 4y$ is parabolic cylinder parallel to the x-axis.

27. $3x+3y-6z = 12$ is a plane.

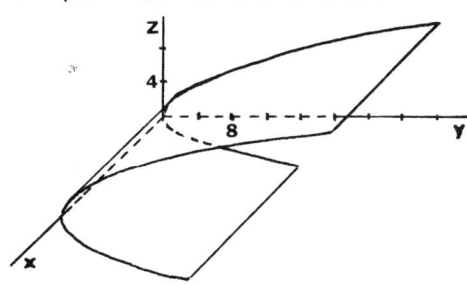
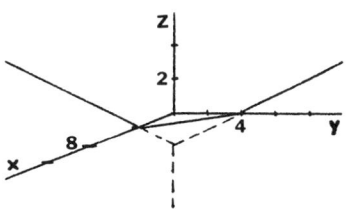

Problem Set 14.8

30. The graph of $3x^2+4y^2+9z^2 = -36$ is the empty set.

33. (a) $\rho^2 = 4$; $\rho = 2$.

(b) $x^2+y^2+z^2 - 2z^2 = 0$; $\rho^2 - 2\rho^2\cos^2\phi = 0$; $\rho^2(1-2\cos^2\phi) = 0$;

$1-2\cos^2\phi = 0$; $\cos^2\phi = 1/2$; $\phi = \pi/4$ or $\phi = 3\pi/4$.

Any of the following (as well as others) would be acceptable:

$$(\phi - \pi/4)(\phi - 3\pi/4) = 0.$$

$$\cos^2\phi = 1/2$$

$$\sec^2\phi = 2$$

$$\tan^2\phi = 1$$

(c) $2x^2 - (x^2+y^2+z^2) = 1$; $2\rho^2\sin^2\phi\cos^2\theta - \rho^2 = 1$; $\rho^2 = \dfrac{1}{2\sin^2\phi\cos^2\theta - 1}$.

(d) $x^2+y^2 = z$; $\rho^2\sin^2\phi\cos^2\theta + \rho^2\sin^2\phi\sin^2\theta = \rho\cos\phi$;

$\rho^2\sin^2\phi(\cos^2\theta+\sin^2\theta) = \rho\cos\phi$; $\rho\sin^2\phi = \cos\phi$; $\rho = \cot\phi\csc\phi$.

[Note that when we divided through by ρ in (c) and (d) we did not lose the pole since it is also a solution of the resulting equations.]

36. $\langle 2,-4,1\rangle$ and $\langle 3,2,-5\rangle$ are normal to the respective planes. The acute angle between the two planes is the same as the acute angle Ω between the lines containing the normal vectors.

$$\cos\Omega = \frac{|6-8-5|}{\sqrt{21}\,\sqrt{38}} = \frac{7}{\sqrt{798}}, \text{ so } \Omega \approx 1.3204 \text{ rad } [75.65°].$$

CHAPTER 15 THE DERIVATIVE IN n-SPACE

Problem Set 15.1 Functions of Two or More Variables

3. (a) $\sin(2\pi) = 0$
   (b) $4\sin(\pi/6) = 2$
   (c) $16\sin(\pi/2) = 16$
   (d) $\pi^2\sin(\pi^2) \approx -4.2469$
   (e) $1.44\sin[(3.1)(4.2)] \approx 0.6311$

6. $F(f(t),g(t)) = F(\ell nt^2, e^{t/2}) = \exp(\ell nt^2) + (e^{t/2})^2 = t^2 + e^t \quad [t \neq 0]$

9. $x+2y+z = 6$ is a plane.

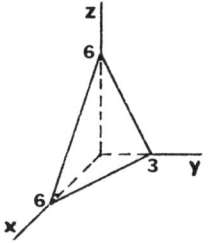

12. $\dfrac{x^2}{4} + \dfrac{y^2}{16} + \dfrac{z^2}{16} = 1, \; z \geq 0$
    is a hemi-ellipsoid.

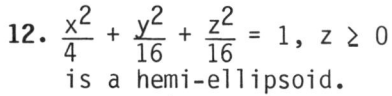

15. $z = \exp[-(x^2+y^2)]$

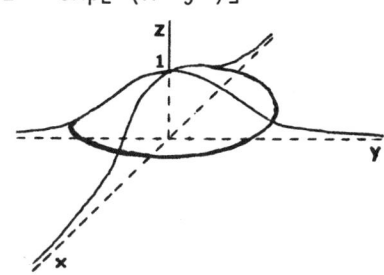

18. $x = zy; \; x = ky$.

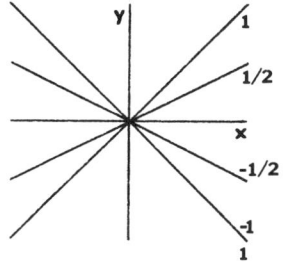

21. $z = \dfrac{x^2+y}{x+y^2}$

   $k = 0$: $y = -x^2$
   [parabola except $(0,0)$]

   $k = 1$: $x^2+y = x+y^2$
   $(x-1/2)^2 - (y-1/2)^2 = 0$
   $y = x$ or $y = -x+1$
   [intersecting lines except $(0,0)$]

   $k = 2$: $x^2+y = 2x+2y^2$
   $\dfrac{(x-1)^2}{7/8} - \dfrac{(y-1/4)^2}{7/16} = 1$
   [hyperbola except $(0,0)$]

   $k = 4$: $x^2+y = 4x+4y^2$
   $\dfrac{(x-2)^2}{63/16} - \dfrac{(y-1/8)^2}{63/64} = 1$
   [hyperbola except $(0,0)$]

Problem Set 15.1

24. $(x-2)^2 + (y+3)^2 = 16/V^2$.

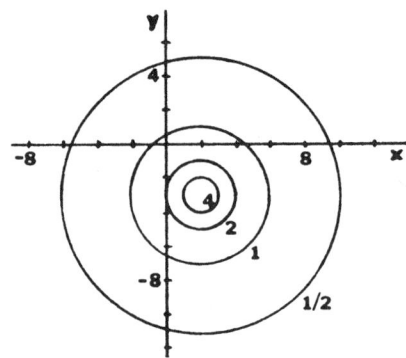

27. $\frac{x^2}{9} + \frac{y^2}{16} + \frac{z^2}{1} \leq 1$. Points inside and on the ellipsoid.

30. $100x^2 + 16y^2 + 25z^2 = k$, $k > 0$; $\frac{x^2}{k/100} + \frac{y^2}{k/16} + \frac{z^2}{k/25} = 1$.

Set of all ellipsoids centered at origin such that their axes have ratio $(1/10):(1/4):(1/5)$ or $2:5:4$.

33. $4x^2 - 9y^2 = k$, k in R; $\frac{x^2}{k/4} - \frac{y^2}{k/9} = 1$, if $k \neq 0$.

Planes $y = \pm 2x/3$ [for k=0] and all hyperbolic cylinders parallel to the z-axis such that the ratio a:b is $(1/2):(1/3)$ or $3:2$ [where a is associated with the x-term].

Problem Set 15.2 Partial Derivatives

3. $f_x(x,y) = \frac{(xy)(2x) - (x^2-y^2)(y)}{(xy)^2} = \frac{x^2+y^2}{x^2y}$.

$f_y(x,y) = \frac{(xy)(-2y) - (x^2-y^2)(x)}{(xy)^2} = \frac{-(x^2+y^2)}{xy^2}$.

6. $f_x(x,y) = (-1/3)(3x^2+y^2)^{-4/3}(6x) = -2x(3x^2+y^2)^{-4/3}$.

$f_y(x,y) = (-1/3)(3x^2+y^2)^{-4/3}(2y) = (-2y/3)(3x^2+y^2)^{-4/3}$.

9. $g_x(x,y) = -ye^{-xy}$; $g_y(x,y) = -xe^{-xy}$.

12. $F_w(w,z) = w \dfrac{1}{\sqrt{1-(w/z)^2}} (1/z) + \sin^{-1}(w/z) = \dfrac{w/z}{\sqrt{1-(w/z)^2}} + \sin^{-1}(w/z)$.

$F_z(w,z) = w \dfrac{1}{\sqrt{1-(w/z)^2}} (-w/z^2) = \dfrac{-(w/z)^2}{\sqrt{1-(w/z)^2}}$.

15. $F_x(x,y) = 2\cos x \cos y$; $F_y(x,y) = -2\sin x \sin y$.

18. $f_x(x,y) = 5(x^3+y^2)^4(3x^2)$; $f_{xy}(x,y) = 60x^2(x^3+y^2)^3(2y) = 120x^2y(x^3+y^2)^3$.

$f_y(x,y) = 5(x^3+y^2)^4(2y)$; $f_{yx}(x,y) = 40y(x^3+y^2)^3(3x^2) = 120x^2y(x^3+y^2)^3$.

21. $F_x(x,y) = \dfrac{(xy)(2)-(2x-y)(y)}{(xy)^2} = \dfrac{y^2}{x^2y^2} = \dfrac{1}{x^2}$; $F_x(3,-2) = \dfrac{1}{9}$.

$F_y(x,y) = \dfrac{(xy)(-1)-(2x-y)(x)}{(xy)^2} = \dfrac{-2x^2}{x^2y^2} = \dfrac{-2}{y^2}$; $F_y(3,-2) = \dfrac{-1}{2}$.

24. $f_x(x,y) = e^y \sinh x$; $f_x(-1,1) = e \sinh(-1) \approx -3.1945$.

$f_y(x,y) = e^y \cosh x$; $f_y(-1,1) = e \cosh(-1) \approx 4.1945$.

27. Let $z = f(x,y) = (1/2)(9x^2+9y^2-36)^{1/2}$; $f_x(x,y) = \dfrac{9x}{2(9x^2+9y^2-36)^{1/2}}$.

$f_x(2,1) = 3$.

30. $T_y(x,y) = 3y^2$; $T_y(3,2) = 12$ degrees per foot.

33. $f_x(x,y) = 3x^2y-y^3$; $f_{xx}(x,y) = 6xy$; $f_y(x,y) = x^3-3xy^2$; $f_{yy}(x,y) = -6xy$.

Therefore, $f_{xx}(x,y) + f_{yy}(x,y) = 0$.

36. $f_x(x,y) = [-\sin(2x^2-y^2)](4x) = -4x \sin(2x^2-y^2)$.

$f_{xx}(x,y) = (-4x)[\cos(2x^2-y^2)](4x) + [\sin(2x^2-y^2)](-4)$.

$f_{xxy}(x,y) = -16x^2[-\sin(2x^2-y^2)](-2y) - 4[\cos(2x^2-y^2)](-2y)$

$= -32x^2y \sin(2x^2-y^2) + 8y \cos(2x^2-y^2)$.

Problem Set 15.2

39. (a) $f_x(x,y,z) = 6xy-yz$.

(b) $f_y(x,y,z) = 3x^2-xz+2yz^2$; $f_y(0,1,2) = 8$.

(c) Using the result in (a), $f_{xy}(x,y,z) = 6x-z$.

42. $f_x(x,y,z) = (1/2)(xy/z)^{-1/2}(y/z)$; $f_x(-2,-1,8) = (1/2)(1/4)^{-1/2}(\frac{-1}{8}) = \frac{-1}{8}$

Problem Set 15.3 Limits and Continuity

3. $\lim_{(x,y) \to (2,\pi)} [x\cos^2 xy - \sin(xy/3)] = 2\cos^2 2\pi - \sin(2\pi/3) = 2 - \frac{\sqrt{3}}{2} \approx 1.1340$

6. $\lim_{(x,y) \to (0,0)} \frac{\tan(x^2+y^2)}{(x^2+y^2)} = \lim_{(x,y) \to (0,0)} \left(\frac{\sin(x^2+y^2)}{x^2+y^2} \cdot \frac{1}{\cos(x^2+y^2)} \right) = (1)(1) = 1$

9. The entire plane since x^2+y^2+1 is never zero.

12. The only points at which f might be discontinuous are where $xy = 0$.

    $\lim_{(x,y) \to (a,0)} \frac{\sin(xy)}{xy} = 1 = f(a,0)$ for all nonzero a in **R**, and then

    $\lim_{(x,y) \to (0,b)} \frac{\sin(xy)}{xy} = 1 = f(0,b)$ for all b in **R**.

    Therefore, f is continuous on the entire plane.

15. Along x-axis (y=0): $\lim_{(x,y) \to (0,0)} \frac{0}{x^2+0} = 0$.

    Along y=x: $\lim_{(x,y) \to (0,0)} \frac{x^2}{2x^2} = \lim_{(x,y) \to (0,0)} \frac{1}{2} = \frac{1}{2}$

    Hence, the limit does not exist because for some points near the origin f(x,y) is getting closer to 0, but for others it is getting closer to 1/2.

18. $\left|\dfrac{xy^2}{x^2+y^2}\right| \le \sqrt{x^2+y^2} < \varepsilon$ in some δ-neighborhood of $(0,0)$ since

$$\lim_{(x,y)\to(0,0)} \sqrt{x^2+y^2} = 0. \text{ Therefore, } \lim_{(x,y)\to(0,0)}\left(\dfrac{xy^2}{x^2+y^2}\right) = 0.$$

21. The boundary consists of the circle and the origin. The set is neither open [since, for example, (1,0) is not an interior point], nor closed [since (0,0) is not in the set].

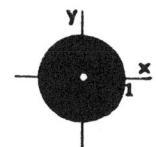

24. The boundary is the set itself along with the origin. The set is neither open (since none of its points are interior points) nor closed (since the origin is not in the set).

Problem Set 15.4 Differentiability

3. $\nabla f(x,y) = \langle (x)(e^{xy}y)+(e^{xy})(1), xe^{xy}x \rangle = e^{xy}\langle xy+1, x^2 \rangle$.

6. $\nabla f(x,y) = \langle 3[\sin^2(x^2y)][\cos(x^2y)](2xy), 3[\sin^2(x^2y)][\cos(x^2y)](x^2)\rangle$
    $= 3x\sin^2(x^2y)\cos(x^2y)\langle 2y, x\rangle$.

9. $\nabla f(x,y) = \langle (x^2y)(e^{x-z})+(e^{x-z})(2xy), x^2e^{x-z}, x^2ye^{x-z}(-1)\rangle$
    $= xe^{x-z}\langle y(x+2), x, -xy\rangle$.

12. $\nabla f(x,y) = \langle e^x\sin y + e^y\cos x, e^x\cos y + e^y\sin x\rangle = e^x\langle \sin y, \cos y\rangle + e^y\langle \cos x, \sin x\rangle$.
    $\nabla f(\pi/6, \pi/3) = e^{\pi/6}\langle\sqrt{3}/2, 1/2\rangle + e^{\pi/3}\langle\sqrt{3}/2, 1/2\rangle = \dfrac{(e^{\pi/6}+e^{\pi/3})}{2}\langle\sqrt{3}, 1\rangle$
    $\approx \langle 3.9298, 2.2689\rangle$.

15. $\nabla f(x,y,z) = \langle 6x, -10y, 2z\rangle$; $\nabla f(1,1-7) = \langle 6, -10, -14\rangle$.

18. $\nabla(f^r) = \langle rf^{r-1}f_x, rf^{r-1}f_y, rf^{r-1}f_z\rangle = rf^{r-1}\langle f_x, f_y, f_z\rangle = rf^{r-1}\nabla f$.

21. $\nabla f(\mathbf{p}) = \nabla g(\mathbf{p}) \Rightarrow \nabla[f(\mathbf{p})-g(\mathbf{p})] = 0 \Rightarrow f(\mathbf{p})-g(\mathbf{p})$ is a constant.

Problem Set 15.5 Directional Derivatives and Gradients

3. $D_\mathbf{u} f(x,y) = \nabla f(x,y) \cdot \mathbf{u}$ [where $\mathbf{u} = \mathbf{a}/|\mathbf{a}|$]

$= \langle 4x+y, x-2y \rangle \cdot \frac{\langle 1,-1 \rangle}{\sqrt{2}}$; $D_\mathbf{u} f(3,-2) = \langle 10,7 \rangle \cdot \frac{\langle 1,-1 \rangle}{\sqrt{2}} = \frac{3}{\sqrt{2}} \approx 2.1213$

6. $D_\mathbf{u} f(x,y) = \langle -ye^{-xy}, -xe^{-xy} \rangle \cdot \frac{\langle -1, \sqrt{3} \rangle}{2}$

$D_\mathbf{u} f(1,-1) = \langle e, -e \rangle \cdot \frac{\langle -1, \sqrt{3} \rangle}{2} = \frac{-e - e\sqrt{3}}{2} \approx -3.7132$

9. f increases most rapidly in the direction of the gradient.

$\nabla f(x,y) = \langle 3x^2, -5y^4 \rangle$; $\nabla f(2,-1) = \langle 12, -5 \rangle$. $\frac{\langle 12,-5 \rangle}{13}$ is the unit vector in that direction. The rate of change of f(x,y) in that direction at that point is the magnitude of the gradient, $|\langle 12,-5 \rangle| = 13$.

12. f increases most rapidly in the direction of the gradient.

$\nabla f(x,y,z) = \langle e^{yz}, xze^{yz}, xye^{yz} \rangle$; $\nabla f(2,0,-4) = \langle 1, -8, 0 \rangle$. $\frac{\langle 1,-8,0 \rangle}{\sqrt{65}}$
is a unit vector in that direction. $|\langle 1,-8,0 \rangle| = \sqrt{65} \approx 8.0623$ is the rate of change of f(x,y,z) in that direction at that point.

15. The level curves are $y/x^2 = k$. For $\mathbf{p} = (1,2)$, $k = 2$, so the level curve through (1,2) is $y/x^2 = 2$ or $y = 2x^2$ [$x \neq 0$].

$\nabla f(x,y) = \langle -2yx^{-3}, x^{-2} \rangle$

$\nabla f(1,2) = \langle -4, 1 \rangle$ which is perpendicular to the parabola at (1,2).

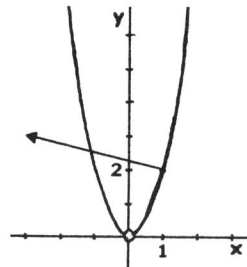

18. $(0, \pi/3)$ is on the y-axis, so the unit vector toward the origin is $-\mathbf{j}$.

$D_\mathbf{u}(x,y) = \langle -e^{-x}\cos y, -e^{-x}\sin y \rangle \cdot \langle 0,-1 \rangle = e^{-x}\sin y$; $D_\mathbf{u}(0, \pi/3) = \sqrt{3}/2$.

21. He should move in the direction of
$-\nabla f(\mathbf{p}) = -\langle f_x(\mathbf{p}), f_y(\mathbf{p})\rangle = -\langle -1/2, -1/4\rangle$
$= (1/4)\langle 2,1\rangle$. Or use $\langle 2,1\rangle$.

The angle α formed with the East is
$\tan^{-1}(1/2) \approx 26.57°$ [N63.43°E]

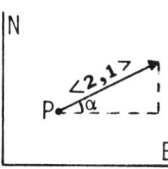

24. $\dfrac{dx/dt}{2x} = \dfrac{dy/dt}{-2y}$; $\dfrac{dx}{x} = \dfrac{dy}{-y}$; $\ln|x| = -\ln|y| + C$.

At $t = 0$: $\ln|-2| = -\ln|1| + C \Rightarrow C = \ln 2$

$\ln|x| = -\ln|y| + \ln 2 = \ln|2/y|$; $|x| = |2/y|$; $|xy| = 2$.

Since the particle starts at $(-2,1)$ and neither x nor y can equal 0, the equation simplifies to $xy = -2$.

$\nabla T(-2,1) = \langle -4,-2\rangle$, so the particle moves downward along the curve.

Problem Set 15.6 The Chain Rule

3. $\dfrac{dw}{dt} = (e^x \sin y + e^y \cos x)(3) + (e^x \cos y + e^y \sin x)(2)$

$= 3e^{3t}\sin 2t + 3e^{2t}\cos 3t + 2e^{3t}\cos 2t + 2e^{2t}\sin 3t$

6. $\dfrac{dw}{dt} = (y+z)(2t) + (x+z)(-2t) + (y+x)(-1)$

$= 2t(2-t-t^2) - 2t(1-t+t^2) - (1)$

$= -4t^3 + 2t - 1$

9. $\dfrac{\partial w}{\partial t} = e^{x^2+y^2}(2x)(s\cos t) + e^{x^2+y^2}(2y)(s\sin s)$

$= 2e^{x^2+y^2}(xs\cos t + y\sin s)$

$= 2(s^2 \sin t \cos t + t \sin^2 s) \exp(s^2\sin^2 t + t^2\sin^2 s)$

Problem Set 15.6

12. $\frac{\partial w}{\partial t} = (e^{xy+z}\,y)(1) + (e^{xy+z}\,x)(-1) + (e^{xy+z})(2t)$

$= e^{xy+z}(y-x+2t)$

$= e^{s^2}(0) = 0$

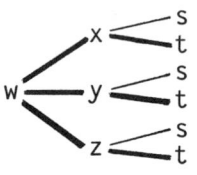

15. $\frac{dw}{dx} = (2u-\tan v)(1) + (-u\sec^2 v)(\pi)$

$= 2x - \tan\pi x - \pi x\sec^2\pi x$

$\left.\frac{dw}{dx}\right|_{x=1/4} = (1/2)-1-(\pi/2) = \frac{1+\pi}{-2}$ -2.0708

18. Let $T = e^{-x-3y}$

$\frac{dT}{dt} = e^{-x-3y}(-1)\frac{dx}{dt} + e^{-x-3y}(-3)\frac{dy}{dt}$

$= e^{-x-3y}(-1)(2) + e^{-x-3y}(-3)(2) = -8e^{-x-3y}$

$\left.\frac{dT}{dt}\right|_{(0,0)} = -8$, so the temperature is decreasing at 8 degrees/minute.

21. Let $F(x,y) = x^3+2x^2y-y^3 = 0$.

Then $\frac{dy}{dx} = \frac{-\partial F/\partial x}{\partial F/\partial y} = \frac{-(3x^2+4xy)}{2x^2-3y^2} = \frac{3x^2+4xy}{3y^2-2x^2}$

24. Let $F(x,y) = x^2\cos y - y^2\sin x = 0$.

Then $\frac{dy}{dx} = \frac{-\partial F/\partial x}{\partial F/\partial y} = \frac{-(2x\cos y - y^2\cos x)}{-x^2\sin y - 2y\sin x} = \frac{2x\cos y - y^2\cos x}{x^2\sin y + 2y\sin x}$

27. $\frac{\partial T}{\partial s} = \frac{\partial T}{\partial x}\frac{\partial x}{\partial s} + \frac{\partial T}{\partial y}\frac{\partial y}{\partial s} + \frac{\partial T}{\partial z}\frac{\partial z}{\partial s} + \frac{\partial T}{\partial w}\frac{\partial w}{\partial s}$

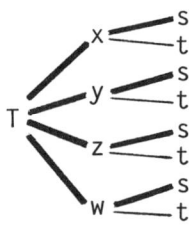

Problem Set 15.7 Tangent Planes, Approximations

3. Let $F(x,y,z) = x^2-y^2+z^2+1 = 0$. $\nabla F(x,y,z) = \langle 2x,-2y,2z\rangle = 2\langle x,-y,z\rangle$. $\nabla F(1,3,\sqrt{7}) = 2\langle 1,-3,\sqrt{7}\rangle$, so $\langle 1,-3,\sqrt{7}\rangle$ is normal to the surface at the point. Then the tangent plane is $1(x-1)-3(y-3)+\sqrt{7}(z-\sqrt{7}) = 0$, or more simply $x-3y+\sqrt{7}z = -1$.

6. Let $f(x,y) = xe^{-2y}$. $\nabla f(x,y) = \langle e^{-2y},-2xe^{-2y}\rangle$. $\nabla f(1,0) = \langle 1,-2\rangle$. Then $\langle 1,-2,-1\rangle$ is normal the the surface at $(1,0,1)$, and the tangent plane is $1(x-1)-2(y-0)-1(z-1) = 0$, or $x-2y-z = 0$.

9. Let $z = f(x,y) = 2x^2y^3$; $dz = 4xy^3dx + 6x^2y^2dy$. For the points given, $dx = -0.01$, $dy = 0.02$, $dz = 4(-0.01) + 6(0.02) = 0.08$.

    $\Delta z = f(0.99,1.02) - f(1,1) = 2(0.99)^2(1.02)^3 - 2(1)^2(1)^3 \approx 0.08018$.

12. Let $z = f(x,y) = \tan^{-1}xy$; $dz = \dfrac{y}{1+x^2y^2}dx + \dfrac{x}{1+x^2y^2}dy = \dfrac{ydx+xdy}{1+x^2y^2}$.

    For the points given, $dx = -0.03$, $dy = -0.01$,
    $dz = \dfrac{(-0.5)(-0.03)+(-2)(-0.01)}{1+(4)(0.25)} = 0.0175$.

    $\Delta z = f(-2.03,-0.51)-f(-2,-0.5) = \tan^{-1}(2.03)(0.51)-\tan^{-1}(1) \approx 0.01734$.

15. For $F(x,y,z) = x^2+4y+z^2 = 0$, $\nabla F(x,y,z) = \langle 2x,4,2z\rangle = 2\langle x,2,z\rangle$, $F(0,-1,2) = 0$, and $\nabla F(0,-1,2) = 2\langle 0,2,2\rangle = 4\langle 0,1,1\rangle$.

    For $G(x,y,z) = x^2+y^2+z^2-6z+7=0$, $\nabla G(x,y,z) = \langle 2x,2y,2z-6\rangle = 2\langle x,y,z-3\rangle$, $G(0,-1,2) = 0$, and $\nabla G(0,-1,2) = 2\langle 0,-1,-1\rangle = -2\langle 0,1,1\rangle$.

    $\langle 0,1,1\rangle$ is normal to both surfaces at $(0,-1,2)$ so the surfaces have the same tangent plane; hence, they are tangent to each other at $(0,-1,2)$.

18. Let $F(x,y,z) = \dfrac{x^2}{a^2} + \dfrac{y^2}{b^2} + \dfrac{z^2}{c^2} = 1$.

    $F(x,y,z) = \langle 2x/a^2, 2y/b^2, 2z/c^2\rangle$. $\nabla F(x_0,y_0,z_0) = 2\langle x_0/a^2, y_0/b^2, z_0/c^2\rangle$.

    The tangent plane at (x_0,y_0,z_0) is $\dfrac{x_0(x-x_0)}{a^2} + \dfrac{y_0(y-y_0)}{b^2} + \dfrac{z_0(z-z_0)}{c^2} = 0$.

    $\dfrac{x_0 x}{a^2} + \dfrac{y_0 y}{b^2} + \dfrac{z_0 z}{c^2} - \left(\dfrac{x_0^2}{a^2} + \dfrac{y_0^2}{b^2} + \dfrac{z_0^2}{c^2}\right) = 0$.

    Therefore, $\dfrac{x_0 x}{a^2} + \dfrac{y_0 y}{b^2} + \dfrac{z_0 z}{c^2} = 1$. since $\dfrac{x_0^2}{a^2} + \dfrac{y_0^2}{b^2} + \dfrac{z_0^2}{c^2} = 1$.

21. $dS = S_A dA + S_W dW = \dfrac{-W}{(A-W)^2} dA + \dfrac{A}{(A-W)^2} dW = \dfrac{-WdA + AdW}{(A-W)^2}$

At $W = 20$, $A = 36$: $dS = \dfrac{-20dA + 36dW}{256} = \dfrac{-5dA + 9dW}{64}$.

Thus, $|dS| \leq \dfrac{5|dA| + 9|dW|}{64} \leq \dfrac{5(0.02) + 9(0.02)}{64} = 0.0004375$

24. Let $F(x,y,z) = xyz = k$; let (a,b,c) be any point on the surface of F. $F(x,y,z) = \langle yz, xz, xy \rangle = \langle k/x, k/y, k/z \rangle = k\langle 1/x, 1/y, 1/z \rangle$. $F(a,b,c) = k\langle 1/a, 1/b, 1/c \rangle$. An equation of the tangent plane at the point is $(1/a)(x-a) + (1/b)(x-b) + (1/c)(x-c) = 0$, or $\dfrac{x}{a} + \dfrac{y}{b} + \dfrac{z}{c} = 3$.

Points of intersection of the tangent plane on the coordinate axes are $(3a,0,0)$, $(0,3b,0)$, and $(0,0,3c)$.

The volume of the tetrahedron is $(1/3)$(area of base)(altitude)

$= \dfrac{1}{3}(\dfrac{1}{2}\;|3a||3b|)(|3c|) = \dfrac{9|abc|}{2} = \dfrac{9|k|}{2}$ (a constant).

Problem Set 15.8 Maxima and Minima

3. $\nabla f(x,y) = \langle 8x^3 - 2x, 6y \rangle = \langle 2x(4x^2-1), 6y \rangle = \langle 0,0 \rangle$ at $(0,0), (0.5,0), (-0.5,0)$, all stationary points.

$f_{xx} = 24x^2 - 2$; $D = f_{xx}f_{yy} - f_{xy}^2 = (24x^2-2)(6) - (0)^2 = 12(12x^2-1)$.

At $(0,0)$: $D = -12$, so $(0,0)$ is a saddle point.

At $(0.5,0)$ and $(-0.5,0)$: $D = 24$ and $f_{xx} = 6$, so local minima occur at at these points.

6. Let $f(x,y) = \langle 3x^2 - 6y, 3y^2 - 6x \rangle = \langle 0,0 \rangle$. Then $3x^2 - 6y = 0$ and $3y^2 - 6x = 0$.

Solving simultaneously, obtain solutions $(0,0)$ and $(2,2)$.

$f_{xx} = 6x$; $D = f_{xx}f_{yy} - f_{xy}^2 = (6x)(6y) - (-6)^2 = 36(xy-1)$.

At $(0,0)$: $D < 0$, so $(0,0)$ is a saddle point.

At $(2,2)$: $D > 0$, $f_{xx} > 0$, so a local minimum occurs here.

9. Let $\nabla f(x,y) = \langle -\sin x - \sin(x+y), -\sin y - \sin(x+y) \rangle = \langle 0,0 \rangle$.

   Then $\begin{bmatrix} -\sin x - \sin(x+y) = 0 \\ \sin y + \sin(x+y) = 0 \end{bmatrix}$. Therefore, $\sin x = \sin y$, so $x = y = \pi/4$.

   However, these values satisfy neither equation. Therefore, the gradient is defined but never zero in its domain, and the boundary of the domain is outside the domain, so there are no critical points.

12. We do not need to use calculus for this one. Each of x^2 and y^2 is minimum at 0 and $(0,0)$ is in S, so x^2+y^2 is minimum at $(0,0)$; the minimum value is 0. Similarly, x^2 and y^2 are maximum at $x=3$ and $y=4$, respectively, and $(3,4)$ is in S, so x^2+y^2 is maximum at $(3,4)$; the maximum value is 25. [Use calculus techniques and compare.]

15. Let x,y,z denote the numbers, so $x+y+z = N$.

    Maximize $P = xyz = xy(N-x-y) = Nxy-x^2y-xy^2$.

    Let $\nabla P(x,y) = \langle Ny-2xy-y^2, Nx-x^2-2xy \rangle = \langle 0,0 \rangle$.

    Then $\begin{bmatrix} Ny-2xy-y^2 = 0 \\ Nx-x^2-2xy = 0 \end{bmatrix}$. $N(x-y) = x^2-y^2 = (x+y)(x-y)$. $x=y$ or $N = x+y$.

    Therefore, $x = y$ [since $N = x+y$ would mean that $P = 0$, certainly not a maximum value].

    Then, substituting into $Nx-x^2-2xy = 0$, we obtain $Nx-x^2-2x^2 = 0$, from which we obtain $x(N-3x) = 0$, so $x = N/3$ [since $x = 0 \Rightarrow P = 0$].

    $P_{xx} = -2y$; $D = P_{xx}P_{yy} - P_{xy}^2 = (-2y)(-2x) - (N-2x-2y)^2 = 4xy - (N-2x-2y)^2$

    At $x = y = N/3$: $D = N^2/3 > 0$, $P_{xx} = -2N/3 < 0$ [so local maximum].

    If $x = y = N/3$, then $z = N/3$.

    Conclusion: Each number is $N/3$. [If the intent is to find three distinct numbers, then there is no maximum value of P that satisfies that condition.]

Problem Set 15.8

18. Let L denote the sum of edge lengths for a box of dimensions x,y,z.
    Minimize $L = 4x+4y+4z$, subject to $V_0 = xyz$.

    $L(x,y) = 4x + 4y + 4V_0/xy$, $x > 0$, $y > 0$.

    Let $\nabla L(x,y) = \left\langle \dfrac{4(x^2y-V_0)}{x^2y}, \dfrac{4(xy^2-V_0)}{xy^2} \right\rangle = \langle 0,0 \rangle$.

    Then $x^2y = V_0$ and $xy^2 = V_0$, from which it follows that $x = y$.
    Therefore, $x = y = z = V_0^{1/3}$.

    $L_{xx} = 8V_0/x^3y$; $D = L_{xx}L_{yy} - L_{xy}^2 = (8V_0/x^3y)(8V_0/xy^3) - (4V_0/x^2y^2)$.

    At $(V_0^{1/3}, V_0^{1/3})$: $D > 0$, $L_{xx} > 0$ [so local minimum].

    There are no other critical points, and as $(x,y) \to$ boundary, $L \to \infty$.

    Conclusion: The optimal box is a cube.

21. Let $\langle x,y,z \rangle$ denote the vector; let S be the sum of its components.

    $x^2+y^2+z^2 = 81$, so $z = (81-x^2-y^2)^{1/2}$.

    Maximize $S(x,y) = x+y+(81-x^2-y^2)^{1/2}$, $0 \leq x^2+y^2 \leq 9$.

    Let $\nabla S(x,y) = \langle 1-x(81-x^2-y^2)^{-1/2}, 1-y(81-x^2-y^2)^{-1/2} \rangle = \langle 0,0 \rangle$.

    Therefore, $x = (81-x^2-y^2)^{1/2}$ and $y = (81-x^2-y^2)^{1/2}$. We then obtain $x = y = 3\sqrt{3}$ as the only stationary point. For these values of x and y, $z = 3\sqrt{3}$ and $S = 9\sqrt{3} \approx 15.59$.

    The boundary needs to be checked. It is fairly easy to check each edge of the boundary separately. The largest value of S at a boundary point occurs at three places and turns out to be $18/\sqrt{2} \approx 12.73$.

    Conclusion: The vector is $3\sqrt{3}\langle 1,1,1 \rangle$.

24. The lines are skew since there are no values of s and t that simultaneously satisfy $t-1 = 3s$, $2t = s+2$, and $t+3 = 2s-1$.

Minimize f, the square of the distance between points on the two lines.

$$f(s,t) = (3s-t+1)^2 + (s+2-2t)^2 + (2s-1-t-3)^2$$

Let $\nabla f(s,t) = \langle 2(3s-t+1)(3) + 2(s-2t+2)(1) + 2(2s-t-4)(2),$
$\qquad\qquad\qquad 2(3s-t+1)(-1) + 2(s-2t+2)(-2) + 2(2s-t-4)(-1)\rangle$

$\qquad\qquad = \langle 28s-14t-6, -14s+12t-2\rangle = \langle 0,0\rangle.$

Solve $\{28s-14t-6 = 0, -14s+12t-2 = 0\}$, obtaining $s = 5/7$, $t = 1$.

$D = f_{ss}f_{tt} - f_{st}^2 = (28)(12) - (-14)^2 > 0$; $f_{ss} = 28 > 0$. [local min.]

The nature of the problem indicates the global minimum occurs here.

$f(5/7,1) = (15/7)^2 + (5/7)^2 + (-25/7)^2 = 875/49.$

Conclusion: The minimum distance between the lines is $\sqrt{875}/7 \approx 4.2258$.

[For another way of doing this problem see Problem 21 of Section 14.4.]

27. $z(x,y) = y^2-x^2$. $\nabla z(x,y) = \langle -2x, 2y\rangle = \langle 0,0\rangle$ at $(0,0)$.

There are no stationary points and no singular points, so consider boundary points.

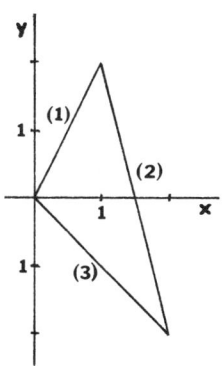

On side (1): $y = 2x$, so $z = 4x^2-x^2 = 3x^2$.
$\qquad\qquad z'(x) = 6x = 0$ if $x = 0$.
$\qquad\qquad$ Therefore, $(0,0)$ is a candidate.

On side (2): $y = -4x+6$, so $z = (-4x+6)^2-x^2$
$\qquad\qquad\qquad = 15x^2-48x+36$.

$\qquad\qquad z'(x) = 30x-48 = 0$ if $x = 1.6$
$\qquad\qquad$ Therefore, $(1.6,-0.4)$ is a candidate.

On side (3): $y = -x$, so $z = (-x)^2-x^2 = 0$.

Also, all vertices are candidates.

x	y	z	
0	0	0	
1.6	-0.4	-2.4	← minimum value of -2.4
2	-2	0	
1	2	3	← maximum value of 3

Problem Set 15.8

Problem Set 15.9 Lagrange's Method

3. Let $\nabla f(x,y) = \lambda \nabla g(x,y)$, where $g(x,y) = x^2+y^2-1 = 0$.

$\langle 8x-4y, -4x+2y \rangle = \lambda \langle 2x, 2y \rangle$.

   (1) $4x-2y = \lambda x$.
   (2) $-2x+y = \lambda y$.
   (3) $x^2+y^2 = 1$.

   (4) $0 = \lambda x + 2\lambda y$.   [From equations (1) and (2)]
   (5) $\lambda = 0$ or $x+2y = 0$. [(4)]

$\lambda=0$: (6) $y = 2x$.   [(1)]
        (7) $x = \pm 1/\sqrt{5}$.   [(6),(3)]
        (8) $y = \pm 2/\sqrt{5}$.   [(7),(6)]

$x+2y=0$: (9) $x = -2y$.
           (10) $y = \pm 1/\sqrt{5}$.   [(9),(3)]
           (11) $x = \mp 2/\sqrt{5}$.   [(10),(9)]

Critical Points: $(1/\sqrt{5}, 2/\sqrt{5})$, $(-1/\sqrt{5}, -2/\sqrt{5})$, $(2/\sqrt{5}, -1/\sqrt{5})$, $(-2/\sqrt{5}, 1/\sqrt{5})$.

$f(x,y)$ is 0 at the first two critical points and 5 at the last two. Therefore, the maximum value of $f(x,y)$ is 5.

6. Let $\nabla f(x,y,z) = \lambda \nabla g(x,y,z)$, where $g(x,y,z) = 2x^2+y^2-3z = 0$.

$\langle 4, -2, 3 \rangle = \lambda \langle 4x, 2y, -3 \rangle$.

   (1) $4 = 4\lambda x$.
   (2) $-2 = 2\lambda y$.
   (3) $3 = -3\lambda$.
   (4) $2x^2+y^2-3z = 0$.

   (5) $\lambda = -1$.   [(3)]
   (6) $x = -1$, $y = 1$. [(5),(1),(2)]
   (7) $z = 1$.   [(6),(4)]

Therefore, $(-1,1,1)$ is a critical point, and $f(-1,1,1) = -3$.

-3 is the minimum rather than the maximum since other points satisfying g have larger values of f. Example: $g(1,1,1) = 0$ and $f(1,1,1) = 5$.

9. Let ℓ and w denote the dimensions of the base, h the depth.

   Maximize $V(\ell,w,h) = \ell wh$ subject to $0.60\ell w + 0.20(\ell w + 2\ell h + 2wh) = 12$ which simplifies to $2\ell w + \ell h + wh = 30$, or $g(\ell,w,h) = 2\ell w + \ell h + wh - 30$.

   Let $\nabla V(\ell,w,h) = \lambda \nabla g(\ell,w,h)$; $\langle wh, \ell h, \ell w \rangle = \lambda \langle 2w+h, 2\ell+h, \ell+w \rangle$

   (1) $wh = \lambda(2w+h)$
   (2) $\ell h = \lambda(2\ell+h)$
   (3) $\ell w = \lambda(\ell+w)$
   (4) $2\ell w + \ell h + wh = 30$

   (5) $(w-\ell)h = 2\lambda(w-\ell)$ [(1),(2)]
   (6) $w = \ell$ or $h = 2\lambda$

   $w=\ell$:   (7) $\ell = 2\lambda = w$ [(3)]   Note: $w \neq 0$, for then $V = 0$.
           (8) $h = 4\lambda$ [(7),(2)]
           (9) $\lambda = \sqrt{5}/2$ [(7),(8),(4)]
           (10) $\ell = w = \sqrt{5}$, $h = 2\sqrt{5}$ [(9),(7),(8)]

   $h=2\lambda$: (11) $\lambda = 0$ [(2)]
              (12) $\ell = w = h = 0$ [(11),(1)-(3)]
              (not possible since this does not satisfy (4).)

   $(\sqrt{5}, \sqrt{5}, 2\sqrt{5})$ is a critical point and $V(\sqrt{5}, \sqrt{5}, 2\sqrt{5}) = 10\sqrt{5} \approx 22.36$ ft^3 is the maximum volume [rather than the minimum volume since, for example, $g(1,1,14) = 30$ and $V(1,1,14) = 14$ which is less than 22.36].

12. Let (x,y,z) denote a point of intersection. Let $f(x,y,z)$ be the square of the distance to the origin. Minimize $f(x,y,z) = x^2+y^2+z^2$ subject to $g(x,y,z) = x+y+z-8 = 0$ and $h(x,y,z) = 2x-y+3z-28 = 0$.

   Let $\nabla f(x,y,z) = \lambda \nabla g(x,y,z) + \mu \nabla h(x,y,z)$
        $\langle 2x, 2y, 2z \rangle = \lambda \langle 1,1,1 \rangle + \mu \langle 2,-1,3 \rangle$

   (1) $2x = \lambda + 2\mu$
   (2) $2y = \lambda - \mu$
   (3) $2z = \lambda + 3\mu$
   (4) $x+y+z = 8$
   (5) $2x-y+3z = 28$

   (6) $3\lambda + 4\mu = 16$ [(1),(2),(3),(4)]
   (7) $2\lambda + 7\mu = 28$ [(1),(2),(3),(5)]
   (8) $\lambda = 0$, $\mu = 4$ [(6),(7)]
   (9) $x = 4$, $y = -2$, $z = 6$ [(8),(1)-(3)]

   $f(4,-2,6) = 56$, and the nature of the problem indicates this is the minimum rather than the maximum.

   Conclusion: The least distance is $\sqrt{56} \approx 7.4833$.

Problem Set 15.9

Problem Set 15.10   Chapter Review Problems

True-False Quiz

3. True. Since $g'(0) = f_x(0,0)$.

6. True. Straight forward calculation of partial derivatives.

9. True. Since $\langle 0,0,-1 \rangle$ is normal to the tangent plane.

12. True. It is nonnegative for all x,y, and it has a value of 0 at (0,0).

15. True. $-D_{\mathbf{u}}f(x,y) = -[\nabla f(x,y) \cdot \mathbf{u}] = \nabla f(x,y) \cdot (-\mathbf{u}) = D_{-\mathbf{u}}f(x,y)$

18. False. (x_0,y_0) could be a singular point.

Miscellaneous Problems

3. $f_x(x,y) = 12x^3y^2 + 14xy^7$
   $f_{xx}(x,y) = 36x^2y^2 + 14y^7$
   $f_{xy}(x,y) = 24x^3y + 98xy^6$

6. $f_x(x,y) = -e^{-x}\sin y$
   $f_{xx}(x,y) = e^{-x}\sin y$
   $f_{xy}(x,y) = -e^{-x}\cos y$

9. $z_y(x,y) = y/2$; $z_y(2,2) = 2/2 = 1$.

12. (a) $\lim\limits_{(x,y) \to (2,2)} \dfrac{x^2-2y}{x^2+2y} = \dfrac{4-4}{4+4} = 0$

(b) Doesn't exist since $\begin{bmatrix} \to 4 \\ \to 0 \end{bmatrix}$.

(c) $\lim\limits_{(x,y) \to (0,0)} \dfrac{(x^2+2y^2)(x^2-2y^2)}{x^2+2y^2} = \lim\limits_{(x,y) \to (0,0)} (x^2-2y^2) = 0$

15. $z = f(x,y) = x^2+y^2$.

⟨1,-√3,0⟩ is horizontal and is normal to the vertical plane that is given. By inspection, ⟨√3,1,0⟩ is also a horizontal vector and is perpendicular to ⟨1,-√3,0⟩, and therefore is parallel to the vertical plane. Then **u** = ⟨√3/2, 1/2⟩ is the corresponding 2-dimensional unit vector.

$D_u f(x,y) = \nabla f(x,y) \cdot \mathbf{u} = \langle 2x, 2y \rangle \cdot \langle \sqrt{3}/2, 1/2 \rangle = \sqrt{3}x + y$.

$D_u f(1,2) = \sqrt{3} + 2 \approx 3.7321$ is the slope of the tangent to the curve.

18. $F_x = F_u u_x + F_v v_x$

$= \dfrac{v}{1+u^2v^2} \dfrac{y}{2\sqrt{xy}} + \dfrac{u}{1+u^2v^2} \dfrac{1}{2\sqrt{x}}$

$= \dfrac{v\sqrt{y} + u}{2(1+u^2v^2)\sqrt{x}}$

$F_y = F_u u_y + F_v v_y$

$= \dfrac{v}{1+u^2v^2} \dfrac{x}{2\sqrt{xy}} + \dfrac{u}{1+u^2v^2} \dfrac{-1}{2\sqrt{y}}$

$= \dfrac{v\sqrt{x} - u}{2(1+u^2v^2)\sqrt{y}}$

21. $F_t = F_x x_t + F_y y_t + F_z z_t$

$= \dfrac{10xy}{z^3} \dfrac{3t^{1/2}}{2} + \dfrac{5x^2}{z^3} \dfrac{1}{t} + \dfrac{-15x^2 y}{z^4} 3e^{3t}$

$= \dfrac{15xy\sqrt{t}}{z^3} + \dfrac{5x^2}{z^3 t} - \dfrac{45x^2 y e^{3t}}{z^4}$

Problem Set 15.10

24. $V = \pi r^2 h$; $dV = V_r dr + V_h dh = 2\pi rh\, dr + \pi r^2 dh$

If $r = 10$, $|dr| \leq 0.02$, $h = 6$, $|dh| \leq 0.01$,

then $|dV| \leq 2\pi rh|dr| + \pi r^2|dh|$

$$\leq 2\pi(10)(6)(0.02) + \pi(100)(0.01) = 3.4\pi$$

$V(10,6) = \pi(100)(6) = 600\pi$. Volume is $600\pi \pm 3.4\pi \approx 1884.96 \pm 10.68$.

27. Let (x,y,z) denote the coordinates of the 1st octant vertex of the box.

Maximize $f(x,y,z) = xyz$ subject to $g(x,y,z) = 36x^2 + 4y^2 + 9z^2 - 36 = 0$

[where $x,y,z > 0$ and the box's volume is $V(x,y,z) = 8f(x,y,z)$.]

Let $\nabla f(x,y,z) = \lambda \nabla g(x,y,z)$
$\langle yz, xz, xy \rangle = \lambda \langle 72x, 8y, 18z \rangle$

(1) $yz = 72\lambda x$
(2) $xz = 8\lambda y$
(3) $xy = 18\lambda z$
(4) $36x^2 + 4y^2 + 9z^2 = 36$

(5) $\dfrac{yz}{xz} = \dfrac{72\lambda x}{8\lambda y}$, so $y^2 = 9x^2$ [(1),(2)]

(6) $\dfrac{yz}{xy} = \dfrac{72\lambda x}{18\lambda z}$, so $z^2 = 4x^2$ [(1),(3)]

(7) $36x^2 + 36x^2 + 36x^2 = 36$, so $x = 1/\sqrt{3}$ [(5),(6),(4)]

(8) $y = 3/\sqrt{3}$, $z = 2/\sqrt{3}$ [(7),(5),(6)]

$V(1/\sqrt{3}, 3/\sqrt{3}, 2/\sqrt{3}) = 8(1/\sqrt{3})(3/\sqrt{3})(2/\sqrt{3}) = 16/\sqrt{3} \approx 9.2376$

The nature of the problem indicates that the critical point yields a maximum value rather than a minimum value.

[For a generalization of this problem, see Problem 11 of Section 15.9.]

CHAPTER 16 THE INTEGRAL IN n-SPACE

Problem Set 16.1 Double Integrals over Rectangles

3. $\iint_R f(x,y)\,dA = \iint_{R_1} 2\,dA + \iint_{R_2} 1\,dA + \iint_{R_3} 3\,dA$

   $= 2A(R_1) + 1A(R_2) + 3A(R_3)$

   $= 2(2) + 1(2) + 3(2) = 12$

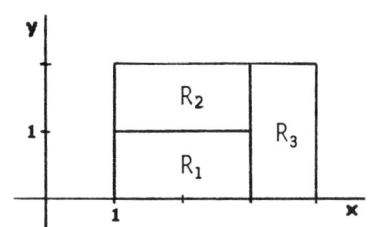

6. $2\iint_R f(x,y)\,dA + 5\iint_R g(x,y)\,dA = 2(3) + 5(5) = 31$

9. $[f(1,1) + f(3,1) + f(5,1) + f(1,3) + f(3,3) + f(5,3)](4)$

   $= [(10) + (8) + (6) + (8) + (6) + (4)](4)$

   $= 168$

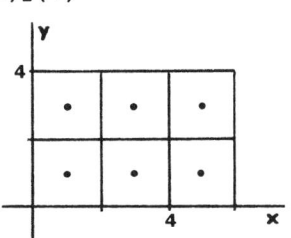

12. $[(41/6) + (33/6) + (25/6) + (35/6) + (27/6) + (19/6)](4) = 120$

15. $z = 6-y$ is a plane parallel to the x-axis. Let T be the area of the front trapezoidal face; let D be the distance between the front and back faces.

    $\iint_R (6-y)\,dA$ = volume of solid

    $= (T)(D) = [(1/2)(6+5)](1) = 5.5$

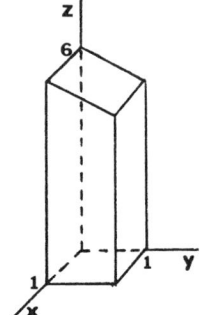

18. $\iint_R m\,dA \leq \iint_R f(x,y)\,dA \leq \iint_R M\,dA$ [Comparison Property]

    Therefore, $m\,A(R) \leq \iint_R f(x,y)\,dA \leq M\,A(R)$

Problem Set 16.2 Iterated Integrals

3. $\displaystyle\int_1^2 \left[\frac{x^2 y}{2} + xy^2\right]_{x=0}^{3} dy = \int_1^2 \left(\frac{9y}{2} + 3y^2\right) dy = \left[\frac{9y^2}{4} + y^3\right]_1^2 = 17 - \frac{13}{4} = 13.75$

6. $\displaystyle\int_0^{\ell n 3}\int_0^{\ell n 2} e^x e^y\, dy\, dx = \int_0^{\ell n 3}\left[e^x e^y\right]_{y=0}^{\ell n 2} dx = \int_0^{\ell n 3}[e^x(2) - e^x(1)]dx$

$\displaystyle = \int_0^{\ell n 3} e^x\, dx = \left[e^x\right]_0^{\ell n 3} = 3-1 = 2$

9. $\displaystyle\int_0^3\left[\frac{2(x^2+y)^{3/2}}{3}\right]_{x=0}^{1} dy = \int_0^3 \frac{2[(1+y)^{3/2} - y^{3/2}]}{3}\, dy = \left[\frac{4[(1+y)^{5/2} - y^{5/2}]}{15}\right]_0^3$

$\displaystyle = \frac{4(32 - 9\sqrt{3}) - 4}{15} = \frac{4(31 - 9\sqrt{3})}{15} \approx 4.1097$

12. $\displaystyle\int_0^1 \left[\frac{y^2}{2(1+x^2)}\right]_{y=0}^{2} dx = \int_0^1 \frac{2}{1+x^2} dx = \left[2\tan^{-1}x\right]_0^1 = 2(\pi/4) - 0 = \pi/2$

15. $\displaystyle\int_0^{\pi/2}\int_0^{\pi/2} \sin(x+y)\, dx\, dy = \int_0^{\pi/2}\left[-\cos(x+y)\right]_{x=0}^{\pi/2} dy$

$\displaystyle = \int_0^{\pi/2}[-\cos(\pi/2 + y) + \cos y]\, dy = \int_0^{\pi/2}(\sin y + \cos y)\, dy$

$\displaystyle = \left[-\cos y + \sin y\right]_0^{\pi/2} = (0+1) - (-1+0) = 2$

18. $z = 2 - x - y$ is a plane.

$x + y + z = 2$

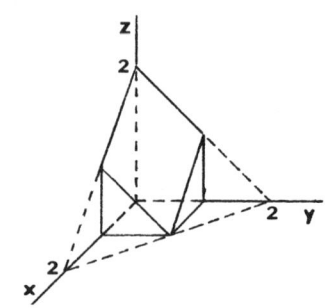

21. $\int_1^3 \int_0^1 (x+y+1) dx dy = \int_1^3 \left[\frac{x^2}{2} + yx + x\right]_{x=0}^1 dy = \int_1^3 (\frac{1}{2} + y + 1) dy$

$= \left[\frac{y}{2} + \frac{y^2}{2} + y\right]_1^3 = 9-2 = 7.$

24. $\int_0^2 \int_0^2 (4-x^2) dx dy = \int_0^2 \left[4x - \frac{x^3}{3}\right]_0^2 dy$

$= \int_0^2 \frac{16}{3} dy = \left[\frac{16y}{3}\right]_0^2 = \frac{32}{3}.$

27. $\int_0^1 \int_0^1 xy e^{x^2} e^{y^2} dy dx = \left(\int_0^1 x e^{x^2} dx\right)\left(\int_0^1 y e^{y^2} dy\right) = \left(\int_0^1 x e^{x^2} dx\right)^2$ [Changed the dummy variable y to the dummy variable x]

$= \left[\frac{e^{x^2}}{2}\right]_0^{1\ 2} = \left(\frac{e-1}{2}\right)^2 \approx 0.7381.$

Problem Set 16.3 Double Integrals over Nonrectangular Regions

3. $\int_{-1}^3 \left[\frac{x^3}{3} + y^2 x\right]_{x=0}^{3y} dy = \int_{-1}^3 (9y^3 + 3y^3) dy = \left[3y^4\right]_{-1}^3 = 243 - 3 = 240.$

6. $\int_1^5 \left[\frac{3}{x} \tan^{-1}(\frac{y}{x})\right]_{y=0}^x dx = \int_1^5 \frac{3}{x} \frac{\pi}{4} dx = \left[\frac{3\pi \ln x}{4}\right]_1^5 = \frac{3\pi \ln 5}{4} \approx 3.7921.$

9. $\int_0^{\pi/9} \left[\tan\theta\right]_{\theta=\pi/4}^{3r} dr = \int_0^{\pi/9} (\tan 3r - 1) dr = \left[\frac{-\ln|\cos 3r|}{3} - r\right]_0^{\pi/9}$

$= \left(\frac{-\ln(1/2)}{3} - \frac{\pi}{9}\right) - \left(\frac{-\ln(1)}{3} - 0\right) = \frac{3\ln 2 - \pi}{9} \approx -0.1180.$

12. $\int_{\pi/6}^{\pi/2} \left[3r^2\cos\theta\right]_{r=0}^{\sin\theta} d\theta = \int_{\pi/6}^{\pi/2} 3\sin^2\theta\cos\theta\, d\theta = \left[\sin^3\theta\right]_{\pi/6}^{\pi/2} = \frac{7}{8} = 0.875$

15. $\int_0^1 \int_{x^2}^{\sqrt{x}} (x^2+2y)\, dy\, dx = \int_0^1 \left[x^2 y + y^2\right]_{y=x^2}^{\sqrt{x}} dx$

$= \int_0^1 [(x^{5/2}+x) - (x^4+x^4)]\, dx = \left[\frac{2x^{7/2}}{7} + \frac{x^2}{2} - \frac{2x^5}{5}\right]_0^1$

$= \frac{2}{7} + \frac{1}{2} - \frac{2}{5} = \frac{27}{70} \approx 0.3857$

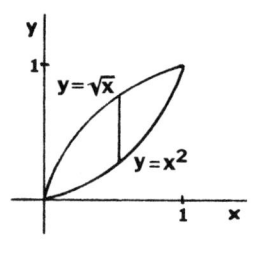

18. Since S is symmetric with respect to the origin and the integrand is an odd function in x, the value of the integral is 0.

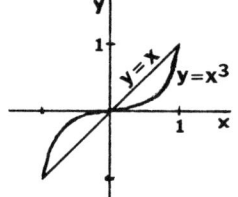

21. $\int_0^5 \int_0^4 \frac{4-y}{2}\, dy\, dx = \left(\int_0^5 1\, dx\right)\left(\int_0^4 \frac{4-y}{2}\, dy\right)$

$= 5\left[2y - \frac{y^2}{4}\right]_0^4 = 5(8-4) = 20$

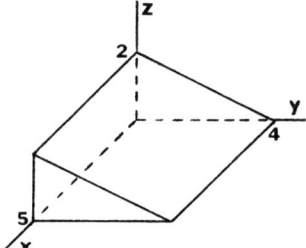

24. $\int_0^3 \int_0^{\sqrt{9-x^2}} (9-x^2-y^2)\, dy\, dx = \int_0^3 \left[(9-x^2)y - \frac{y^3}{3}\right]_{y=0}^{\sqrt{9-x^2}} dx$

$= \int_0^3 \frac{2(9-x^2)^{3/2}}{3}\, dx = \int_0^{\pi/2} 18\cos^3 t \cdot 3\cos t\, dt$

$= \int_0^{\pi/2} 54\cos^4 t\, dt = \int_0^{\pi/2} \left(\frac{81}{4} + 27\cos 2t + \frac{27\cos 4t}{4}\right) dt$

$= \left[\frac{81t}{4} + \frac{27\sin 2t}{2} + \frac{27\sin 4t}{16}\right]_0^{\pi/2} = \frac{81\pi}{8} \approx 31.8086$

[At the 3rd step the substitution, x = 3sint, was used. At the 5th step the identity $\cos^2 A = (1+\cos 2A)/2$ was used a few times.]

27. $\int_0^1 \int_0^x \tan x^2 \, dy \, dx = \int_0^1 \left[y \tan x^2 \right]_{y=0}^x dx$

$= \int_0^1 x \tan x^2 \, dx = \left[\frac{-\ln|\cos x^2|}{2} \right]_0^1$

$= \frac{-\ln(\cos 1)}{2} \approx 0.3078$

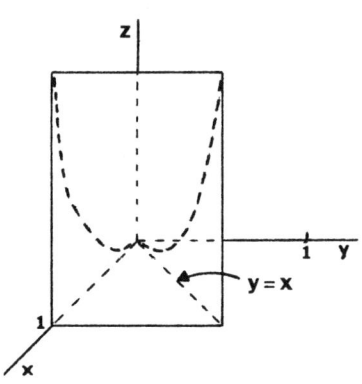

30. Making use of symmetry, the volume is

$2 \iint_{R_1} (16-x^2)^{1/2} dA = 2 \int_0^4 \int_0^x (16-x^2)^{1/2} dy \, dx$

$= 2 \int_0^4 \left[(16-x^2)^{1/2} y \right]_{y=0}^x dx$

$= 2 \int_0^4 (16-x^2)^{1/2} x \, dx = \left[\frac{-2(16-x^2)^{3/2}}{3} \right]_0^4$

$= 0 + \frac{2(64)}{3} = \frac{128}{3} \approx 42.6667$

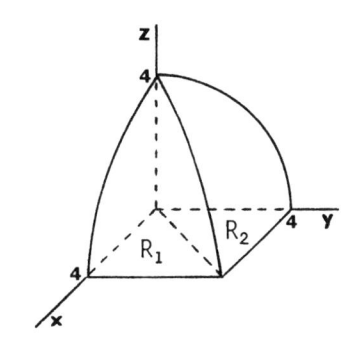

33. $\int_0^1 \int_{y^4}^{\sqrt{y}} f(x,y) \, dx \, dy$

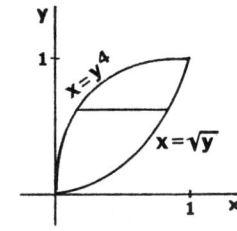

36. $\int_{-1}^1 \int_{x^2-1}^0 f(x,y) \, dy \, dx$

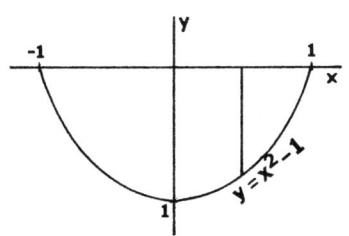

Problem Set 16.3

39. $\int_0^2 \int_0^{y^2} \sin(y^3) \, dx \, dy = \int_0^2 \left[x \sin(y^3) \right]_{x=0}^{y^2} dy$

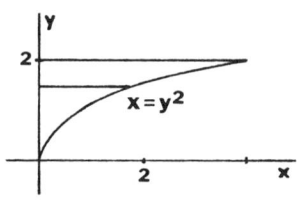

$= \int_0^2 y^2 \sin(y^3) \, dy = \left[\frac{-\cos(y^3)}{3} \right]_0^2$

$= \frac{1 - \cos 8}{3} \approx 0.3818$

Problem Set 16.4 Double Integrals in Polar Coordinates

3. $\int_0^\pi \left[\frac{r^3}{3} \right]_{r=0}^{\sin\theta} d\theta = \int_0^\pi \frac{\sin^3\theta}{3} d\theta = \int_0^\pi \frac{(1-\cos^2\theta)\sin\theta}{3} d\theta = \left[\frac{-\cos\theta}{3} + \frac{\cos^3\theta}{9} \right]_0^\pi$

$= \left(\frac{1}{3} - \frac{1}{9} \right) - \left(\frac{-1}{3} + \frac{1}{9} \right) = \frac{4}{9}$

6. $\int_0^{\pi/6} \int_0^{4\sin\theta} r \, dr \, d\theta = \int_0^{\pi/6} \left[\frac{r^2}{2} \right]_0^{4\sin\theta} d\theta$

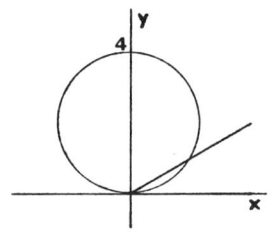

$= \int_0^{\pi/6} 8\sin^2\theta \, d\theta = \int_0^{\pi/6} 4(1-\cos 2\theta) \, d\theta$

$= \left[4\theta - 2\sin 2\theta \right]_0^{\pi/6} = \frac{2\pi}{3} - \sqrt{3} \approx 0.3623$

9. $2\int_{5\pi/6}^{3\pi/2} \int_0^{2-4\sin\theta} r \, dr \, d\theta = 2\int_{5\pi/6}^{3\pi/2} \left[\frac{r^2}{2} \right]_0^{2-4\sin\theta} d\theta$

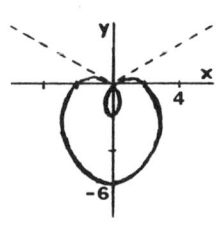

$= 2\int_{5\pi/6}^{3\pi/2} (6 - 8\sin\theta - 4\cos 2\theta) \, d\theta$

$= 2\left[6\theta + 8\cos\theta - 2\sin 2\theta \right]_{5\pi/6}^{3\pi/2} = 2(4\pi + 3\sqrt{3}) \approx 35.5250$

12. $\int_0^{\pi/4}\int_0^2 (4-r^2)^{1/2} r\,dr\,d\theta = \int_0^{\pi/4}\left[\frac{(4-r^2)^{3/2}}{-3}\right]_0^2 d\theta$

$= \int_0^{\pi/4} (8/3)\,d\theta = \left[\frac{8\theta}{3}\right]_0^{\pi/4} = 2\pi/3 \approx 2.0944$

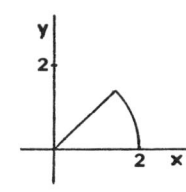

15. $\int_0^{\pi/2}\int_0^1 (4-r^2)^{-1/2} r\,dr\,d\theta = \int_0^{\pi/2}\left[-(4-r^2)^{1/2}\right]_0^1 d\theta$

$= \int_0^{\pi/2} (-\sqrt{3} + 2)\,d\theta = (-\sqrt{3} + 2)(\pi/2) \approx 0.4209$

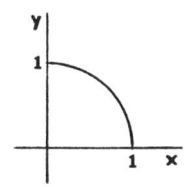

18. $\int_0^{\pi/4}\int_{\sec\theta}^{2\cos\theta} r^{-1} r\,dr\,d\theta = \int_0^{\pi/4}\left[r\right]_{\sec\theta}^{2\cos\theta} d\theta$

$= \int_0^{\pi/4} (2\cos\theta - \sec\theta)\,d\theta$

$= \left[2\sin\theta - \ln|\sec\theta + \tan\theta|\right]_0^{\pi/4}$

$= [\sqrt{2} - \ln(\sqrt{2} + 1)] - [0 - \ln(1+0)] = \sqrt{2} - \ln(\sqrt{2} + 1) \approx 0.5328$

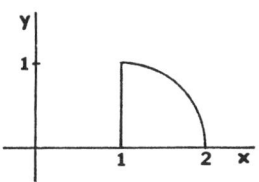

21. $\int_{-5}^0 \int_{\sqrt{3}x}^{-x} (y^2)\,dy\,dx = \int_{-5}^0 \left[\frac{y^3}{3}\right]_{\sqrt{3}x}^{-x} dx$

$= \int_{-5}^0 \frac{-1-3\sqrt{3}}{3} x^3\,dx = \left[\frac{(-1-3\sqrt{3})x^4}{12}\right]_{-5}^0$

$= \frac{(1+3\sqrt{3})625}{12} \approx 322.7163$

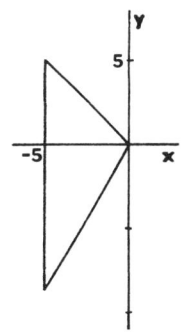

Problem Set 16.4

Problem Set 16.5 Applications of Double Integrals

3. $m = \int_0^\pi \int_0^{\sin x} y \, dy \, dx = \int_0^\pi \left[\frac{y^2}{2}\right]_0^{\sin x} dx$

$= \int_0^\pi \frac{\sin^2 x}{2} dx = \int_0^\pi \frac{1-\cos 2x}{4} dx = \left[\frac{x}{4} - \frac{\sin 2x}{8}\right]_0^\pi = \frac{\pi}{4}$

$M_x = \int_0^\pi \int_0^{\sin x} y \cdot y \, dy \, dx = \int_0^\pi \left[\frac{y^3}{3}\right]_0^{\sin x} dx = \int_0^\pi \frac{\sin^3 x}{3} dx$

$= \frac{1}{3}\int_0^\pi (1-\cos^2 x)\sin x \, dx = \frac{1}{3}\left[-\cos x + \frac{\cos^3 x}{3}\right]_0^\pi = \frac{4}{9}$

$\bar{y} = \frac{M_x}{m} = \frac{4/9}{\pi/4} = \frac{16}{9\pi} \approx 0.5659; \quad \bar{x} = \pi/2 \text{ (by symmetry)}$

6. $m = \int_0^1 \int_0^{e^x} (2-x+y) \, dy \, dx = \int_0^1 \left[(2-x)y + \frac{y^2}{2}\right]_{y=0}^{e^x} dx$

$= \int_0^1 \left(2e^x - xe^x + \frac{e^{2x}}{2}\right) dx = \left[2e^x - (xe^x - e^x) + \frac{e^{2x}}{4}\right]_0^1 = \frac{e^2+8e-13}{4}$

$M_x = \int_0^1 \int_0^{e^x} (2-x+y) y \, dy \, dx = \int_0^1 \left[y^2 - \frac{xy^2}{2} + \frac{y^3}{3}\right]_{y=0}^{e^x} dx$

$= \int_0^1 \left(e^{2x} - \frac{xe^{2x}}{2} + \frac{e^{3x}}{3}\right) dx = \left[\frac{e^{2x}}{2} - \left(\frac{xe^{2x}}{4} - \frac{e^{2x}}{8}\right) + \frac{e^{3x}}{9}\right]_0^1$

$= \left(\frac{e^2}{2} - \frac{e^2}{4} + \frac{e^2}{8} + \frac{e^3}{9}\right) - \left(\frac{1}{2} - 0 + \frac{1}{8} + \frac{1}{9}\right) = \frac{8e^3+27e^2-53}{72}$

$M_y = \int_0^1 \int_0^{e^x} (2-x+y) x \, dy \, dx = \int_0^1 \left[2xy - x^2 y + \frac{xy^2}{2}\right]_{y=0}^{e^x} dx$

$= \int_0^1 \left(2xe^x - x^2 e^x + \frac{xe^{2x}}{2}\right) dx$

$= \left[(2xe^x - 2e^x) - (x^2 e^x - 2xe^x + 2e^x) + \left(\frac{xe^{2x}}{4} - \frac{e^{2x}}{8}\right)\right]_0^1 = \frac{e^2-8e+33}{8}$

6. (continued)

$$\bar{x} \equiv \frac{M_y}{m} = \frac{e^2-8e+33}{2(e^2+8e-13)} \approx 0.5777; \quad \bar{y} \equiv \frac{M_x}{m} = \frac{8e^3+27e^2-53}{18(e^2+8e-13)} \approx 1.0577$$

9. $I_x = \int_0^3 \int_{y^2}^9 y^2(x+y)\,dx\,dy = \int_0^3 \left(\frac{81y^2}{2} + 9y^3 - \frac{y^6}{2} + y^5\right)dy = \frac{7533}{28} \approx 269$

$I_y = \int_0^9 \int_0^{\sqrt{x}} x^2(x+y)\,dy\,dx = \int_0^9 \left(x^{7/2} + \frac{x^3}{2}\right)dx = \frac{41553}{8} \approx 5194$

$I_z = I_x + I_y = \frac{305937}{56} \approx 5463$

12. $I_x = \int_0^a \int_0^{a-y} (x^2+y^2)y^2\,dx\,dy = \frac{1}{3}\int_0^a (a^3-3a^2y+3ay^2-y^3+3ay^4-3y^5)dy = \frac{7a^6}{180}$

$I_y = \frac{7a^6}{180}; \quad I_z = \frac{7a^6}{90}$.

15. $m = \delta\pi a^2$. The moment of inertia about diameter AB is

$$I = I_x = \int_0^{2\pi}\int_0^a \delta r^2 \sin^2\theta\, r\,dr\,d\theta = \int_0^{2\pi} \frac{\delta a^4 \sin^2\theta}{4}\,d\theta$$

$$= \frac{\delta a^4}{8}\int_0^{2\pi}(1-\cos 2\theta)\,d\theta = \frac{\delta a^4}{8}\left[\theta - \frac{\sin 2\theta}{2}\right]_0^{2\pi} = \frac{\delta a^4 \pi}{4}$$

$$\bar{r} = (I/m)^{1/2} = \left(\frac{\delta a^4 \pi/4}{\delta \pi a^2}\right)^{1/2} = \frac{a}{2}$$

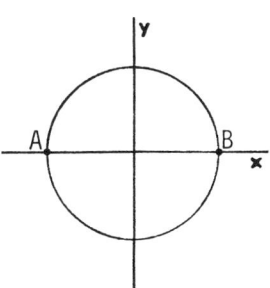

Problem Set 16.5

Problem Set 16.6 Surface Area

3. $z = f(x,y) = (4-y^2)^{1/2}$; $f_x(x,y) = 0$; $f_y(x,y) = -y(4-y^2)^{-1/2}$.

$$A(G) = \int_0^1\int_1^2 \sqrt{y^2(4-y^2)^{-1} + 1}\ dxdy = \int_0^1\int_1^2 \frac{2}{\sqrt{4-y^2}}\ dxdy$$

$$= \int_0^1 \frac{2}{\sqrt{4-y^2}}\ dy = \left[2\sin^{-1}(y/2)\right]_0^1\ 2(\pi/6) - 2(0)$$

$$= \pi/3 \approx 1.0472$$

6. Let $z = f(x,y) = x^2+y^2$; $f_x(x,y) = 2x$; $f_y(x,y) = 2y$.

$$A(G) = 4\int_0^2\int_0^{\sqrt{4-x^2}} \sqrt{4x^2+4y^2+1}\ dydx$$

$$= 4\int_0^{\pi/2}\int_0^2 (4r^2+1)^{1/2} r\, drd\theta$$

$$= 4\int_0^{\pi/2} \left[\frac{(4r^2+1)^{3/2}}{12}\right]_0^2\ dr = \frac{(17^{3/2}-1)}{3}\cdot\frac{\pi}{2} \approx 36.1769$$

9. $z=(25-x^2-y^2)^{1/2}$; $f_x(x,y) = -x(25-x^2-y^2)^{-1/2}$; $f_y(x,y) = -y(25-x^2-y^2)^{-1/2}$

$$A(G) = \int_0^4\int_0^{\sqrt{16-x^2}} [(x^2+y^2)(25-x^2-y^2)^{-1}+1]^{1/2}\ dydx$$

$$= \int_0^{\pi/2}\int_0^4 [r^2(25-r^2)^{-1}+1]^{1/2} r\, drd\theta$$

$$= \int_0^{\pi/2}\int_0^4 5r(25-r^2)^{-1/2}\, drd\theta$$

$$= \int_0^{\pi/2} \left[-5(25-r^2)^{1/2}\right]_0^4 d\theta = \int_0^{\pi/2} 10\ d\theta = 10(\pi/2) = 5\pi \approx 15.7080$$

Problem Set 16.7 Triple Integrals (Cartesian Coordinates)

3. $\int_1^4 \int_{z-1}^{2z} \int_0^{y+2z} dx\,dy\,dz = \int_1^4 \int_{z-1}^{2z} (y+2z)\,dy\,dz = \int_1^4 \left[\frac{y^2}{2} + 2yz\right]_{y=z-1}^{2z} dx$

   $= \int_1^4 \left(\frac{7z^2}{2} + 3z - \frac{1}{2}\right) dz = \left[\frac{7z^3}{6} + \frac{3z^2}{2} - \frac{z}{2}\right]_1^4 = \frac{189}{2} = 94.5$

6. $\int_0^{\pi/2} \int_0^z \int_0^y \sin(x+y+z)\,dx\,dy\,dz = \int_0^{\pi/2} \int_0^z [-\cos(2y+z) + \cos(y+z)]\,dy\,dz$

   $= \int_0^{\pi/2} \left(\frac{-\sin 3z}{2} + \sin 2z - \frac{\sin z}{2}\right) dz = \frac{1}{3}$

9. $3x+2y+6z = 12$

   $\int_0^1 \int_0^3 \int_0^{(12-3x-2y)/6} f(x,y,z)\,dz\,dy\,dx$

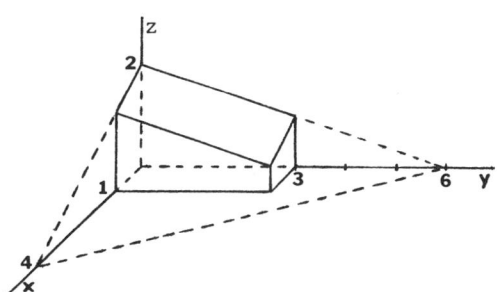

12. $\int_0^4 \int_0^{\sqrt{y}} \int_0^{3x/2} f(x,y,z)\,dz\,dx\,dy$

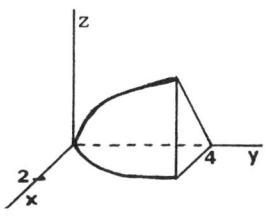

15. Using the cross product of vectors along edges it is easy to show that $\langle 2,6,9\rangle$ is normal to the upward face. Then obtain that its equation is $2x+6y+9z = 18$.

    $\int_0^3 \int_{2x/3}^{(9-x)/3} \int_0^{(18-2x-6y)/9} f(x,y,z)\,dz\,dy\,dx$

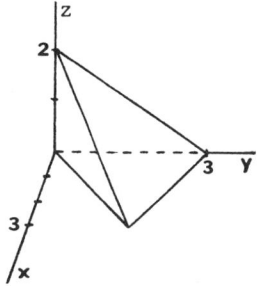

18. $\int_0^3 \int_0^1 \int_y^{\sqrt{2y-y^2}} f(x,y,z)\, dxdydz$

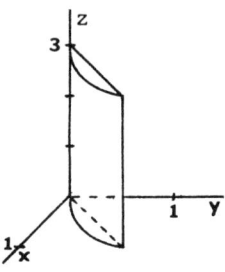

21. $V = 4\int_0^1 \int_0^{\sqrt{y}} \int_0^{\sqrt{y}} 1\, dzdxdy = 4\int_0^1 \int_0^{\sqrt{y}} \sqrt{y}\, dxdy$

$= 4\int_0^1 \sqrt{y}\sqrt{y}\, dy = \left[2y^2\right]_0^1 = 2$

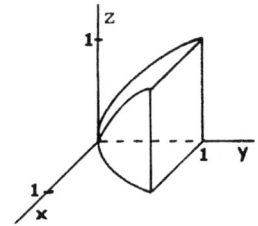

24. $\delta(x,y,z) = k(x^2+y^2+z^2)$. In evaluating the coordinates of the center of mass, k is a factor of the numerator and denominator and so may be cancelled. Hence, for sake of convenience we may just let k = 1 when determining the center of mass. Note that this is not valid if we are concerned with the values of moments or mass.

$m = 4\int_0^3 \int_0^{\sqrt{9-x^2}} \int_0^4 (x^2+y^2+z^2)\, dzdydx$

$= 4\int_0^3 \int_0^{\sqrt{9-x^2}} [4(x^2+y^2) + \tfrac{64}{3}]\, dydx$

$= 4\int_0^{\pi/2} \int_0^3 (4r^2 + \tfrac{64}{3})r\, drd\theta$ [changing to polar]

$= 4\int_0^{\pi/2} \left[r^4 + \tfrac{32r^2}{3}\right]_0^3 d\theta = 4\int_0^{\pi/2} 177\, d\theta = 354\pi$

$M_{xy} = 4\int_0^{\pi/2} \int_0^3 \int_0^4 z(r^2+z^2)r\, dzdrd\theta$ [polar coordinates]

$= 4\int_0^{\pi/2} \int_0^3 (8r^3+64r)\, drd\theta = 4\int_0^{\pi/2} 450\, d\theta = 900\pi$

$\bar{z} = \dfrac{900\pi}{354\pi} = \dfrac{150}{59} \approx 2.5425$; $\bar{x} = \bar{y} = 0$ (by symmetry).

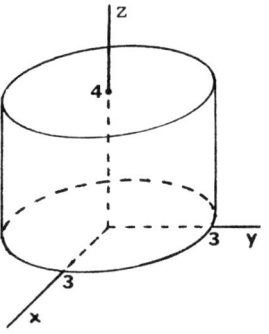

27. The limits of integration are those for the first octant part of a sphere of radius 1.

$$\int_0^1 \int_0^{\sqrt{1-x^2}} \int_0^{\sqrt{1-x^2-y^2}} f(x,y,z) \, dzdydx.$$

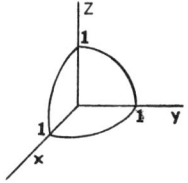

30. $\int_0^5 \int_0^2 \int_0^{2-x} f(x,y,z) \, dzdxdy + \int_5^9 \int_0^{\sqrt{9-y}} \int_0^{2-x} f(x,y,z) \, dzdxdy.$

Problem Set 16.8 Triple Integrals (Cylindrical and Spherical Coordinates)

3. $\int_0^{2\pi} \int_0^2 \int_{r^2/4}^{\sqrt{5-r^2}} r \, dzdrd\theta = \int_0^{2\pi} \int_0^2 [r(5-r^2)^{1/2} - \frac{r^3}{4}] drd\theta$

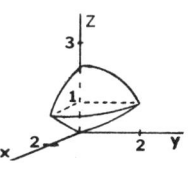

$= \int_0^{2\pi} \frac{5^{3/2}-4}{3} d\theta = \frac{2\pi(5^{3/2}-4)}{3} \approx 15.0385.$

6. Let $(x,y,z) = 1$ [See comment at beginning of write-up of Problem 24 of the previous section.]

$m = \int_0^{2\pi} \int_1^2 \int_0^{12-r^2} 1 \, rdzdrd\theta = \int_0^{2\pi} \int_1^2 r(12-r^2) drd\theta$

$= \int_0^{2\pi} \left[6r^2 - \frac{r^4}{4}\right]_1^2 d\theta = \int_0^{2\pi} (57/4) d\theta = 57\pi/2.$

$M_{xy} = \int_0^{2\pi} \int_1^2 \int_0^{12-r^2} z \, rdzdrd\theta = \int_0^{2\pi} \int_1^2 \frac{(12-r^2)^2(-2r)}{-4} drd\theta$

$= \int_0^{2\pi} \frac{11^3-8^3}{12} d\theta = \frac{273\pi}{2}.$ Therefore, $\bar{z} = \frac{273\pi/2}{57\pi/2} = \frac{91}{19} \approx 4.7895.$

$\bar{x} = \bar{y} = 0$ (by symmetry).

9. Let $\delta(x,y,z) = \rho$. [Letting k = 1 -- see comment at the beginning of the write-up of Problem 24 of the previous section.]

$$m = \int_0^{\pi/2}\int_0^{2\pi}\int_0^a \rho^3\sin\phi\, d\rho d\theta d\phi = \int_0^{\pi/2}\int_0^{2\pi} \frac{a^4\sin\phi}{4}\, d\theta d\phi$$

$$= \int_0^{\pi/2} \frac{\pi a^4\sin\phi}{2}\, d\phi = \frac{\pi a^4}{2}$$

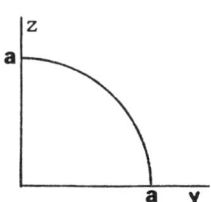

$$M_{xy} = \int_0^{\pi/2}\int_0^{2\pi}\int_0^a \rho^4\sin\phi\cos\phi\, d\rho d\theta d\phi \quad [z = \rho\cos\phi]$$

$$= \int_0^{\pi/2}\int_0^{2\pi} \frac{a^5\sin 2\phi}{10}\, d\theta d\phi = \int_0^{\pi/2} \frac{\pi a^5\sin 2\phi}{5}\, d\phi$$

$$= \frac{\pi a^5}{5};\ \bar{z} = \frac{\pi a^5/5}{\pi a^4/2} = 0.4a;\ \bar{x} = \bar{y} = 0 \text{ (by symmetry)}.$$

12. Volume $= \int_{\pi/4}^{\pi/2}\int_0^{2\pi}\int_0^4 \rho^2\sin\phi\, d\rho d\theta d\phi$

$$= \int_{\pi/4}^{\pi/2}\int_0^{2\pi} \frac{64\sin\phi}{3}\, d\theta d\phi = \int_{\pi/4}^{\pi/2} \frac{128\pi\sin\phi}{3}\, d\phi$$

$$= 64\sqrt{2}\pi/3 \approx 94.7815$$

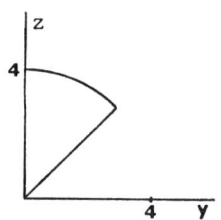

15. Volume $= \int_0^{\pi}\int_0^{\sin\theta}\int_{r^2}^{r\sin\theta} 1\, rdzdrd\theta$

$$= \int_0^{\pi}\int_0^{\sin\theta} r(r\sin\theta - r^2)drd\theta = \int_0^{\pi} \frac{\sin^4\theta}{12}\, d\theta$$

$$= \frac{1}{48}\int_0^{\pi} [1 - 2\cos 2\theta + \frac{1+\cos 4\theta}{2}]d\theta = \frac{\pi}{32} \approx 0.0982$$

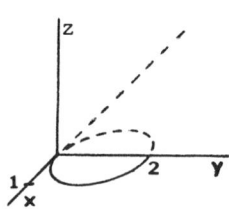

18. $x = \rho\sin\phi\cos\theta$, $y = \rho\sin\phi\sin\theta$, $z = \rho\cos\phi$

$$J(\rho,\phi,\theta) = \begin{vmatrix} x_\rho & x_\phi & x_\theta \\ y_\rho & y_\phi & y_\theta \\ z_\rho & z_\phi & z_\theta \end{vmatrix} = \begin{vmatrix} \sin\phi\cos\theta & \rho\cos\phi\cos\theta & -\rho\sin\phi\sin\theta \\ \sin\phi\sin\theta & \rho\cos\phi\sin\theta & \rho\sin\phi\cos\theta \\ \cos\phi & -\rho\sin\phi & 0 \end{vmatrix}$$

$$= (\cos\phi)(\rho^2\cos\phi\sin\phi)(\cos^2\theta+\sin^2\theta) + (\rho\sin\phi)(\rho\sin^2\phi)(\cos^2\theta+\sin^2\theta)$$

$$= \rho^2\sin\phi(\cos^2\phi+\sin^2\phi) = \rho^2\sin\phi.$$

[Expansion was along the third row of the determinant.]

Problem Set 16.9 Chapter Review Problems

True-False Quiz

3. **True.** Inside integral is 0 since $\sin(x^3y^2)$ is an odd function in x.

6. **True.** $f(x,y) \geq f(x_0,y_0)/2$ in some neighborhood N of (x_0,y_0) due to the continuity. Then

$$\iint_R f(x,y)\,dA \geq \iint_N (1/2)f(x_0,y_0)\,dA = (1/2)f(x_0,y_0)(\text{Area N}) > 0$$

9. **True.** See the comment at the beginning of the write-up of Problem 24, Section 16.7.

12. **True.** See Page 705 of the text. $A(T) = (\text{Area of base})(\sec 30°)$

$$= \pi(1)^2(2/\sqrt{3}) = 2\sqrt{3}\pi/3.$$

15. **True.** $|\nabla f|$ is the magnitude of the greatest increase in f.

$$|D_u f| = |\nabla f \cdot u| = |\langle f_x, f_y \rangle \cdot u| = \sqrt{f_x^2 + f_y^2}\,(1)\cos\theta \leq \sqrt{4+4} = \sqrt{8}$$

Therefore, Area(G) $\leq$ Area(R) max$\{\sec\gamma\} \leq (1)\sec(\tan^{-1}\sqrt{8}) = 3$

Miscellaneous Problems

3. $\int_0^{\pi/2} \left[\frac{r^2 \cos\theta}{2} \right]_{r=0}^{2\sin\theta} d\theta = \int_0^{\pi/2} 2\sin^2\theta \cos\theta \, d\theta = \left[\frac{2\sin^3\theta}{3} \right]_0^{\pi/2} = \frac{2}{3}$

6. $\int_0^{\pi/2} \int_0^{\cos x} f(x,y) \, dx\, dy$

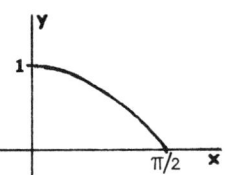

9. (a) $8 \int_0^a \int_0^{\sqrt{a^2-z^2}} \int_0^{\sqrt{a^2-y^2-z^2}} dx\, dy\, dz$

(b) $8 \int_0^{\pi/2} \int_0^a \int_0^{\sqrt{a^2-r^2}} r \, dz\, dr\, d\theta$

(c) $8 \int_0^{\pi/2} \int_0^{\pi/2} \int_0^a \rho^2 \sin\phi \, d\rho\, d\phi\, d\theta$

12. $\int_0^{2\pi} \int_2^3 (r^{-2}) r \, dr\, d\theta = \int_0^{2\pi} \left[\ell n\, r \right]_2^3 = \int_0^{2\pi} \ell n(3/2) \, d\theta = 2\pi \ell n(3/2) \quad 2.5476$

15. $z = f(x,y) = (9-y^2)^{1/2}$; $f_x(x,y) = 0$; $f_y(x,y) = -y(9-y^2)^{-1/2}$.

Area $= \int_0^3 \int_{y/3}^y \sqrt{y^2(9-y^2)^{-1}+1} \, dx\, dy$

$= \int_0^3 \int_{y/3}^y 3(9-y^2)^{-1/2} dx\, dy$

$= \int_0^3 (9-y^2)^{-1/2}(2y) dy = \left[-2(9-y^2)^{1/2} \right]_0^3 = 6$

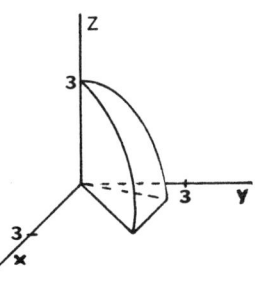

18. $m = \iint_R 1 \, dA = \int_0^{2\pi} \int_0^{4(1+\sin\theta)} r \, dr\, d\theta = \int_0^{2\pi} (1+2\sin\theta + \frac{1-\cos 2\theta}{2}) d\theta = 24\pi$

$M_x = \iint_R y \, dA = \int_0^{2\pi} \int_0^{4(1+\sin\theta)} r\sin\theta \, r\, dr\, d\theta = 80\pi$

$\bar{y} = \frac{80\pi}{24\pi} = \frac{10}{3}$; $\bar{x} = 0$ (by symmetry).

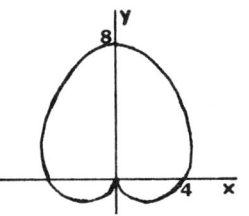

CHAPTER 17   VECTOR CALCULUS

Problem Set 17.1   Vector Fields

3. $F(x,y) = \langle -x, 2y \rangle$.

6. $F(x,y,z) = \langle 0, 0, -z \rangle$.

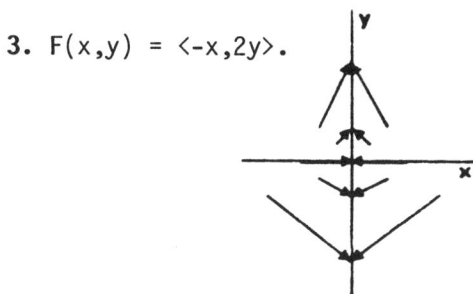
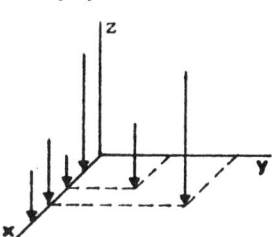

9. $f(x,y,z) = \ln|x| + \ln|y| + \ln|z|$; $f(x,y,z) = \langle x^{-1}, y^{-1}, z^{-1} \rangle$.

12. $\nabla f(x,y,z) = \langle 0, 2ye^{-2z}, -2y^2 e^{-2z} \rangle = 2ye^{-2z}\langle 0, 1, -y \rangle$.

15. divF = $\nabla \cdot F$ = 0 + 0 + 0 = 0.
    curlF = $\nabla \times F$ = $\langle x-x, y-y, z-z \rangle$ = **0**.

18. divF = $\nabla \cdot F$ = 0+0+0 = 0.
    curlF = $\nabla \times F$ = $\langle 1-1, 1-1, 1-1 \rangle$ = **0**.

21. $F(x,y,z) = -cr|r|^{-3} = -c\langle x,y,z \rangle |r|^{-3}$, where $|r| = (x^2+y^2+z^2)^{1/2}$.

   1st component of curlF: $\frac{\partial}{\partial y}\left(-cz|r|^{-3}\right) - \frac{\partial}{\partial z}\left(-cy|r|^{-3}\right)$

   $= -cz(-3)|r|^{-4} \frac{\partial |r|}{\partial y} - (-y)(-3)|r|^{-4} \frac{\partial |r|}{\partial z}$

   $= 3c|r|^{-4}\left(z[y(x^2+y^2+z^2)^{-1/2}] - y[z(x^2+y^2+z^2)^{-1/2}]\right) = 0$.

   The 2nd component is exactly the same except that z replaces y and x replaces z, and the 3rd component is also the same except that x replaces y and y replaces z, in the work shown for the 1st component.

   Therefore, curlF = $\langle 0,0,0 \rangle$ = **0**.

21. (continued)

$$\text{div} F = \nabla \cdot \langle -cx|r|^{-3}, -cy|r|^{-3}, -cz|r|^{-3} \rangle$$
$$= [-cx(-3)|r|^{-4}x(x^2+y^2+z^2)^{-1/2} - c|r|^{-3}]$$
$$+ [-cy(-3)|r|^{-4}y(x^2+y^2+z^2)^{-1/2} - c|r|^{-3}]$$
$$+ [-cz(-3)|r|^{-4}z(x^2+y^2+z^2)^{-1/2} - c|r|^{-3}]$$
$$= 3c|r|^{-4}(x^2+y^2+z^2)^{-1/2}(x^2+y^2+z^2) - 3c|r|^{-3} = 0$$

24. $\lim_{(x,y,z) \to (a,b,c)} F(x,y,z) = F(a,b,c)$

Problem Set 17.2 Line Integrals

3. Let $x = t$, $y = 2t$, t in $[0,\pi]$.

Then $\int_C (\sin x + \cos y) ds = \int_0^\pi (\sin t + \cos 2t)\sqrt{1+4}\, dt = 2\sqrt{5} \approx 4.4721$

6. $\int_0^{2\pi} (16\cos^2 t + 16\sin^2 t + 9t^2)(16\sin^2 t + 16\cos^2 t + 9)^{1/2} dt$

$= \int_0^{2\pi} (16+9t^2)(5) dt = \left[80t+15t^3\right]_0^{2\pi} = 160\pi + 120\pi^3 \approx 4878.11$

9. $\int_C y^3 dx + x^3 dy = \int_{C_1} y^3 dx + x^3 dy + \int_{C_2} y^3 dx + x^3 dy$

$= \int_1^{-2} (-4)^3 dy + \int_{-4}^2 (-2)^3 dx = 192 + (-48) = 144$

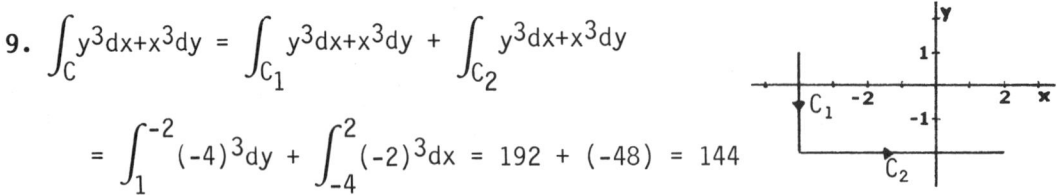

12. $\int_0^1 [x^2 + (x)2x]dx$ [letting x be the parameter; i.e., $x = x$, $y = x^2$]

$= \int_0^1 3x^2 dx = 1$

15. On C_1: $y = z = dy = dz = 0$
On C_2: $x = 2$, $z = dx = dz = 0$
On C_3: $x = 2$, $y = 3$, $dx = dy = 0$

$\int_0^2 x\,dx + \int_0^3 (2-2y)dy + \int_0^4 (4+3-z)dz$

$= \left[\frac{x^2}{2}\right]_0^2 + \left[2y-y^2\right]_0^3 + \left[7z - \frac{z^2}{2}\right]_0^4 = 2 + (-3) + 20 = 19$

18. Let $\delta(x,y,z) = k$ (a constant)

$m = k\int_C 1\,ds = k\int_0^{3\pi} 1(a^2\sin^2 t + a^2\cos^2 t + b^2)^{1/2}dt = 3\pi k(a^2+b^2)^{1/2}$

$M_{xy} = k\int_C z\,ds = k(a^2+b^2)^{1/2}\int_0^{3\pi} bt\,dt = \frac{9\pi^2 bk(a^2+b^2)^{1/2}}{2}$

$M_{xz} = k\int_C y\,ds = k(a^2+b^2)^{1/2}\int_0^{3\pi} a\sin t\,dt = ak(a^2+b^2)^{1/2}(2) = 2ak(a^2+b^2)^{1/2}$

$M_{yz} = k\int_C x\,ds = k(a^2+b^2)^{1/2}\int_0^{3\pi} a\cos t\,dt = ak(a^2+b^2)^{1/2}(0) = 0$

Therefore, $\bar{x} = \frac{M_{yz}}{m} = 0$; $\bar{y} = \frac{M_{xz}}{m} = \frac{2a}{3\pi}$; $\bar{z} = \frac{M_{xy}}{m} = \frac{3\pi b}{2}$.

21. $W = \int_C \mathbf{F}\cdot d\mathbf{r} = \int_C (x+y)dx + (x-y)dy$

$= \int_0^{\pi/2} [(a\cos t + b\sin t)(-a\sin t) + (a\cos t - b\sin t)(b\cos t)]dt$

Problem Set 17.2

21. (continued)

$$= \int_0^{\pi/2} [-(a^2+b^2)\sin t \cos t + ab(\cos^2 t - \sin^2 t)]dt$$

$$= \int_0^{\pi/2}\left(\frac{-(a^2+b^2)\sin 2t}{2} + ab\cos 2t\right)dt = \left[\frac{(a^2+b^2)\cos 2t}{4} + \frac{ab\sin 2t}{2}\right]_0^{\pi/2} = \frac{a^2+b^2}{-2}.$$

24. $W = \int_C \mathbf{F} \cdot d\mathbf{r} = \int_C y\,dx + z\,dy + x\,dz = \int_0^2 [(t^2)(1) + (t^3)(2t) + (t)(3t^2)]dt$

$$= \int_0^2 (2t^4 + 3t^3 + t^2)dt = \frac{64}{5} + 12 + \frac{8}{3} = \frac{412}{15} \approx 27.4667.$$

Problem Set 17.3   Independence of Path

3. $M_y = 90x^4y - 36y^5 \neq N_x$ since $N_x = 90x^4y - 12y^5$, so $\mathbf{F}$ is not conservative.

6. $M_y = (4y^2)(-2xy\sin xy^2) + (8y)(\cos xy^2) \neq N_x$ since

$N_x = (8x)(-y^2\sin xy^2) + (8)(\cos xy^2)$, so $\mathbf{F}$ is not conservative.

9. $M_y = 0 = N_x$, $M_z = 0 = P_x$, and $N_z = 0 = P_y$, so $\mathbf{F}$ is conservative.
f satisfies $f_x(x,y,z) = 3x^2$, $f_y(x,y,z) = 6y^2$, and $f_z(x,y,z) = 9z^2$.
Therefore, f satisfies (1) $f(x,y,z) = x^3 + C_1(y,z)$,

(2) $f(x,y,z) = 2y^3 + C_2(x,z)$,

(3) $f(x,y,z) = 3z^3 + C_3(x,y)$.

A function with an arbitrary constant that satisfies (1), (2), and (3)
is $f(x,y,z) = x^3 + 2y^3 + 3z^3 + C$.

12. $M_y = e^x\cos y = N_x$, so the line integral is independent of the path.

Let $f_x(x,y) = e^x\sin y$ and $f_y(x,y) = e^x\cos y$

Then $f(x,y) = e^x\sin y + C_1(y)$ and $f(x,y) = e^x\sin y + C_2(x)$.

Choose $f(x,y) = e^x\sin y$.

By Theorem A, $\int_{(0,0)}^{(1,\pi/2)} e^x\sin y\, dx + e^x\cos y\, dy = \left[e^x\sin y\right]_{(0,0)}^{(1,\pi/2)} = e$.

[Or use line segments $(0,0)$ to $(1,0)$, then $(1,0)$ to $(1,\pi/2)$.]

15. $M_y = 1 = N_x$, $M_z = 1 = P_x$, $N_z = 1 = P_y$ (so path independent).

From inspection observe that $f(x,y,z) = xy+xz+yz$ satisfies

$\nabla f = \langle y+z, x+z, x+y\rangle$, so the integral equals $\left[xy+xz+yz\right]_{(0,0,0)}^{(-1,0,\pi)} = -\pi$

[Or use line segments $(0,1,0)$ to $(1,1,0)$, then $(1,1,0)$ to $(1,1,1)$.]

18. $f_x(x,y,z) = -kx(x^2+y^2+z^2)^{-1/2}$, so $f(x,y,z) = -k(x^2+y^2+z^2)^{1/2} + C_1(y,z)$.

Similarly, using f_y, obtain $f(x,y,z) = -k(x^2+y^2+z^2)^{1/2} + C_2(x,z)$, and

using f_z, obtain $f(x,y,z) = -k(x^2+y^2+z^2)^{1/2} + C_2(x,z)$.

Thus, one potential function for F is $f(x,y,z) = -k(x^2+y^2+z^2)^{1/2}$.

21. $\int_C \mathbf{F}\cdot d\mathbf{r} = \int_a^b (m\mathbf{r}''\cdot\mathbf{r}')\, dt = m\int_a^b (x''x' + y''y' + z''z')\, dt$

$= m\left[\frac{(x')^2}{2} + \frac{(y')^2}{2} + \frac{(z')^2}{2}\right]_a^b = \frac{m}{2}\left[|\mathbf{r}'(t)|^2\right]_a^b = \frac{m}{2}\left[|\mathbf{r}'(b)| - |\mathbf{r}'(a)|\right]$

Problem Set 17.3

Problem Set 17.4 Green's Theorem in the Plane

3. $\oint_C (2x+y^2)dx + (x^2+2y)dy = \iint_S (2x-2y)dA$

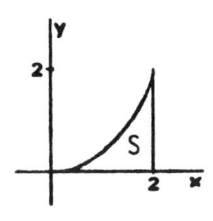

$= \int_0^2 \int_0^{x^3/4} (2x-2y)dydx = \int_0^2 \left(\frac{x^4}{2} - \frac{x^6}{16}\right)dx$

$= \frac{16}{5} - \frac{8}{7} = \frac{72}{35} \approx 2.0571.$

6. $\oint_C (e^{3x}+2y)dx + (x^2+\sin y)dy = \iint_S (2x-2)dA$

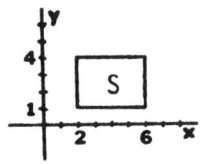

$= \int_1^4 \int_2^6 (2x-2)dxdy = \int_1^4 24\,dy = 24(3) = 72.$

9. (a) $\iint_S \text{div}\,F\,dA = \iint_S (M_x+N_y)dA = \iint_S (0+0)dA = 0.$

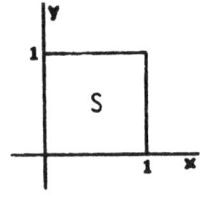

(b) $\iint_S (\text{curl}\,F)\cdot k\,dA = \iint_S (N_x-M_y)dA = \iint_S (2x-2y)dA$

$= \int_0^1 \int_0^1 (2x-2y)dxdy = \int_0^1 (1-2y)dy = 0.$

12. (a) $\iint_S \text{div}\,F\,dA = \iint_S (M_x+N_y)dA = \iint_S (1+1)dA = 2[A(S)] = 2\pi.$

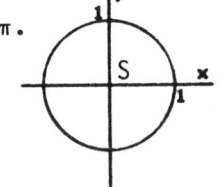

(b) $\iint_S (\text{curl}\,F)\cdot k\,dA = \iint_S (N_x-M_y)dA = \iint_S (0-0)dA = 0.$

15. $W = \oint_C F\cdot T\,ds = \iint_S (N_x-M_y)dA = \iint_S (-2y-2y)dA$

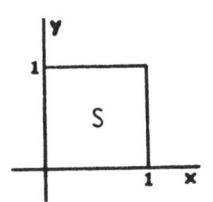

$= \int_0^1 \int_0^1 -4y\,dxdy = \int_0^1 -4y\,dy = -2.$

18. $\oint_C Mdx+Ndy = \iint_S (N_x-M_y)dA = 0$

Therefore, $\int_C F \cdot dr$ is independent of path since

$\int_{C_1} F \cdot dr - \int_{C_2} F \cdot dr = \int_C F \cdot dr = 0$

[where C is the loop C_1 followed by $-C_2$]

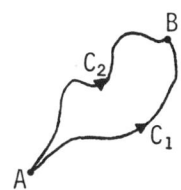

Therefore, $\int_{C_1} F \cdot dr = \int_{C_2} F \cdot dr$, so F is conservative.

21. Use Green's Theorem with $M(x,y) = -y$ and $N(x,y) = 0$.

$\oint_C (-y)dx = \iint_S [0-(-1)]dA = A(S)$.

Now use Green's Theorem with $M(x,y) = 0$ and $N(x,y) = x$.

$\oint_C xdy = \iint_S (1-0)dA = A(S)$.

24. $W = \oint_C F \cdot T ds = \iint_S (curl F) \cdot k \, dA = \iint_S (N_x-M_y)dA = \iint_S (-3-2)dA$

$= -5[A(S)] = -5(3a^2\pi/8) = -15a^2\pi/15$ [using result of Problem 23]

Problem Set 17.5 Surface Integrals

3. $\iint_R (x+y)\sqrt{[-x(4-x^2)^{-1/2}]^2 + 0 + 1} \, dA = \int_0^{\sqrt{3}}\int_0^1 \frac{2(x+y)}{(4-x^2)^{1/2}} dy\,dx$

$= \int_0^{\sqrt{3}} \frac{2x+1}{(4-x^2)^{1/2}} dx = \left[-2(4-x^2)^{1/2} + \sin^{-1}(x/2)\right]_0^{\sqrt{3}} = \frac{\pi+6}{3} \approx 3.0472$

6. $\iint_R y(4y^2+1)^{1/2} dA = \int_0^3 \int_0^2 (4y^2+1)^{1/2} y \, dy\, dx$

$= \int_0^3 \frac{(17^{3/2}-1)}{12} dx = \frac{17^{3/2}-1}{4} \approx 17.2732$

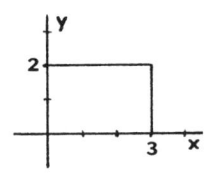

9. $\iint_G \mathbf{F} \cdot \mathbf{n} \, dS = \iint_R (-Mf_x - Nf_y + P) dA = \int_0^1 \int_0^{1-y} (8y+4x+0) dx\, dy$

$= \int_0^1 [8(1-y)y + 2(1-y)^2] dy = \int_0^1 (-6y^2+4y+2) dy = 2$

12. $\iint_R [-Mf_x - Nf_y + P] dA = \iint_R [-2x(x^2+y^2)^{-1/2} - 5y(x^2+y^2)^{-1/2} + 3] dA$

$= \int_0^{2\pi} \int_0^1 [(-2r\cos\theta - 5r\sin\theta) r^{-1} + 3] r\, dr\, d\theta$

$= \int_0^{2\pi} (-2\cos\theta - 5\sin\theta + 3) d\theta \int_0^1 r\, dr = (6\pi)(1/2) = 3\pi$

15. [Let $\delta = 1$.] $m = \iint_S 1\, dS = \iint_R (1+1+1)^{1/2} dA$

$= \sqrt{3} \int_0^a \int_0^{a-y} dx\, dy = \sqrt{3} \int_0^a (a-y) dy = \frac{a^2\sqrt{3}}{2}$

$M_{xy} = \iint_S z\, dS = \iint_R (a-x-y)\sqrt{3}\, dA$

$= \sqrt{3} \int_0^a \int_0^{a-y} (a-x-y) dx\, dy = \sqrt{3} \int_0^a [a(a-y) - \frac{(a-y)^2}{2} - y(a-y)] dy$

$= \sqrt{3} \int_0^a \left(\frac{a^2}{2} - ay + \frac{y^2}{2}\right) dy = \frac{a^3\sqrt{3}}{6}$

$\bar{z} = \frac{M_{xy}}{m} = \frac{a}{3}$; then $\bar{x} = \bar{y} = \frac{a}{3}$ [by symmetry].

Problem Set 17.6 Gauss's Divergence Theorem

3. $\iint_{\partial S} \mathbf{F} \cdot \mathbf{r} \, dS = \iiint_S (M_x + N_y + P_z) \, dV = \int_0^c \int_0^b \int_0^a (2xyz + 2xyz + 2xyz) \, dx \, dy \, dz$

$= \int_0^c \int_0^b 3a^2 yz \, dy \, dz = \int_0^c \frac{3a^2 b^2 z}{2} \, dz = \frac{3a^2 b^2 c^2}{4}.$

6. $\iiint_S (M_x + N_y + P_z) \, dV = \iiint_S (2x + 1 + 2z) \, dV$

$= \int_0^{2\pi} \int_0^2 \int_0^{2-r\cos\theta} (2r\cos\theta + 1 + 2z) r \, dz \, dr \, d\theta$

$= \int_0^{2\pi} \int_0^2 [(2r^2\cos\theta + r)(2 - r\cos\theta) + r(2 - r\cos\theta)^2] \, dr \, d\theta$

$= \int_0^{2\pi} \int_0^2 (6r - r^3\cos^2\theta - r^2\cos\theta) \, dr \, d\theta = \int_0^{2\pi} (12 - 4\cos^2\theta - \frac{8\cos\theta}{3}) \, d\theta = 20\pi.$

9. $\iiint_S (M_x + N_y + P_z) \, dV = \iiint_S (2 + 3 + 4) \, dV = 9(\text{Volume of spherical shell})$

$= 9(4\pi/3)(5^3 - 3^3) = 1176\pi \approx 3694.51.$

12. $V(S) = \frac{1}{3} \iint_{\partial S} \mathbf{F} \cdot \mathbf{n} \, dS \quad \text{for } \mathbf{F} = \langle x, y, z \rangle$

$= \frac{1}{3} \iiint_S 3 \, dV = \int_0^{2\pi} \int_0^a \int_0^h r \, dz \, dr \, d\theta$

$= \int_0^{2\pi} \int_0^a rh \, dr \, d\theta = \int_0^{2\pi} \frac{a^2 h}{2} \, d\theta = 2\pi \frac{a^2 h}{2} = \pi a^2 h.$

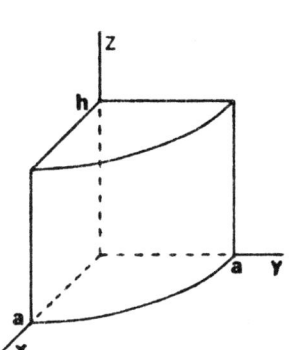

15. $\iint_{\partial S} D_n f \, dS = \iint_{\partial S} \nabla f \cdot \mathbf{n} \, dS = \iiint_S \text{div}(\nabla f) \, dV = \iiint_S (f_{xx} + f_{yy} + f_{zz}) \, dV$

$= \iiint_S \nabla^2 f \, dV$

18. $\iint_{\partial S} f D_n g \, dS = \iint_{\partial S} f(\nabla g \cdot \mathbf{n}) \, dS = \iint_{\partial S} (f \nabla g) \cdot \mathbf{n} \, dS = \iiint_S \text{div}(f \nabla g) \, dV$ [Gauss]

$= \iiint_S [f(\text{div} \nabla g) + (\nabla f) \cdot (\nabla g)] \, dV$ [See Problem 20(c) of Section 17.1]

$= \iiint_S (f \nabla^2 g + \nabla f \cdot \nabla g) \, dV$

Problem Set 17.7   Stokes's Theorem

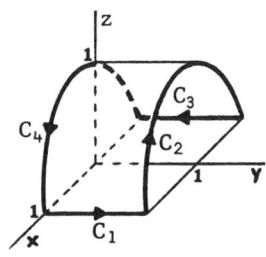

3. $\iint_S (\text{curl} \mathbf{F}) \cdot \mathbf{n} \, dS = \oint_{\partial S} \mathbf{F} \cdot \mathbf{T} \, ds$

$= \oint_{\partial S} (y+z) \, dx + (x^2 + z^2) \, dy + y \, dz$

$[*] = \int_0^1 1 \, dt + \int_0^\pi [(1 + \sin t)(-\sin t) + \cos t] \, dt + \int_0^1 -1 \, dt + \int_0^\pi \sin^2 t \, dt$

$= \int_0^\pi (-\sin t + \cos t) \, dt = -2$

The result at [*] was obtained by integrating along S by doing so along C_1, C_2, C_3, C_4 in that order.

Along C_1: x=1, y=t, z=0, dx=dz=0, dy=dt, t in [0,1]
Along C_2: x=cost, y=1, z=sint, dx=-sintdt, dy=0, dz=costdt, t in [0,π]
Along C_3: x=-1, y=1-t, z=0, dx=dz=0, dy=-dt, t in [0,1]
Along C_4: x=-cost, y=0, z=sint, dx=sintdt, dy=0, dz=costdt, t in [0,π]

6. ∂S is the circle $x^2+y^2 = 1$, $z = 0$

$$\iint_S (\text{curl} F) \cdot n \, dS = \oint_{\partial S} F \cdot T \, ds$$

$$= \oint_{\partial S} (z-y)dx + (z+x)dy + (-x-y)dz$$

$$= \int_0^{2\pi} [(-\sin t)(-\sin t) + (\cos t)(\cos t)]dt = 2\pi$$

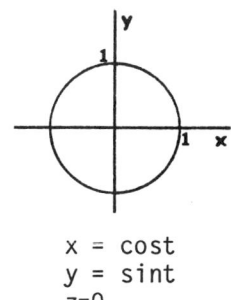

$x = \cos t$
$y = \sin t$
$z = 0$
t in $[0, 2\pi]$

9. $\text{curl} F = \langle -1+1, 0-1, 1-1 \rangle = \langle 0, -1, 0 \rangle$. The unit normal vector that is needed to apply Stokes's Theorem points downward. It is

$$n = \frac{\langle -1, -2, -1 \rangle}{\sqrt{6}}$$

$$\oint_C F \cdot T \, ds = \iint_S (\text{curl} F) \cdot n \, dS = \iint_S (2/\sqrt{6}) dS$$

$$= \iint_R (2/\sqrt{6})[(1+4+1)]^{1/2} dA = \iint_R 2 \, dA$$

$$= 2(\text{Area of triangle in xy-plane}) = 2(1) = 2$$

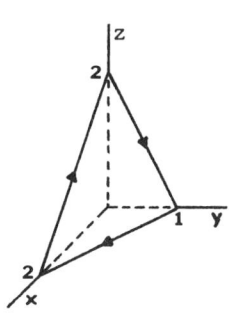

12. $\text{curl} F = \langle -1-1, -1-1, -1-1 \rangle = -2\langle 1,1,1 \rangle$, $n = \frac{\langle 1, 0, 1 \rangle}{\sqrt{2}}$, so $(\text{curl} F) \cdot n = -2\sqrt{2}$.

$$\oint_C F \cdot T \, ds = \iint_S (\text{curl} F) \cdot n \, dS = -2\sqrt{2} \iint_S 1 \, dS$$

$$= -2\sqrt{2} \iint_R \sec 45° \, dA = -2\sqrt{2}\sqrt{2}[A(R)] = -4\pi$$

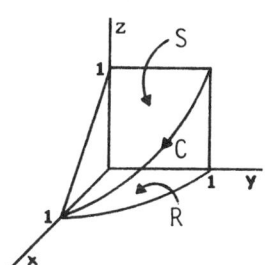

Problem Set 17.7

Problem Set 17.8 Chapter Review Problems

True-False Quiz

3. False. grad(curlF) is not defined since curlF is not a scalar field.

6. True. See the three equivalent conditions on Page 739 of the text.

9. True. It is the case in which the surface is in a plane.

12. True. divF = 0, so by Gauss's Divergence Theorem, the integral given equals $\iiint_D 0\,dV$ where D is the solid sphere for which S = ∂D.

Miscellaneous Problems

3. curl(f∇f) = (f)(curl∇f) + (∇f × ∇f) = (f)(0) + 0 = 0

6. $M_x = 2y = N_y$ so the integral in independent of the path.
   Find any function f(x,y) such that $f_x(x,y) = y^2$ and $f_y(x,y) = 2xy$.
   $f(x,y) = xy^2 + C_1(y)$ and $f(x,y) = xy^2 + C_2(x)$, so let $f(x,y) = xy^2$.
   Then the given integral equals $\left[xy^2\right]_{(0,0)}^{(1,2)} = 4$.

9. (a) $\int_0^1 0\,dx + \int_0^1 (1+y^2)\,dy + \int_1^0 x\,dx + \int_1^0 y^2\,dy$

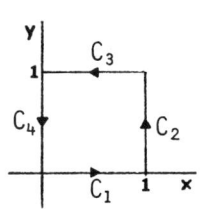

$= 0 + \frac{4}{3} - \frac{1}{2} - \frac{1}{3} = \frac{1}{2}$

9. (continued)

(b) A vector equation of C_3 is
$\langle x,y \rangle = \langle 2,1 \rangle + t\langle -2,-1 \rangle$ for
t in [0,1], so let $x = 2-2t$,
$y = 1-t$ for t in [0,1] be
parametric equations of C_3.

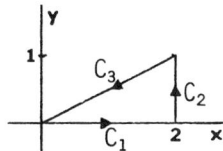

$$\int_0^2 0\,dx + \int_0^1 (4+y^2)\,dy + \int_0^1 [2(1-t)^2(-2) + 5(1-t)^2(-1)]\,dt$$

$$= 0 + \frac{13}{3} - 3 = \frac{4}{3}$$

(c) $\int_0^{2\pi} [(\cos t)(\sin t)(-\sin t) + (\cos^2 t + \sin^2 t)(\cos t)]\,dt$

$$= \int_0^{2\pi} \cos^3 t\,dt = \int_0^{2\pi} (1-\sin^2 t)\cos t\,dt$$

$$= \left[\sin t - \frac{\sin^3 t}{3}\right]_0^{2\pi} = 0$$

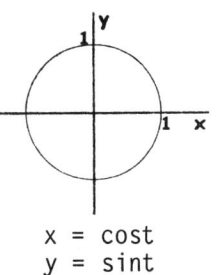

$x = \cos t$
$y = \sin t$
t in $[0, 2\pi]$

12. $\iint_G xyz\,dS = \iint_R xy(x+y)(\sec\gamma)\,dA$

$$= \sqrt{3}\int_0^1 \int_0^{-2x+2} (x^2 y + xy^2)\,dy\,dx$$

$$= \sqrt{3}\int_0^1 \left(\frac{4x^2(1-x)^2}{2} + \frac{8x(1-x)^3}{3}\right)dx$$

$$= \frac{-2\sqrt{3}}{3}\int_0^1 (x^4 - 6x^3 + 9x^2 - 4x)\,dx$$

$$= \frac{-2\sqrt{3}}{3}\left(\frac{1}{5} - \frac{3}{2} + 3 - 2\right) = \frac{\sqrt{3}}{5} \approx 0.3464$$

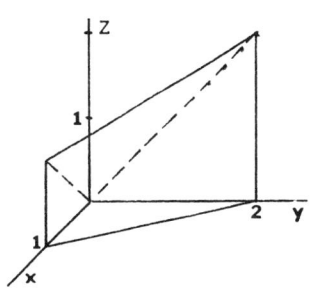

$\cos\gamma = \dfrac{\langle -1,-1,1 \rangle \cdot \langle 0,0,1 \rangle}{\sqrt{3}}$

Therefore, $\sec\gamma = \sqrt{3}$

Problem Set 17.8

15. $\text{curl}\,\mathbf{F} = \langle 3-0, 0-0, -1-1 \rangle = \langle 3, 0, -2 \rangle$, $\mathbf{n} = \dfrac{\langle a,b,1 \rangle}{\sqrt{a^2+b^2+1}}$

$$\oint_C \mathbf{F}\cdot\mathbf{T}\,ds = \iint_S (\text{curl}\,\mathbf{F})\cdot\mathbf{n}\,dS$$

$$= \iint_S \frac{3a-2}{\sqrt{a^2+b^2+1}}\,dS = \frac{3a-2}{\sqrt{a^2+b^2+1}}\,[A(S)]$$

$$= \frac{3a-2}{\sqrt{a^2+b^2+1}}\,(9\pi) \qquad [S \text{ is a circle of radius 3.}]$$

$$= \frac{9\pi(3a-2)}{\sqrt{a^2+b^2+1}}$$

CHAPTER 18 DIFFERENTIAL EQUATIONS

Problem Set 18.1 Linear First-Order Equations

Notes: (1) "IF" denotes "Integrating Factor."

(2) After multiplying each side of $y' + f(x)y = g(x)$ by an integrating factor, the equation simplifies to $D[y(IF)] = f(x)(IF)$.

3. $y' + \dfrac{x}{1-x^2} y = \dfrac{ax}{1-x^2}$ IF: $\exp\left(\int \dfrac{x}{1-x^2}\right)dx = \exp[\ell n(1-x^2)^{-1/2}] = (1-x^2)^{-1/2}$

$D[y(1-x^2)^{-1/2}] = ax(1-x^2)^{-3/2}$ [See note (2) above.]

Then $y(1-x^2)^{-1/2} = a(1-x^2)^{-1/2} + C$, so $y = a + C(1-x^2)^{1/2}$.

6. $y' - ay = f(x)$ IF: $e^{\int -adx} = e^{-ax}$

$D[ye^{-ax}] = e^{-ax}f(x)$

Then $ye^{-ax} = \int e^{-ax}f(x)dx$, so $y = e^{ax}\int e^{-ax}f(x)dx$.

9. $y' + f(x)y = f(x)$ IF: $e^{\int f(x)dx}$

$D[ye^{\int f(x)dx}] = f(x) e^{\int f(x)dx}$

Then $ye^{\int f(x)dx} = e^{\int f(x)dx} + C$, so $y = 1 + Ce^{-\int f(x)dx}$.

12. $y' + 3y = e^{2x}$ IF: $e^{\int 3dx} = e^{3x}$

$D[ye^{3x}] = e^{5x}$

Then $ye^{3x} = \dfrac{e^{5x}}{5} + C$. $x=0, y=1 \Rightarrow C = \dfrac{4}{5}$, so $ye^{3x} = \dfrac{e^{5x}}{5} + \dfrac{4}{5}$.

Therefore, $y = \dfrac{e^{2x} + 4e^{-3x}}{5}$ is the particular solution through (0,1).

15. Let Q denote the number of pounds of chemical A after t minutes.

$\frac{dQ}{dt}$ = (2 lbs/gal)(3gal/min) - (Q lbs/20gal)(3gal/min) = 6 - $\frac{3Q}{20}$ lbs/min

$Q' + \frac{3}{20}Q = 6$ IF: $e^{\int(3/20)dx} = e^{3t/20}$

$D[Qe^{3t/20}] = 6e^{3t/20}$

Then $Qe^{3t/20} = 40e^{3t/20} + C$. t=0, Q=10 => C = -30.

Therefore, $Q(t) = 40 - 30e^{-3t/20}$, so $Q(20) = 40 - 30e^{-3} \approx 38.506$ lbs.

18. $\frac{dQ}{dt} = \frac{-2Q}{50+t}$ or $Q' + \frac{2}{50+t}Q = 0$ IF: $\exp\left(\int\frac{2}{50+t}dt\right) = e^{2\ell n(50+t)} = (50+t)^2$

$D[Q(50+t)^2] = 0$

Then $Q(50+t)^2 = C$. t=0, Q=30 => C=75000. Thus, $Q(50+t)^2 = 75000$.

If Q=25, $25(50+t)^2 = 75000$, so t = $\sqrt{3000} - 50 \approx 4.772$ minutes.

21. LI' + RI = E, so 3.5I' + 100I = 120sin377t

    I' + 285.7143I ≈ 34.2857sin377t   IF: $e^{285.7143t}$

    $D[Ie^{285.7143t}] \approx 34.2857e^{285.7143t}\sin 377t$

    Then, using integration formula 67,

    $Ie^{285.7143t} \approx 34.2857\left[\frac{e^{285.7143t}}{223761.6612}(285.7143\sin 377t - 377\cos 377t)\right] + C$

    $\approx 0.0001532e^{285.7143t}(285.7143\sin 377t - 377\cos 377t) + C$

    t=0, I=0 => C ≈ 0.05776

    (a) I ≈ 0.0001532(285.7143sin377t - 377cos377t) + 0.05776$e^{-285.7143t}$

        ≈ 0.04377sin377t - 0.05776cos377t + 0.05776$e^{-285.7t}$

    (b) As t→∞, I → 0.04377sin377t - 0.05776cos377t

Problem Set 18.2 Second-Order Homogeneous Equations

3. Auxiliary Equation: $r^2+6r-7 = 0$, $(r+7)(r-1) = 0$ has roots $-7, 1$.

General Solution: $y = C_1 e^{-7x} + C_2 e^x$

$y' = -7C_1 e^{-7x} + C_2 e^x$

If $x=0$, $y=0$, $y'=4$, then $0 = C_1 + C_2$ and $4 = -7C_1 + C_2$, so $C_1 = -1/2$ and $C_2 = 1/2$.

Therefore, $y = \dfrac{e^x - e^{-7x}}{2}$

6. Auxiliary Equation: $r^2+10r+25 = 0$, $(r+5)^2 = 0$ has roots $5, 5$.

General Solution: $y = C_1 e^{-5x} + C_2 x e^{-5x}$ or $y = (C_1 + C_2 x) e^{-5x}$

9. Auxiliary Equation: $r^2+4 = 0$ has roots $\pm 2i$.

General Solution: $y = C_1 \cos 2x + C_2 \sin 2x$

If $x=0$ and $y=2$, then $2 = C_1$; if $x=\pi/4$ and $y=3$, then $3 = C_2$.

Therefore, $y = 2\cos 2x + 3\sin 2x$

12. Auxiliary Equation: $r^2+r+1 = 0$ has roots $\dfrac{-1}{2} \pm \dfrac{\sqrt{3}}{2} i$.

General Solution: $y = C_1 e^{(-1/2)x} \cos(\sqrt{3}/2)x + C_2 e^{(-1/2)x} \sin(\sqrt{3}/2)x$

$y = e^{-x/2}[C_1 \cos(\sqrt{3}/2)x + C_2 \sin(\sqrt{3}/2)x]$

15. Auxiliary Equation: $r^4+3r^2-4=0$, $(r+1)(r-1)(r^2+4)=0$ has roots $-1, 1, \pm 2i$.

General Solution: $y = C_1 e^{-x} + C_2 e^x + C_3 \cos 2x + C_4 \sin 2x$

18. $e^u = \cosh u + \sinh u$ and $e^{-u} = \cosh u - \sinh u$ [See Problems 1 and 2 of Problem Set 7.8.]

Auxiliary Equation: $r^2 - 2bx - c^2 = 0$

Roots of Auxiliary Equation: $\dfrac{2b \pm \sqrt{4b^2+4c^2}}{2} = b \pm \sqrt{b^2+c^2}$

18. (continued)

General Solution: $y = C_1 e^{b+\sqrt{b^2+c^2}} + C_2 e^{b-\sqrt{b^2+c^2}}$

$= e^b[C_1(\cosh\sqrt{b^2+c^2} + \sinh\sqrt{b^2+c^2}) + C_2(\cosh\sqrt{b^2+c^2} - \sinh\sqrt{b^2+c^2})]$

$= e^b[(C_1+C_2)\cosh\sqrt{b^2+c^2} + (C_1-C_2)\sinh\sqrt{b^2+c^2}]$

$= e^b[D_1\cosh\sqrt{b^2+c^2} + D_2\sinh\sqrt{b^2+c^2}]$

21. (*) $x^2 y'' + 5xy' + 4y = 0$

Let $x = e^z$. Then $z = \ln x$; $y' = \dfrac{dy}{dx} = \dfrac{dy}{dz}\dfrac{dz}{dx} = \dfrac{dy}{dz}\dfrac{1}{x}$;

$y'' = \dfrac{dy'}{dx} = \dfrac{d}{dx}\left(\dfrac{dy}{dz}\dfrac{1}{x}\right) = \dfrac{dy}{dz}\dfrac{-1}{x^2} + \dfrac{1}{x}\dfrac{d^2y}{dz^2}\dfrac{dz}{dx} = \dfrac{dy}{dz}\dfrac{-1}{x^2} + \dfrac{1}{x}\dfrac{d^2y}{dz^2}\dfrac{1}{x}$

$\left(-\dfrac{dy}{dz} + \dfrac{d^2y}{dz^2}\right) + \left(5\dfrac{dy}{dz}\right) + 4y = 0$ [Substituting y' and y'' into (*)]

$\dfrac{d^2y}{dz^2} + 4\dfrac{dy}{dz} + 4y = 0$

Auxiliary Equation: $r^2+4r+4 = 0$, $(r+2)^2 = 0$ has roots $-2,-2$.

General solution: $y = (C_1+C_2 z)e^{-2z}$, $y = (C_1+C_2\ln x)e^{-2\ln x}$

$y = (C_1+C_2\ln x)x^{-2}$

Problem Set 18.3 The Nonhomogeneous Equation

3. Auxliary Equation: $r^2-2r+1 = 0$ has roots $1,1$.

$y_h = (C_1+C_2 x)e^x$

Let $y_p = B_2 x^2 + B_1 x + B_0$; $y_p' = 2B_2 x + B_1$; $y_p'' = 2B_2$.

Then $(2B_2) - 2(2B_2 x + B_1) + (B_2 x^2 + B_1 x + B_0) = x^2 + x$

$B_2 x^2 + (-4B_2+B_1)x + (2B_2-2B_1+B_0) = x^2 + x$

Thus, $B_2 = 1$, $-4B_2+B_1 = 1$, $2B_2-2B_1+B_0 = 0$, so $B_2 = 1$, $B_1 = 5$, $B_0 = 8$.

General Solution: $y = x^2 + 5x + 8 + (C_1+C_2 x)e^x$

6. Auxiliary Equation: $r^2+6r+9 = 0$, $(r+3)^2 = 0$ has roots $-3,-3$.

   $y_h = (C_1+C_2x)e^{-3x}$

   Let $y_p = Be^{-x}$; $y_p' = -Be^{-x}$; $y_p'' = Be^{-x}$

   Then $(Be^{-x}) + 6(-Be^{-x}) + 9(Be^{-x}) = 2e^{-x}$; $4Be^{-x} = 2e^{-x}$; $B = 1/2$.

   General Solution: $y = (1/2)e^{-x} + (C_1+C_2x)e^{-3x}$

9. Auxiliary Equation: $r^2-r-2 = 0$, $(r+1)(r-2) = 0$ has roots $-1,2$.

   $y_h = C_1e^{-x} + C_2e^{2x}$

   Let $y_p = B\cos x + C\sin x$; $y_p' = -B\sin x + C\cos x$; $y_p'' = -B\cos x - C\sin x$

   Then $(-B\cos x - C\sin x) - (-B\sin x + C\cos x) - 2(B\cos x + C\sin x) = 2\sin x$.

   $(-3B-C)\cos x + (B-3C)\sin x = 2\sin x$, so $-3B-C=0$ and $B-3C=2$; $B = \frac{1}{5}$; $C = \frac{-3}{5}$

   General Solution: $(1/5)\cos x - (3/5)\sin x + C_1e^{2x} + C_2e^{-x}$

12. Auxiliary Equation: $r^2+9 = 0$ has roots $\pm 3i$, so $y_h = C_1\cos 3x + C_2\sin 3x$.

    Let $y_p = Bx\cos 3x + Cx\sin 3x$; $y_p' = (-3bx+C)\sin 3x + (B+3Cx)\cos 3x$

    $y_p'' = (-9Bx+6C)\cos 3x + (-9Cx-6B)\sin 3x$

    Then substituting into the original equation and simplifying, obtain

    $6C\cos 3x - 6B\sin 3x = \sin 3x$, so $C = 0$ and $B = -1/6$.

    General Solution: $y = (-1/6)x\cos 3x + C_1\cos 3x + C_2\sin 3x$

15. Auxiliary Equation: $r^2-5r+6 = 0$ has roots $2,3$, so $y_h = C_1e^{2x} + C_2e^{3x}$.

    Let $y_p = Be^x$; $y_p' = Be^x$; $y_p'' = Be^x$.

    Then $(Be^x) - 5(Be^x) + 6(Be^x) = 2e^x$; $2Be^x = 2e^x$; $B = 1$.

    General Solution: $y = e^x + C_1e^{2x} + C_2e^{3x}$

    $y' = e^x + 2C_1e^{2x} + 3C_2e^{3x}$

    If $x=0$, $y=1$, $y'=0$, then $1 = 1+C_1+C_2$ and $0 = 1+2C_1+3C_2$; $C_1=1$, $C_2=-1$.

    Therefore, $y = e^x + e^{2x} - e^{3x}$.

Problem Set 18.3

18. Auxiliary Equation: $r^2-4 = 0$ has roots 2,-2, so $y_h = C_1e^{2x} + C_2e^{-2x}$.
Let $y_p = v_1e^{2x} + v_2e^{-2x}$, subject to $v_1'e^{2x} + v_2'e^{-2x} = 0$,

$$\text{and } v_1'(2e^{2x}) + v_2'(-2e^{-2x}) = e^{2x}$$

Then $v_1'(4e^{2x}) = e^{2x}$ and $v_2'(-4e^{-2x}) = e^{2x}$;

$v_1' = 1/4$ and $v_2' = -e^{4x}/4$; $v_1 = x/4$ and $v_2 = -e^{4x}/16$.

General Solution: $y = \frac{xe^{2x}}{4} - \frac{e^{2x}}{16} + C_1e^{2x} + C_2e^{-2x}$

21. Auxiliary Equation: $r^2-3r+2 = 0$ has roots 1,2, so $y_h = C_1e^x + C_2e^{2x}$.
Let $y_p = v_1e^x + v_2e^{2x}$ subject to $v_1'e^x + v_2'e^{2x} = 0$,

$$\text{and } v_1'(e^x) + v_2'(2e^{2x}) = \frac{e^x}{e^x+1}$$

Then $v_1' = \frac{-e^x}{e^x(e^x+1)}$ so $v_1 = \int \frac{-e^x}{e^x(e^x+1)} dx = \int \frac{-1}{u(u+1)} du = \int \left(\frac{-1}{u} + \frac{1}{u+1}\right) du$

$= -\ell n u + \ell n(u+1) = \ell n\left(\frac{u+1}{u}\right) = \ell n \frac{e^x+1}{e^x} = \ell n(1+e^{-x})$

$v_2' = \frac{e^x}{e^{2x}(e^x+1)}$ so $v_2 = -e^{-x} + \ell n(1+e^{-x})$ [similar to finding v_1]

General Solution: $y = e^x \ell n(1+e^{-x}) - e^x + e^{2x}\ell n(1+e^{-x}) + C_1e^x + C_2e^{2x}$

$y = (e^x+e^{2x})\ell n(1+e^{-x}) + D_1e^x + D_2e^{2x}$

Problem Set 18.4 Applications of Second-Order Equations

3. Equilibrium position is where $y=0$; $0 = -\cos 8t$; $8t = \pi/2, 3\pi/2, \ldots$

   $t = \pi/16, 3\pi/16, \ldots$

   At each of these values of t, $|y'(t)| = |8\sin t| = 8$ ft/sec.

6. $k = 20$ lbs/ft; $w = 10$ lbs; $y_0 = 1$ ft; $q = 4$ sec-lb/ft.

$B = \sqrt{\frac{(20)(32)}{10}} = 8$; $E = \frac{(4)(32)}{10} = 12.8$; $E^2 - 4B^2 < 0$, so damped motion.

Roots of auxiliary equation are $\frac{-E \pm \sqrt{E^2 - 4B^2}}{2} = -6.4 \pm 4.8i$

General Solution is $y = e^{-6.4t}(C_1\cos 4.8t + C_2\sin 4.8t)$.

$y' = e^{-6.4t}(-4.8C_1\sin 4.8t + 4.8C_2\cos 4.8t) - 6.4e^{-6.4t}(C_1\cos 4.8t + C_2\sin 4.8t)$

If $t=0$, $y=1$, $y'=0$, then $1=C_1$ and $0=4.8C_2-6.4C_1$, so $C_1 = 1$ and $C_2 = 4/3$.

Therefore, $y = e^{-6.4t}[\cos 4.8t + (4/3)\sin 4.8t]$.

9. $LQ'' + RQ' + \frac{Q}{C} = E(t)$; $10^6 Q' + 10^6 Q = 1$; $Q' + Q = 10^{-6}$ IF: e^t

$D[Qe^t] = 10^{-6}e^t$; $Qe^t = 10^{-6}e^t + C$; $Q = 10^{-6} + Ce^{-t}$

If $t=0$, $Q=0$, then $C = -10^{-6}$.

Therefore, $Q(t) = 10^{-6} - 10^{-6}e^{-t} = 10^{-6}(1-e^{-t})$.

12. $LQ'' + RQ' + \frac{Q}{C} = E$; $10^{-2}Q'' + \frac{Q}{10^{-7}} = 20$; $Q'' + 10^9 Q = 2000$.

The auxiliary equation, $r^2 + 10^9 = 0$, has roots $\pm 10^{9/2}i$.

$Q_h = C_1\cos 10^{9/2}t + C_2\sin 10^{9/2}t$

$Q_p = 2000(10^{-9}) = 2(10^{-6})$ is a particular solution. [by inspection]

General Solution: $Q(t) = 2(10^{-6}) + C_1\cos 10^{9/2}t + C_2\sin 10^{9/2}t$

Then $I(t) = Q'(t) = -10^{9/2}C_1\sin 10^{9/2}t + 10^{9/2}C_2\cos 10^{9/2}t$

If $t=0$, $Q=0$, $I=0$, then $0 = 2(10^{-6}) + C_1$ and $0 = C_2$.

Therefore, $I(t) = -10^{9/2}(-2[10^{-6}])\sin 10^{9/2}t = 2(10^{-3/2})\sin 10^{9/2}t$.

15. $C\sin(\beta t + \gamma) = C(\sin\beta t\cos\gamma + \cos\beta t\sin\gamma) = (C\cos\gamma)\sin\beta t + (C\sin\gamma)\cos\beta t$

$= C_1\sin\beta t + C_2\cos\beta t$, where $C_1 = C\cos\gamma$ and $C_2 = C\sin\gamma$.

[Note that $C_1^2 + C_2^2 = C^2\cos^2\gamma + C^2\sin^2\gamma = C^2$.]

Problem Set 18.4

Problem Set 18.5 Chapter Review Problems

True-False Quiz

3. True. $y' = \sec^2 x + \sec x \tan x$

$2y' - y^2 = (2\sec^2 x + 2\sec x \tan x) - (\tan^2 x + 2\sec x \tan x + \sec^2 x)$

$= \sec^2 x - \tan^2 x = 1$

6. False. Replacing y by $C_1 u_1(x) + C_2 u_2(x)$ would yield, on the left side,

$C_1 f(x) + C_2 f(x) = (C_1 + C_2) f(x)$ which is $f(x)$ only if $C_1 + C_2 = 1$
or $f(x) = 0$.

9. False. That is the form of y_h.

y_p should have the form $Bx\cos 3x + Cx\sin 3x$.

Miscellaneous Problems

3. (Linear first-order) $y' + 2xy = 2x$ 　　　　IF: $e^{\int 2x\, dx} = e^{x^2}$

$D[ye^{x^2}] = 2xe^{x^2}$; $ye^{x^2} = e^{x^2} + C$; $y = 1 + Ce^{-x^2}$.

If $x=0$, $y=3$, then $3 = 1+C$, so $C = 2$.

Therefore, $y = 1 + 2e^{-x^2}$.

6. (Linear first-order) $y' + (\tan x)y = 2\sec x$

IF: $e^{\int \tan x\, dx} = e^{\ln \sec x} = \sec x$ [if $\sec x > 0$]

$D[y \sec x] = 2\sec^2 x$; $y \sec x = 2\tan x + C$; $y = 2\sin x + C\cos x$.

[Same result is obtained if $\sec x < 0$.]

9. (Second-order homogeneous)

The auxiliary equation, $r^2-3r+2 = 0$, has roots 1,2.

The general solution is $y = C_1e^x + C_2e^{2x}$

$y' = C_1e^x + 2C_2e^{2x}$

If $x=0$, $y=0$, $y'=3$, then $0 = C_1+C_2$ and $3 = C_1+2C_2$, so $C_1=-3$, $C_2= 3$.

Therefore, $y = -3e^x+3e^{2x}$.

12. (Second-order nonhomogeneous)

The auxiliary equation, $r^2+4r+4 = 0$, has roots -2,-2.

$y_h = C_1e^{-2x} + C_2xe^{-2x} = (C_1+C_2x)e^{-2x}$.

Let $y_p = Be^x$; $y_p' = Be^x$; $y_p'' = Be^x$.

$(Be^x) + 4(Be^x) + 4(Be^x) = 3e^x$, so $B = 1/3$.

General Solution: $y = e^x/3 + (C_1+C_2x)e^{-2x}$

15. (Second-order homogeneous)

The auxiliary equation, $r^2+6r+25 = 0$, has roots $-3\pm 4i$.

General Solution: $y = e^{-3x}(C_1\cos 4x + C_2\sin 4x)$

18. (Fourth-order homogeneous) The auxiliary equation, $r^4-3r^2-10 = 0$ or $(r^2-5)(r^2+2) = 0$, has roots $-\sqrt{5}, \sqrt{5}, \pm\sqrt{2}i$.

General Solution: $y = C_1e^{\sqrt{5}x} + C_2e^{-\sqrt{5}x} + C_3\cos\sqrt{2}x + C_4\sin\sqrt{2}x$

21. (Simple harmonic motion)

$k = 5$; $w = 10$; $y_0 = -1$.

$B = \sqrt{\frac{(5)(32)}{10}} = 4$

Then the equation of motion is $y = -\cos 4t$.

The amplitude is $|-1| = 1$; the period is $2\pi/4 = \pi/2$.

Problem Set 18.5